Autodesk Inventor 2024 and Engineering Graphics
An Integrated Approach

Randy H. Shih
Oregon Institute of Technology

SDC
PUBLICATIONS

SDC Publications
P.O. Box 1334
Mission, KS 66222
913-262-2664
www.SDCpublications.com
Publisher: Stephen Schroff

ISBN-13: 978-1-63057-583-0
ISBN-10: 1-63057-583-6

Printed and bound in the United States of America.

Preface

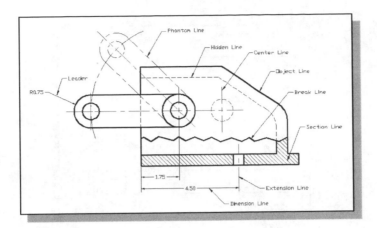

The primary goal of *Autodesk Inventor 2024 and Engineering Graphics: An Integrated Approach* is to introduce Engineering Graphics with the use of a modern Computer Aided Design package – Autodesk Inventor® 2024. This text is intended to be used as a training guide for students and professionals. The chapters in this text proceed in a pedagogical fashion to guide you from constructing basic shapes to making complete sets of engineering drawings. This text takes a hands-on, exercise-intensive approach to all the important concepts of Engineering Graphics, as well as in-depth discussions of parametric feature-based CAD techniques. This textbook contains a series of fifteen chapters, with detailed step-by-step tutorial style lessons, designed to introduce beginning CAD users to the graphic language used in all branches of technical industry. Emphases in lessons on Engineering Graphics are placed on graphical analysis, principles of orthographic projection, auxiliary views, pictorial drawings, dimensioning methods, sectioning, creating assembly and working drawings set with adherence to ANSI/ASME drafting standards. This book does not attempt to cover all of Autodesk Inventor 2024's features, only to provide an introduction to the software. It is intended to help you establish a good basis for exploring and growing in the exciting field of Computer Aided Engineering.

Acknowledgments

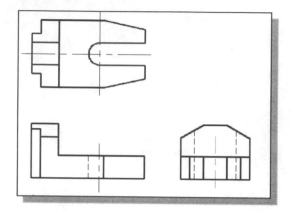

This book would not have been possible without a great deal of support. First, special thanks to two great teachers, Prof. George R. Schade of University of Nebraska-Lincoln and Mr. Denwu Lee from Taiwan, who taught me the fundamentals, the intrigue and the sheer fun of Computer Aided Engineering.

The effort and support of the editorial and production staff of SDC Publications is gratefully acknowledged.

I am grateful that the Department of Mechanical and Manufacturing Engineering and Technology at Oregon Institute of Technology has provided me with an excellent environment in which to pursue my interests in teaching and research. I would also like to thank Emeritus Professor Charles Hermach for many helpful comments and encouragement.

Finally, truly unbounded thanks are due to my wife Hsiu-Ling and our daughter Casandra for their understanding and encouragement throughout this project.

Randy H. Shih
Klamath Falls, Oregon
Spring, 2023

Table of Contents

Chapter 1
Introduction

Chapter 2
Parametric Modeling Fundamentals

Chapter 3
Constructive Solid Geometry Concepts

Chapter 4
Geometric Constructions

Chapter 5
Model History Tree

Chapter 6
Geometric Construction Tools

Chapter 7
Orthographic Projection and Multiview Constructions

Chapter 8
Dimensioning and Notes

Chapter 9
Tolerancing and Fits

Chapter 10
Pictorials and Sketching

Chapter 11
Auxiliary Views and Reference Geometry

Chapter 12
Section Views & Symmetrical Features in Designs

Chapter 15
Introduction to Stress Analysis

Appendix

Index

Notes:

Autodesk Inventor Certified User Examination Overview

The Autodesk Inventor Certified User examination is a performance-based exam. The examination is comprised of approximately 30 questions to be completed in 50 minutes. The test items will require you to use the Autodesk Inventor software to perform specific tasks and then answer questions about the tasks.

Performance-based testing is defined as ***Testing by Doing***. This means you actually perform the given task then answer the questions regarding the task. Performance-based testing is widely accepted as a better way of ensuring the users have the skills needed, rather than just recalling information.

The Autodesk Inventor Certified User examination is designed to test specific performance tasks in the following 6 sections:

➢ Every effort has been made to cover the exam objectives included in the Autodesk Inventor Certified User Examination. However, the format and topics covered by the examination are constantly changing. Students planning to take the Certified User Examination are advised to visit the Autodesk website and obtain information regarding the format and details about the Autodesk Inventor Certified User Examination.

Section 1: User Interface

Objectives: Primary Environments, UI Navigation/Interaction, Graphics Window Display, Navigation Control

Section 2: File Management

Objectives: Project Files

Section 3: Sketches

Objectives: Creating 2D Sketches, Draw Tools, Sketch Constraints, Pattern Sketches, Modify Sketches, Format Sketches, Sketch Doctor, Shared Sketches, Sketch Parameters

Certification Examination Performance Task	Covered in this book on Chapter – Page

Section 4: Parts
Objectives: Creating parts, Work Features, Pattern Features, Part Properties

Section 5: Assemblies

Objectives: Creating Assemblies, Viewing Assemblies, Animation Assemblies, Adaptive Features, Parts, and Subassemblies

Certification Examination Performance Task	Covered in this book on Chapter – Page

Section 6: Drawings
Objectives: Create drawings

Certification Examination Performance Task	Covered in this book on Chapter – Page

Tips for Taking the Autodesk Certified User Examination

1. **Study:** The first step to maximize your potential on an exam is to sufficiently prepare for it. You need to be familiar with the Autodesk Inventor package, and this can only be achieved by doing designs and exploring the different commands available. The Autodesk Inventor Certified User exam is designed to measure your familiarity with the Autodesk Inventor software. You must be able to perform the given task and answer the exam questions correctly and quickly.

2. **Make Notes**: Take notes on what you learn either while attending classroom sessions or going through study material. Use these notes as a review guide before taking the actual test.

3. **Time Management**: Manage the time you spent on each question. Always remember you do not need to score 100% to pass the exam. Also keep in mind that some questions are weighed more heavily and may take more time to answer.

4. **Use Common Sense**: If you are unable to get the correct answer and unable to eliminate all distracters, then you need to select the best answer from the remaining selections. This may be a task of selecting the best answer from amongst several correct answers, or it may be selecting the least incorrect answer from amongst several poor answers.

5. **Use the Autodesk Inventor Help system**: If you get confused and can't think of the answer, remember the Autodesk Inventor help system is a great tool to confirm your considerations.

6. **Take Your Time:** The examination has a time limit. If you encounter a question you cannot answer in a reasonable amount of time, use the Save As feature to save a copy of the data file, and mark the question for review. When you review the question, open your copy of the data file and complete the performance task. After you verify that you have entered the answer correctly, unmark the question so it no longer appears as marked for review.

7. **Be Cautious and Don't Act in Haste:** Devote some time to ponder and think of the correct answer. Ensure that you interpret all the options correctly before selecting from available choices. Don't go into panic mode while taking a test. Use the *Review* screen to ensure you have reviewed all the questions you may have marked for review. When you are confident that you have answered all questions, end the examination to submit your answers for scoring. You will receive a score report once you have submitted your answers.

8. **Relax before the exam:** In order to avoid last minute stress, make sure that you arrive 10 to 15 minutes early and relax before taking the exam.

Chapter 1
Introduction

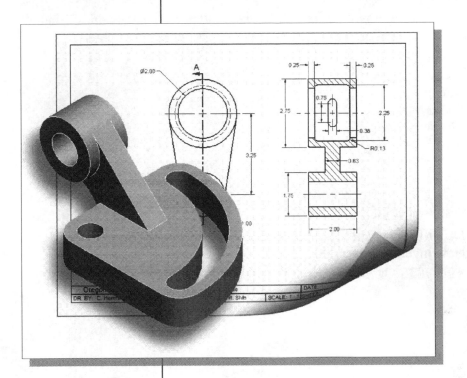

Learning Objectives

- ◆ **Introduction to Engineering Graphics**
- ◆ **Development of Computer Geometric Modeling**
- ◆ **Feature-Based Parametric Modeling**
- ◆ **Startup Options and Units Setup**
- ◆ **Autodesk Inventor Screen Layout**
- ◆ **User Interface & Mouse Buttons**
- ◆ **Autodesk Inventor Online Help**

Autodesk Inventor Certified User Exam Objectives Coverage

Section 1: User Interface

Objectives: Primary Environments, UI Navigation/Interaction, Graphics Window Display, Navigation Control

Section 2: File Management

Objectives: Project Files

Introduction

Engineering Graphics, also known as **Technical Drawing**, is the technique of creating accurate representations of designs, an *engineering drawing*, for construction and manufacturing. An **engineering drawing** is a type of drawing that is technical in nature, used to fully and clearly define requirements for engineered items, and is usually created in accordance with standardized conventions for layout, nomenclature, interpretation, appearance, size, etc. A skilled practitioner of the art of engineering drawing is known as a *draftsman* or *draftsperson*.

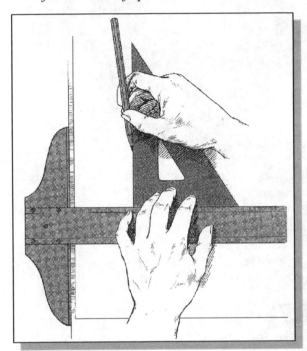

The basic mechanics of drafting is to use a pencil and draw on a piece of paper. For engineering drawings, papers are generally placed on a drafting table and additional tools are used. A T-square is one of the standard tools commonly used with a drafting table.

A T-square is also known as a *sliding straightedge*; parallel lines can be drawn simply by moving the T-square and running a pencil along the T-square's straightedge. The T-square is more typically used as a tool to hold other tools such as triangles. For example, one or more triangles can be placed on the T-square and lines can be drawn at chosen angles on the paper.

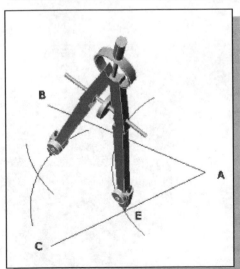

In addition to the triangles, other tools are used to draw curves and circles. Primary among these are the compass, used for drawing simple arcs and circles, and the French curve, typically a piece of plastic with a complex curve on it. A spline is a rubber coated articulated metal that can be manually bent to almost any curve.

This basic drafting system requires an accurate table and constant attention to the positioning of the tools. A common error draftspersons faced was allowing the triangles to push the top of the T-square down slightly, thereby throwing off all angles. Drafting in general was already a time-consuming process.

A solution to these problems was the introduction of the "**drafting machine**," a device that allowed the draftsperson to have an accurate right angle at any point on the page quite quickly. These machines often included the ability to change the angle, thereby removing the need for the triangles as well.

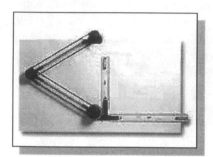

In addition to the mechanics of drawing the lines onto a piece of paper, drafting requires an understanding of geometry and the professional skills of the specific designer. At one time, drafting was a sought-after job, considered one of the more demanding and highly-skilled positions of the trade. Today, the mechanics of the drafting task have been largely automated and greatly accelerated through the use of **computer aided design** (CAD) systems. Proficiency in using CAD systems has also become one of the more important requirements for engineers and designers.

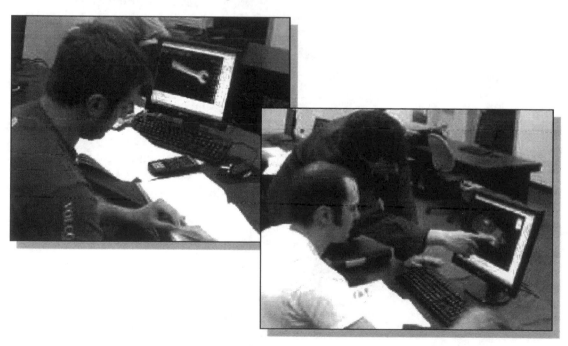

Drawing in CAD Systems

Computer Aided Design (CAD) is the process of creating designs with the aid of computers. This includes the generation of computer models, analysis of design data and the creation of necessary drawings. **Autodesk Inventor** is a CAD software package developed by *Autodesk, Inc*. The **Autodesk Inventor** software is a tool that can be used for design and drafting activities. Two-dimensional and three-dimensional models created in **Autodesk Inventor** can be transferred to other computer programs for further analysis and testing. The computer models can also be used in manufacturing equipment such as machining centers, lathes, mills or rapid prototyping machines to manufacture the product.

Rapid changes in the field of **computer aided engineering** (CAE) have brought exciting advances to the industry. Recent advances have made the long-sought goal of reducing design time, producing prototypes faster, and achieving higher product quality closer to a reality.

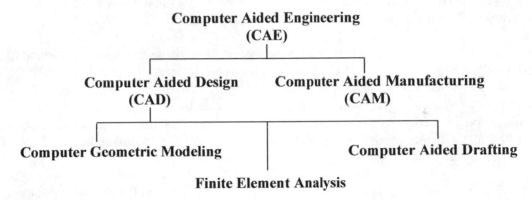

Computer Aided Engineering (CAE)

Computer Aided Design (CAD) — **Computer Aided Manufacturing (CAM)**

Computer Geometric Modeling — **Computer Aided Drafting**

Finite Element Analysis

Development of Computer Aided Design

Computer Aided Design is a relatively new technology, and its rapid expansion in the last fifty years is truly amazing. Computer-modeling technology has advanced along with the development of computer hardware. The first-generation CAD programs, developed in the 1950s, were mostly non-interactive; CAD users were required to create program-codes to generate the desired two-dimensional (2D) geometric shapes. Initially, the development of CAD technology occurred mostly in academic research facilities. The Massachusetts Institute of Technology, Carnegie-Mellon University, and Cambridge University were the leading pioneers at that time. The interest in CAD technology spread quickly and several major industry companies, such as General Motors, Lockheed, McDonnell, IBM, and Ford Motor Co., participated in the development of interactive CAD programs in the 1960s. Usage of CAD systems was primarily in the automotive industry, aerospace industry, and government agencies that developed their own programs for their specific needs. The 1960s also marked the beginning of the development of finite element analysis methods for computer stress analysis and computer aided manufacturing for generating machine tool paths.

The 1970s are generally viewed as the years of the most significant progress in the development of computer hardware, namely the invention and development of **microprocessors**. With the improvement in computing power, new types of 3D CAD programs that were user-friendly and interactive became reality. CAD technology quickly expanded from very simple **computer aided drafting** to very complex **computer aided design**. The use of 2D and 3D wireframe modelers was accepted as the leading-edge technology that could increase productivity in industry. The developments of surface modeling and solid modeling technologies were taking shape by the late 1970s, but the high cost of computer hardware and programming slowed the development of such technology. During this period, the available CAD systems all required room-sized mainframe computers that were extremely expensive.

In the 1980s, improvements in computer hardware brought the power of mainframes to the desktop at less cost and with more accessibility to the general public. By the mid-1980s, CAD technology had become the main focus of a variety of manufacturing industries and was very competitive with traditional design/drafting methods. It was during this period of time that 3D solid modeling technology had major advancements, which boosted the usage of CAE technology in industry.

The introduction of the *feature-based parametric solid modeling* approach, at the end of the 1980s, elevated CAD/CAM/CAE technology to a new level. In the 1990s, CAD programs evolved into powerful design/manufacturing/management tools. CAD technology has come a long way, and during these years of development, modeling schemes progressed from two-dimensional (2D) wireframe to three-dimensional (3D) wireframe, to surface modeling, to solid modeling and, finally, to feature-based parametric solid modeling.

The first-generation CAD packages were simply 2D **computer aided drafting** programs, basically the electronic equivalents of the drafting board. For typical models, the use of this type of program would require that several views of the objects be created individually as they would be on the drafting board. The 3D designs remained in the designer's mind, not in the computer database. Mental translations of 3D objects to 2D views are required throughout the use of these packages. Although such systems have some advantages over traditional board drafting, they are still tedious and labor intensive. The need for the development of 3D modelers came quite naturally, given the limitations of the 2D drafting packages.

The development of three-dimensional modeling schemes started with three-dimensional (3D) wireframes. Wireframe models are models consisting of points and edges, which are straight lines connecting between appropriate points. The edges of wireframe models are used, similar to lines in 2D drawings, to represent transitions of surfaces and features. The use of lines and points is also a very economical way to represent 3D designs.

The development of the 3D wireframe modeler was a major leap in the area of computer geometric modeling. The computer database in the 3D wireframe modeler contains the locations of all the points in space coordinates, and it is typically sufficient to create just one model rather than multiple views of the same model. This single 3D model can then be viewed from any direction as needed. Most 3D wireframe modelers allow the user to create projected lines/edges of 3D wireframe models. In comparison to other types of 3D modelers, the 3D wireframe modelers require very little computing power and generally can be used to achieve reasonably good representations of 3D models. However, because surface definition is not part of a wireframe model, all wireframe images have the inherent problem of ambiguity. Two examples of such ambiguity are illustrated.

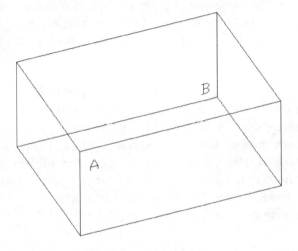

Wireframe Ambiguity: Which corner is in front, A or B?

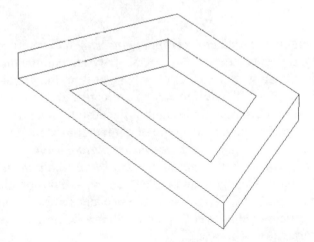

A non-realizable object: Wireframe models contain no surface definitions.

Surface modeling is the logical development in computer geometry modeling to follow the 3D wireframe modeling scheme by organizing and grouping edges that define polygonal surfaces. Surface modeling describes the part's surfaces but not its interiors. Designers are still required to interactively examine surface models to ensure that the various surfaces on a model are contiguous throughout. Many of the concepts used in 3D wireframe and surface modelers are incorporated in the solid modeling scheme, but it is solid modeling that offers the most advantages as a design tool.

In the solid modeling presentation scheme, the solid definitions include nodes, edges, and surfaces, and it is a complete and unambiguous mathematical representation of a precisely enclosed and filled volume. Unlike the surface modeling method, solid modelers start with a solid or use topology rules to guarantee that all of the surfaces are stitched together properly. Two predominant methods for representing solid models are **constructive solid geometry** (CSG) representation and **boundary representation** (B-rep).

The CSG representation method can be defined as the combination of 3D solid primitives. What constitutes a "primitive" varies somewhat with the software but typically includes a rectangular prism, a cylinder, a cone, a wedge, and a sphere. Most solid modelers also allow the user to define additional primitives, which are shapes typically formed by the basic shapes. The underlying concept of the CSG representation method is very straightforward; we simply **add** or **subtract** one primitive from another. The CSG approach is also known as the machinist's approach, as it can be used to simulate the manufacturing procedures for creating the 3D object.

In the B-rep representation method, objects are represented in terms of their spatial boundaries. This method defines the points, edges, and surfaces of a volume, and/or issues commands that sweep or rotate a defined face into a third dimension to form a solid. The object is then made up of the unions of these surfaces that completely and precisely enclose a volume.

By the 1980s, a new paradigm called *concurrent engineering* had emerged. With concurrent engineering, designers, design engineers, analysts, manufacturing engineers, and management engineers all work together closely right from the initial stages of the design. In this way, all aspects of the design can be evaluated, and any potential problems can be identified right from the start and throughout the design process. Using the principles of concurrent engineering, a new type of computer modeling technique appeared. The technique is known as the *feature-based parametric modeling technique*. The key advantage of the *feature-based parametric modeling technique* is its capability to produce very flexible designs. Changes can be made easily, and design alternatives can be evaluated with minimum effort. Various software packages offer different approaches to feature-based parametric modeling, yet the end result is a flexible design defined by its design variables and parametric features.

Feature-Based Parametric Modeling

One of the key elements in the Autodesk Inventor solid modeling software is its use of the **feature-based parametric modeling technique**. The feature-based parametric modeling approach has elevated solid modeling technology to the level of a very powerful design tool. Parametric modeling automates the design and revision procedures by the use of parametric features. Parametric features control the model geometry by the use of design variables. The word *parametric* means that the geometric definitions of the design, such as dimensions, can be varied at any time during the design process. Features are predefined parts or construction tools for which users define the key parameters. A part is described as a sequence of engineering features, which can be modified and/or changed at any time. The concept of parametric features makes modeling more closely match the actual design-manufacturing process than the mathematics of a solid modeling program. In parametric modeling, models and drawings are updated automatically when the design is refined.

Parametric modeling offers many benefits:

- **We begin with simple, conceptual models with minimal detail; this approach conforms to the design philosophy of "shape before size."**

- **Geometric constraints, dimensional constraints, and relational parametric equations can be used to capture design intent.**

- **The ability to update an entire system, including parts, assemblies and drawings after changing one parameter of complex designs.**

- **We can quickly explore and evaluate different design variations and alternatives to determine the best design.**

- **Existing design data can be reused to create new designs.**

- **Quick design turn-around.**

One of the key features of Autodesk Inventor is the use of an assembly-centric paradigm, which enables users to concentrate on the design without depending on the associated parameters or constraints. Users can specify how parts fit together and the Autodesk Inventor *assembly-based fit function* automatically determines the parts' sizes and positions. This unique approach is known as the **Direct Adaptive Assembly approach**, which defines part relationships directly with no order dependency.

The *Adaptive Assembly approach* is a unique design methodology that can only be found in Autodesk Inventor. The goal of this methodology is to improve the design process and allows you, the designer, to **Design the Way You Think**.

Getting Started with Autodesk Inventor

Autodesk Inventor is composed of several application software modules (these modules are called *applications*), all sharing a common database. In this text, the main concentration is placed on the solid modeling modules used for part design. The general procedures required in creating solid models, engineering drawings, and assemblies are illustrated.

How to start Autodesk Inventor depends on the type of workstation and the particular software configuration you are using. With most *Windows* systems, you may select **Autodesk Inventor** on the *Start* menu or select the **Autodesk Inventor** icon on the desktop. Consult your instructor or technical support personnel if you have difficulty starting the software. The program takes a while to load, so be patient.

The tutorials in this text are based on the assumption that you are using Autodesk Inventor's default settings. If your system has been customized for other uses, contact your technical support personnel to restore the default software configuration.

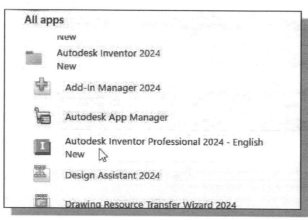

The Screen Layout and Getting Started Toolbar

Once the program is loaded into the memory, the Inventor window appears on the screen with the *Get Started* toolbar options activated.

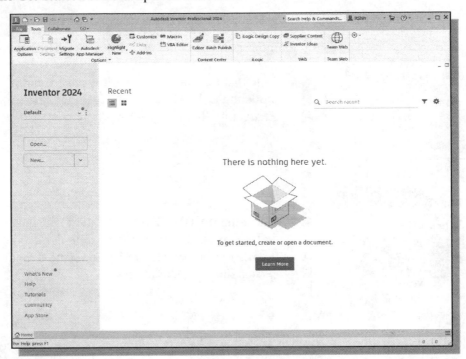

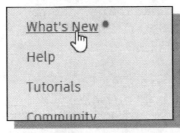

❖ Note that the *Get Started* toolbar contains helpful information in regard to using the Inventor software. For example, clicking the **What's New** option will bring up the *Internet Browser*, which contains the list of new features that are included in this release of Autodesk Inventor.

❖ You are encouraged to browse through the different information available in the *Getting Started Toolbar* section.

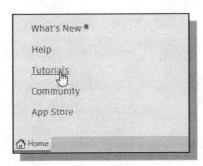

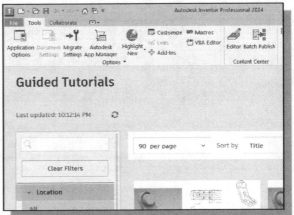

The New File Dialog Box and Units Setup

When starting a new CAD file, the first thing we should do is to choose the units we would like to use. We will use the English (feet and inches) setting for this example.

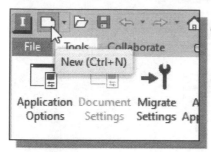

- Select the **New** icon with a single click of the left-mouse-button in the *Quick Access* toolbar.

- ❖ Note that the **New** option allows us to start a new modeling task, which can be creating a new model or several other modeling tasks.

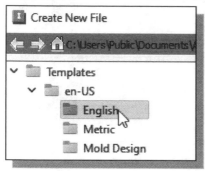

- Select the **en-US->English** tab in the *New File* dialog box as shown. Note the default tab contains the file options which are based on the default units chosen during installation.

- Select the **Standard(in).ipt** icon as shown. The different icons are templates for the different modeling tasks. The **idw** file type stands for drawing file, the **iam** file type stands for assembly file, and the **ipt** file type stands for part file. The **ipn** file type stands for assembly presentation.

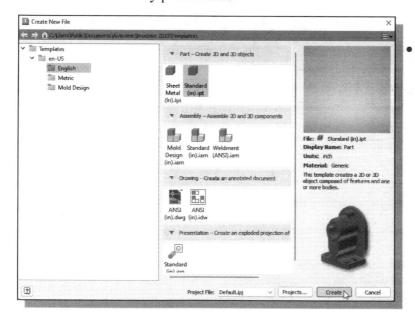

- Click **Create** in the *Create New File* dialog box to accept the selected settings.

The Default Autodesk Inventor Screen Layout

The default Autodesk Inventor drawing screen contains the *pull-down* menus, the *Standard* toolbar, the *Features* toolbar, the *Sketch* toolbar, the *drawing* area, the *browser* area, and the *Status Bar*. A line of quick text appears next to the icon as you move the *mouse cursor* over different icons. You may resize the Autodesk Inventor drawing window by clicking and dragging the edges of the window, or relocate the window by clicking and dragging the window title area.

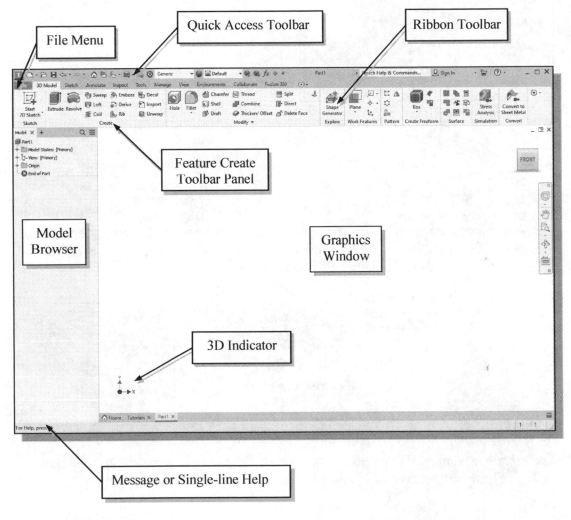

- The **Ribbon Toolbar** is a relatively new feature in Autodesk Inventor; the *Ribbon Toolbar* is composed of a series of tool panels, which are organized into tabs labeled by task. The *Ribbon* provides a compact palette of all of the tools necessary to accomplish the different modeling tasks. The drop-down arrow next to any icon indicates additional commands are available on the expanded panel; access the expanded panel by clicking on the drop-down arrow.

- **File Menu**

The *File* menu at the upper left corner of the main window contains tools for all file-related operations, such as Open, Save, Export, etc.

- **Quick Access Toolbar**

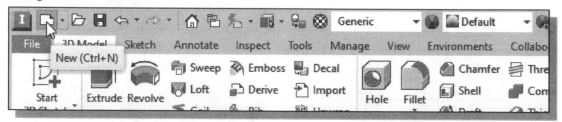

The *Quick Access* toolbar at the top of the *Inventor* window allows us quick access to file-related commands and to Undo/Redo the last operations.

- **Ribbon Tabs and Tool panels**

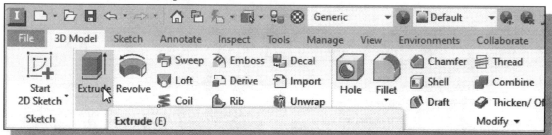

The *Ribbon* is composed of a series of tool panels, which are organized into tabs labeled by task. The assortments of tool panels can be accessed by clicking on the tabs.

- **Online Help Panel**

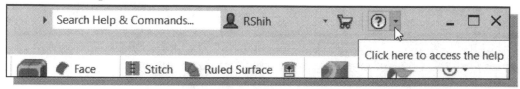

The *Help* options panel provides us with multiple options to access online help for Autodesk Inventor. The **Online Help system** provides general help information, such as command options and command references.

- **3D Model Toolbar**

The *3D Model* toolbar provides tools for creating the different types of 3D features, such as Extrude, Revolve, Sweep, etc.

- **Graphics Window**

The *graphics window* is the area where models and drawings are displayed.

- **Message and Status Bar**

The *Message and Status Bar* area shows a single-line help when the cursor is on top of an icon. This area also displays information pertinent to the active operation. For example, in the figure above, the coordinates and length information of a line are displayed while the *Line* command is activated.

Mouse Buttons

Autodesk Inventor utilizes the mouse buttons extensively. In learning Autodesk Inventor's interactive environment, it is important to understand the basic functions of the mouse buttons. It is highly recommended that you use a mouse or a tablet with Autodesk Inventor since the package uses the buttons for various functions.

- **Left mouse button**
  The **left mouse button** is used for most operations, such as selecting menus and icons, or picking graphic entities. One click of the button is used to select icons, menus and form entries, and to pick graphic items.

- **Right mouse button**
  The **right mouse button** is used to bring up additional available options. The software also utilizes the **right mouse button** the same as the **ENTER** key and is often used to accept the default setting to a prompt or to end a process.

- **Middle mouse button/wheel**
  The middle mouse button/wheel can be used to **Pan** (hold down the wheel button and drag the mouse) or **Zoom** (turn the wheel) realtime.

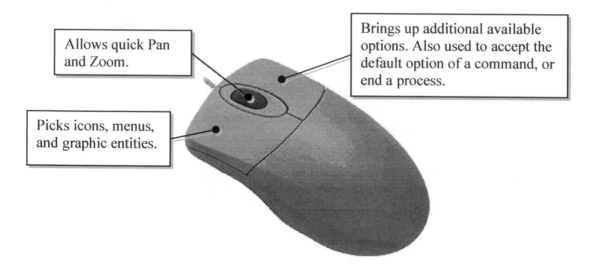

Brings up additional available options. Also used to accept the default option of a command, or end a process.

Allows quick Pan and Zoom.

Picks icons, menus, and graphic entities.

[Esc] – Canceling Commands

The [**Esc**] key is used to cancel a command in Autodesk Inventor. The [**Esc**] key is located near the top-left corner of the keyboard. Sometimes, it may be necessary to press the [**Esc**] key twice to cancel a command; it depends on where we are in the command sequence. For some commands, the [**Esc**] key is used to exit the command.

Autodesk Inventor Help System

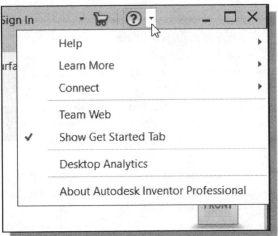

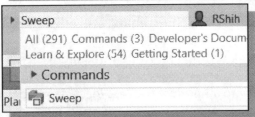

❖ Several types of help are available at any time during an Autodesk Inventor session. Autodesk Inventor provides many help functions, such as:

• Use the **Help** button near the upper right corner of the *Inventor* window.

• Help quick-key: Press the [**F1**] key to access the *Inventor Help* system.

• Use the *Info Center* to get information on a specific topic.

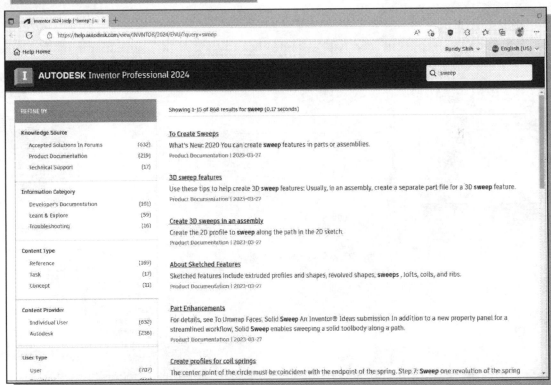

• Use the Internet to access information on the Autodesk community website.

Data Management Using Inventor Project files

With Autodesk Inventor, it is quite feasible to create designs without any regard to using the Autodesk Inventor data management system. Data management becomes critical for projects involving complex designs, especially when multiple team members are involved, or when we are working on integrating multiple design projects, or when it is necessary to share files among the design projects. Autodesk Inventor provides a fairly flexible data management system. It allows one person to use the basic option to help manage the locations of the different design files, or a team of designers can use the data management system to manage their projects stored on a networked computer system.

The Autodesk Inventor data management system organizes files based on **projects**. Each project is identified with a main folder that can contain files and folders associated to the design. In Autodesk Inventor, a *project* file (.ipj) defines the locations of all files associated with the project, including templates and library files.

The Autodesk Inventor data management system uses two types of projects:
➢ **Single-user Project**
➢ **Autodesk Vault Project** (installation of *Autodesk Vault* software is required)

The *single-user project* is for simpler projects where all project files are located on the same computer. The *Autodesk Vault project* is more suitable for projects requiring multiple users using a networked computer system.

• Click on the **Projects** icon with a single click of the left-mouse-button in the *[File]* → *[Manage]* pull-down menu.

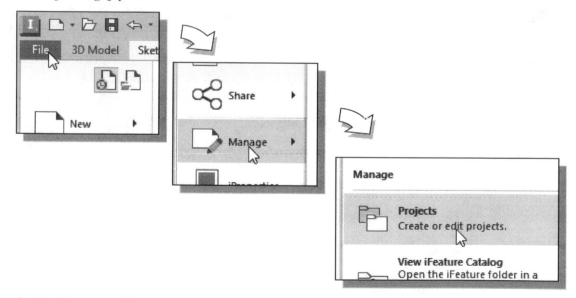

❖ The **Projects Editor** appears on the screen, which allows us to access the settings of new or existing Inventor Single User Project or Autodesk Vault Project.

❖ In the *Projects Editor*: the **Default** project is available.

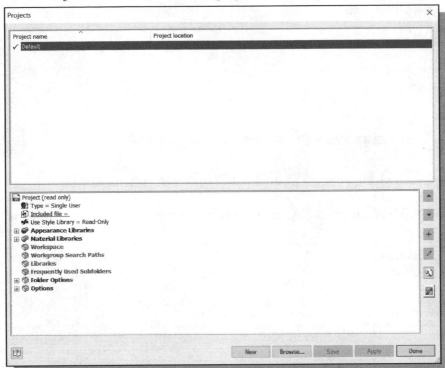

The **Default** project is automatically active by default, and the *default project* does not define any location for files. In other words, the data management system is not used. Using the *default project*, designs can still be created and modified, and any model file can be opened and saved anywhere without regard to project and file management.

Set up of a New Inventor Project

In this section, we will create a new *Inventor* project for the chapters of this book using the *Inventor* built-in **Single User Project** option. Note that it is also feasible to create a separate project for each chapter.

1. Click **New** to begin the setup of a new *project file*.

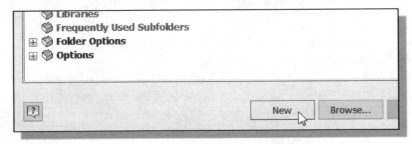

2. The *Inventor project wizard* appears on the screen; select the **New Single User Project** option as shown.

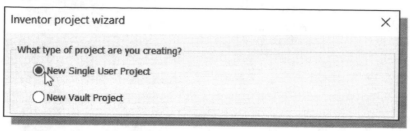

 3. Click **Next** to proceed with the next setup option.

4. In the *Project File Name* input box, enter **Parametric-Modeling** as the name of the new project.

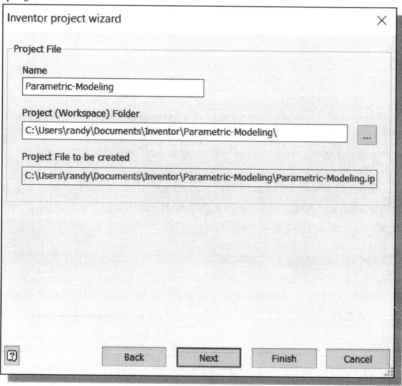

5. In the *Project Folder* input box, note the default folder location, such as **C:\Users\Documents\Inventor \Parametric Modeling\,** and choose a preferred folder name as the folder name of the new project.

 6. Click **Finish** to proceed with the creation of the new project.

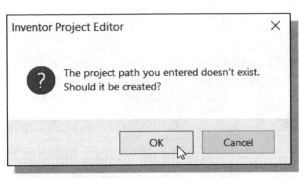

7. A *warning message* appears on the screen, indicating the specified folder does not exist. Click **OK** to create the folder.

8. A second *warning message* appears on the screen, indicating that the newly created project cannot be made active since an *Inventor* file is open. Click **OK** to close the message dialog box.

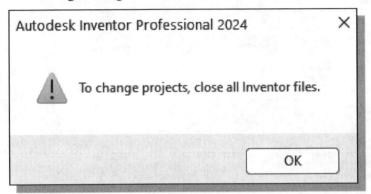

*   The new project has been created and its name appears in the project list area as shown in the figure.

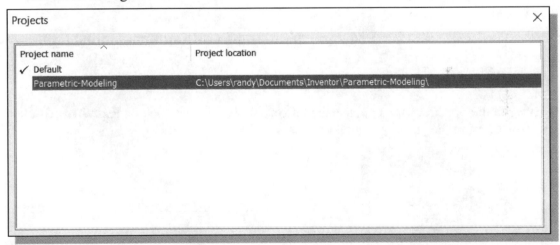

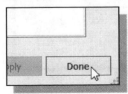

9. Click **Done** to exit the *Inventor Projects Editor* option.

The Content of the Inventor Project File

An *Inventor* project file is actually a text file in .xml format with an **.ipj** extension. The file specifies the paths to the folder containing the files in the project. To assure that links between files work properly, it is advised to add the locations for folders to the project file before working on model files.

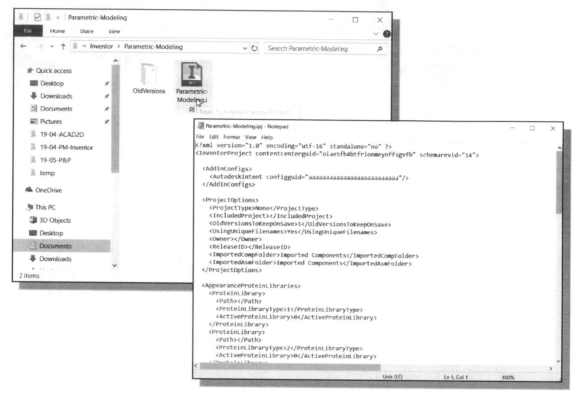

Leaving Autodesk Inventor

To leave the *Application* menu, use the left-mouse-button and click on **Exit Autodesk Inventor** from the pull-down menu.

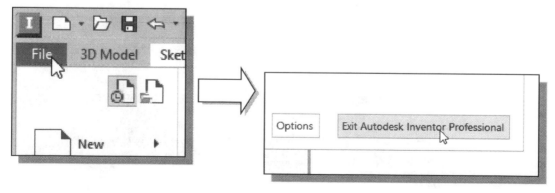

Chapter 2
Parametric Modeling Fundamentals

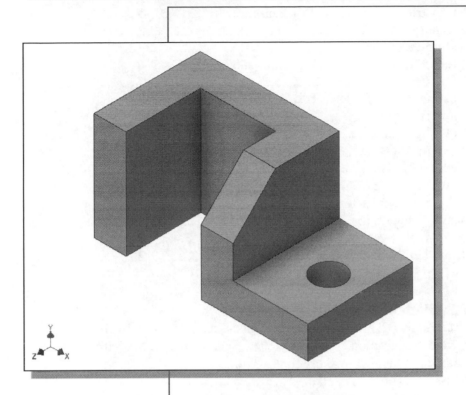

Learning Objectives

- ◆ **Create Simple Extruded Solid Models**
- ◆ **Understand the Basic Parametric Modeling Procedure**
- ◆ **Create 2D Sketches**
- ◆ **Understand the "Shape before Size" Design Approach**
- ◆ **Use the Dynamic Viewing Commands**
- ◆ **Create and Edit Parametric Dimensions**

Autodesk Inventor Certified User Exam Objectives Coverage

Section 3: Sketches

Objectives: Creating 2D Sketches, Draw Tools, Sketch Constraints, Pattern Sketches, Modify Sketches, Format Sketches, Sketch Doctor, Shared Sketches, Sketch Parameters

Section 4: Parts

Objectives: Creating parts, Work Features, Pattern Features, and Part Properties

Introduction

The **feature-based parametric modeling** technique enables the designer to incorporate the original **design intent** into the construction of the model. The word *parametric* means the geometric definitions of the design, such as dimensions, can be varied at any time in the design process. Parametric modeling is accomplished by identifying and creating the key features of the design with the aid of computer software. The design variables, described in the sketches as parametric relations, can then be used to quickly modify/update the design.

In Autodesk Inventor, the parametric part modeling process involves the following steps:

1. **Create a rough two-dimensional sketch of the basic shape of the base feature of the design.**

2. **Apply/modify constraints and dimensions to the two-dimensional sketch.**

3. **Extrude, revolve, or sweep the parametric two-dimensional sketch to create the base solid feature of the design.**

4. **Add additional parametric features by identifying feature relations and complete the design.**

5. **Perform analyses on the computer model and refine the design as needed.**

6. **Create the desired drawing views to document the design.**

The approach of creating two-dimensional sketches of the three-dimensional features is an effective way to construct solid models. Many designs are in fact the same shape in one direction. Computer input and output devices we use today are largely two-dimensional in nature, which makes this modeling technique quite practical. This method also conforms to the design process that helps the designer with conceptual design along with the capability to capture the *design intent*. Most engineers and designers can relate to the experience of making rough sketches on restaurant napkins to convey conceptual design ideas. Autodesk Inventor provides many powerful modeling and design-tools, and there are many different approaches to accomplishing modeling tasks. The basic principle of **feature-based modeling** is to build models by adding simple features one at a time. In this chapter, the general parametric part modeling procedure is illustrated; a very simple solid model with extruded features is used to introduce the Autodesk Inventor user interface. The display viewing functions and the basic two-dimensional sketching tools are also demonstrated.

The Adjuster Design

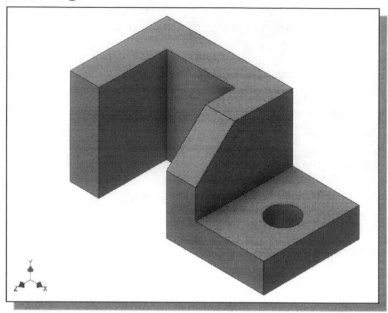

Starting Autodesk Inventor

1. Select the **Autodesk Inventor** option on the *Start* menu or select the **Autodesk Inventor** icon on the desktop to start Autodesk Inventor. The Autodesk Inventor main window will appear on the screen.

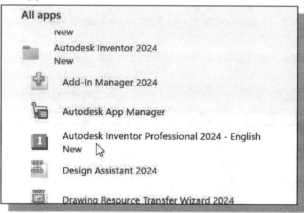

2. Click on the **three dots** next to the *Projects* icon, then select the **Settings** option as shown.

3. In the *Projects List*, **double-click** on the ***Parametric-Modeling*** project name to activate the project as shown.

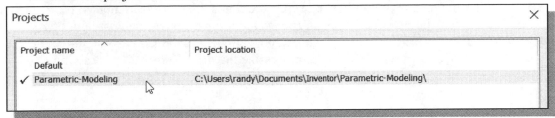

- Note that Autodesk Inventor will keep this activated project as the default project until another project is activated.

4. Click **Done** to accept the setting and end the *Projects Editor*.

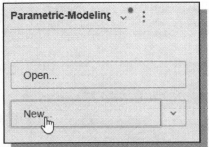

5. Select the **New File** icon with a single click of the left-mouse-button.

- Notice the ***Parametric-Modeling*** project name is displayed as the active project.

6. Select the **en-US → English** tab as shown below. When starting a new CAD file, the first thing we should do is choose the units we would like to use. We will use the English setting (inches) for this example.

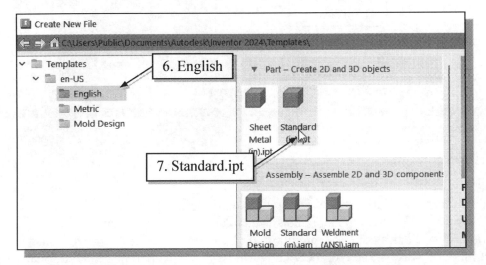

7. Select the **Standard(in).ipt** icon as shown.

8. Pick **Create** in the *New File* dialog box to accept the selected settings.

The Default Autodesk Inventor Screen Layout

The default Autodesk Inventor drawing screen contains the *pull-down* menus, the *Standard* toolbar, the *3D Model* toolbar, the *Sketch* toolbar, the *drawing* area, the *browser* area, and the *Status Bar*. A line of quick text appears next to the icon as you move the *mouse cursor* over different icons. You may resize the *Autodesk Inventor* drawing window by clicking and dragging the edges of the window, or relocate the window by clicking and dragging the window title area.

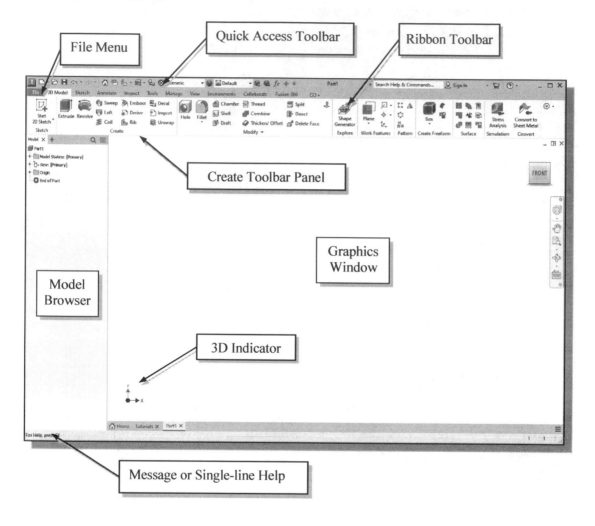

The **Ribbon Toolbar** is a relatively new feature in Autodesk Inventor; the *Ribbon Toolbar* is composed of a series of tool panels, which are organized into tabs labeled by task. The *Ribbon* provides a compact palette of all of the tools necessary to accomplish the different modeling tasks. The drop-down arrow next to any icon indicates additional commands are available on the expanded panel; access the expanded panel by clicking on the drop-down arrow.

Sketch Plane – It is an XY monitor, but an XYZ World

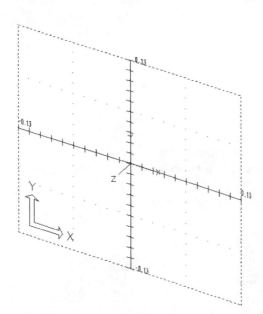

Design modeling software is becoming more powerful and user friendly, yet the system still does only what the user tells it to do. When using a geometric modeler, we therefore need to have a good understanding of what its inherent limitations are.

In most 3D geometric modelers, 3D objects are located and defined in what is usually called **world space** or **global space**. Although a number of different coordinate systems can be used to create and manipulate objects in a 3D modeling system, the objects are typically defined and stored using the world space. The world space is usually a **3D Cartesian coordinate system** that the user cannot change or manipulate.

In engineering designs, models can be very complex, and it would be tedious and confusing if only the world coordinate system were available. Practical 3D modeling systems allow the user to define **Local Coordinate Systems (LCS)** or **User Coordinate Systems (UCS)** relative to the world coordinate system. Once a local coordinate system is defined, we can then create geometry in terms of this more convenient system.

Although objects are created and stored in 3D space coordinates, most of the geometric entities can be referenced using 2D Cartesian coordinate systems. Typical input devices such as a mouse or digitizer are two-dimensional by nature; the movement of the input device is interpreted by the system in a planar sense. The same limitation is true of common output devices, such as displays and plotters. The modeling software performs a series of three-dimensional to two-dimensional transformations to correctly project 3D objects onto the 2D display plane.

The Autodesk Inventor *sketching plane* is a special construction approach that enables the planar nature of the 2D input devices to be directly mapped into the 3D coordinate system. The *sketching plane* is a local coordinate system that can be aligned to an existing face of a part, or a reference plane.

Think of the sketching plane as the surface on which we can sketch the 2D sections of the parts. It is similar to a piece of paper, a whiteboard, or a chalkboard that can be attached to any planar surface. The first sketch we create is usually drawn on one of the established datum planes. Subsequent sketches/features can then be created on sketching planes that are aligned to existing **planar faces of the solid part** or **datum planes.**

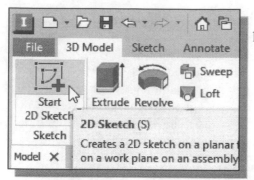

1. Activate the **Start 2D Sketch** icon with a single click of the left-mouse-button.

2. Move the cursor over the edge of the *XY Plane* in the graphics area. When the *XY Plane* is highlighted, click once with the **left-mouse-button** to select the *Plane* as the sketch plane for the new sketch.

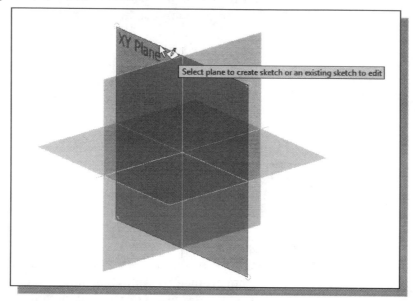

*   The *sketching plane* is a reference location where two-dimensional sketches are created. Note that the *sketching plane* can be any planar part surface or datum plane.

3. Confirm the main *Ribbon* area is switched to the **Sketch** toolbars; this indicates we have entered the 2D Sketching mode.

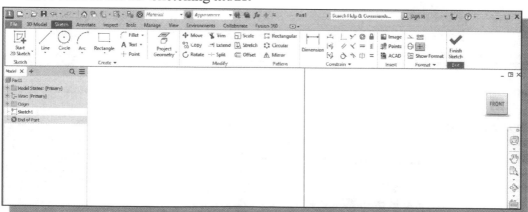

Create A Rough Sketch

Quite often during the early design stage, the shape of a design may not have any precise dimensions. Most conventional CAD systems require the user to input the precise lengths and locations of all geometric entities defining the design, which are not available during the early design stage. With *parametric modeling*, we can use the computer to elaborate and formulate the design idea further during the initial design stage. With Autodesk Inventor, we can use the computer as an electronic sketchpad to help us concentrate on the formulation of forms and shapes for the design. This approach is the main advantage of *parametric modeling* over conventional solid-modeling techniques.

As the name implies, a ***rough sketch*** is not precise at all. When sketching, we simply sketch the geometry so that it closely resembles the desired shape. Precise scale or lengths are not needed. Autodesk Inventor provides many tools to assist us in finalizing sketches. For example, geometric entities such as horizontal and vertical lines are set automatically. However, if the rough sketches are poor, it will require much more work to generate the desired parametric sketches. Here are some general guidelines for creating sketches in Autodesk Inventor:

- **Create a sketch that is proportional to the desired shape.** Concentrate on the shapes and forms of the design.

- **Keep the sketches simple.** Leave out small geometry features such as fillets, rounds and chamfers. They can easily be placed using the Fillet and Chamfer commands after the parametric sketches have been established.

- **Exaggerate the geometric features of the desired shape.** For example, if the desired angle is 85 degrees, create an angle that is 50 or 60 degrees. Otherwise, Autodesk Inventor might assume the intended angle to be a 90-degree angle.

- **Draw the geometry so that it does not overlap.** The geometry should eventually form a closed region. *Self-intersecting* geometry shapes are not allowed.

- **The sketched geometric entities should form a closed region.** To create a solid feature, such as an extruded solid, a closed region is required so that the extruded solid forms a 3D volume.

- ➢ **Note:** The concepts and principles involved in *parametric modeling* are very different, and sometimes they are totally opposite, to those of conventional computer aided drafting. In order to understand and fully utilize Autodesk Inventor's functionality, it will be helpful to take a *Zen* approach to learning the topics presented in this text: **Have an open mind and temporarily forget your experiences using conventional Computer Aided Drafting systems.**

Step 1: Creating a Rough Sketch

The *Sketch* toolbar provides tools for creating the basic geometry that can be used to create features and parts.

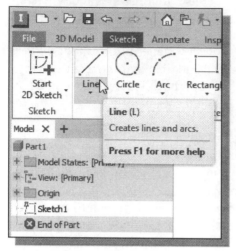

1. Move the graphics cursor to the **Line** icon in the *Sketch* toolbar. A *Help-tip box* appears next to the cursor and a brief description of the command is displayed at the bottom of the drawing screen: "*Creates Straight lines and arcs.*"

2. Select the icon by clicking once with the **left-mouse-button**; this will activate the Line command. Autodesk Inventor expects us to identify the starting location of a straight line.

Graphics Cursors

Notice the cursor changes from an arrow to a crosshair when graphical input is expected.

1. Left-click a starting point for the shape, roughly just below and to the right of the center of the graphics window.

2. As you move the graphics cursor, you will see a digital readout next to the cursor and also in the *Status Bar* area at the bottom of the window. The readout gives you the cursor location, the line length, and the angle of the line measured from horizontal. Move the cursor around and you will notice different symbols appear at different locations.

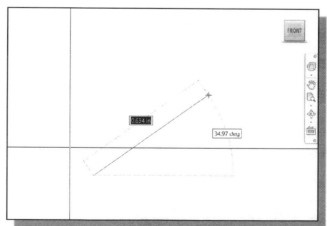

➢ The readout displayed next to the cursor is called the ***Dynamic Input***. This option is part of the **Heads-Up Display** option that is now available in Inventor. Note that *Dynamic Input* can be used for entering precise values, but its usage is somewhat limited in *parametric modeling*.

3. Move the graphics cursor toward the right side of the graphics window and create a horizontal line as shown below (**Point 2**). Notice the geometric constraint symbol, a short horizontal line indicating the geometric property, is displayed.

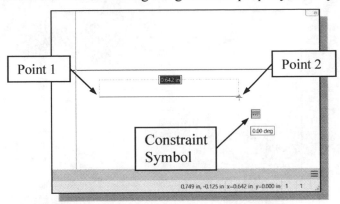

Geometric Constraint Symbols

Autodesk Inventor displays different visual clues, or symbols, to show you alignments, perpendicularities, tangencies, etc. These constraints are used to capture the *design intent* by creating constraints where they are recognized. Autodesk Inventor displays the governing geometric rules as models are built. To prevent constraints from forming, hold down the [**Ctrl**] key while creating an individual sketch curve. For example, while sketching line segments with the Line command, endpoints are joined with a Coincident *constraint*, but when the [**Ctrl**] key is pressed and held, the inferred constraint will not be created.

	Vertical	indicates a line is vertical
	Horizontal	indicates a line is horizontal
	Dashed line	indicates the alignment is to the center point or endpoint of an entity
	Parallel	indicates a line is parallel to other entities
	Perpendicular	indicates a line is perpendicular to other entities
	Coincident	indicates the cursor is at the endpoint of an entity
	Concentric	indicates the cursor is at the center of an entity
	Tangent	indicates the cursor is at tangency points to curves

1. Complete the sketch as shown below, creating a closed region ending at the starting point (**Point 1**). Do not be overly concerned with the actual size of the sketch. Note that all line segments are sketched horizontally or vertically.

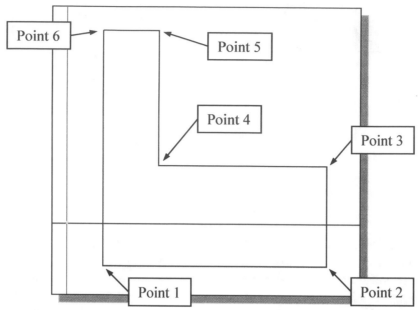

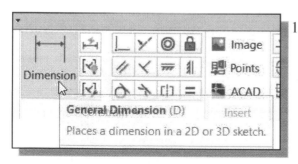

2. Inside the graphics window, click once with the **right-mouse-button** to display the option menu. Select Cancel [Esc] in the pop-up menu, or hit the [**Esc**] key once to end the Sketch Line command.

Step 2: Apply/Modify Constraints and Dimensions

As the sketch is made, Autodesk Inventor automatically applies some of the geometric constraints (such as horizontal, parallel, and perpendicular) to the sketched geometry. We can continue to modify the geometry, apply additional constraints, and/or define the size of the existing geometry. In this example, we will illustrate adding dimensions to describe the sketched entities.

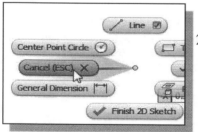

1. Move the cursor to the *Constrain* toolbar area; it is the toolbar next to the *2D Draw* toolbar. Note the first icon in this toolbar is the General Dimension icon. The Dimension command is generally known as **Smart Dimensioning** in parametric modeling.

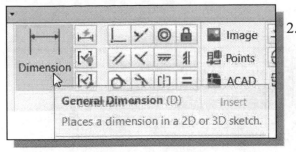

2. Move the cursor on top of the Dimension icon. The **Smart Dimensioning** command allows us to quickly create and modify dimensions. **Left-click** once on the icon to activate the Dimension command.

3. The message "*Select Geometry to Dimension*" is displayed in the *Status Bar* area at the bottom of the *Inventor* window. Select the bottom horizontal line by left-clicking once on the line.

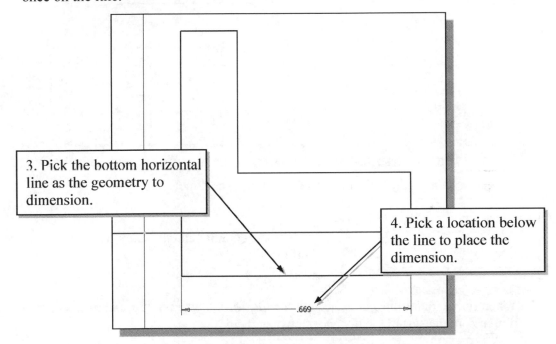

3. Pick the bottom horizontal line as the geometry to dimension.

4. Pick a location below the line to place the dimension.

.669

4. Move the graphics cursor below the selected line and **left-click** to place the dimension. (Note that the value displayed on your screen might be different than what is shown in the figure above.)

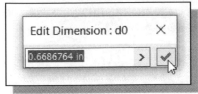

5. Accept the *default value* by clicking on the **Accept** button as shown.

❖ The **General Dimension** command will create a length dimension if a single line is selected.

6. The message "*Select Geometry to Dimension*" is displayed in the *Status Bar* area, located at the bottom of the *Inventor* window. Select the top-horizontal line as shown below.

7. Select the bottom-horizontal line as shown below.

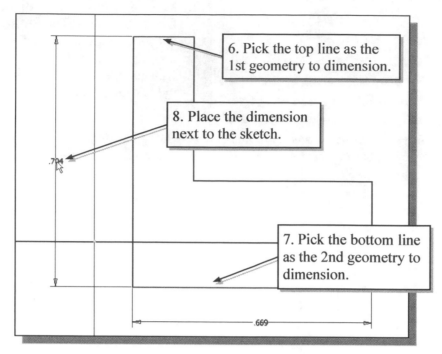

6. Pick the top line as the 1st geometry to dimension.

8. Place the dimension next to the sketch.

7. Pick the bottom line as the 2nd geometry to dimension.

8. Pick a location to the left of the sketch to place the dimension.

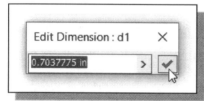

Edit Dimension : d1   ✕

0.7037775 in

9. Accept the default value by clicking on the **Accept** button.

❖ When two parallel lines are selected, the General Dimension command will create a dimension measuring the distance between them.

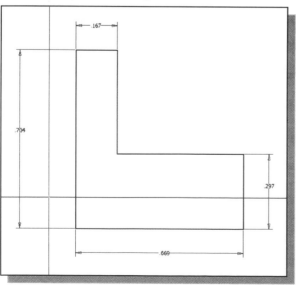

10. On your own, repeat the above steps and create additional dimensions (accepting the default values created by Inventor) so that the sketch appears as shown.

Dynamic Viewing Functions – Zoom and Pan

Autodesk Inventor provides a special user interface called *Dynamic Viewing* that enables convenient viewing of the entities in the graphics window.

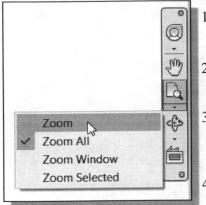

1. Click on the **Zoom** icon located in the *Navigation* bar as shown.

2. Move the cursor near the center of the graphics window.

3. Inside the graphics window, **press and hold down the left-mouse-button**, then move downward to enlarge the current display scale factor.

4. Press the **[Esc]** key once to exit the Zoom command.

5. Click on the **Pan** icon located above the Zoom command in the *Navigation* bar. The icon is the picture of a hand.

➤ The Pan command enables us to move the view to a different position. This function acts as if you are using a video camera.

6. On your own, use the Zoom and Pan options to reposition the sketch near the center of the screen.

Modifying the Dimensions of the Sketch

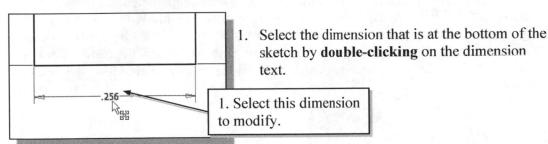

1. Select the dimension that is at the bottom of the sketch by **double-clicking** on the dimension text.

1. Select this dimension to modify.

2. In the *Edit Dimension* window, the current length of the line is displayed. Enter **2.5** to set the length of the line.

3. Click on the **Accept** icon to accept the entered value.

➤ Autodesk Inventor will now update the profile with the new dimension value.

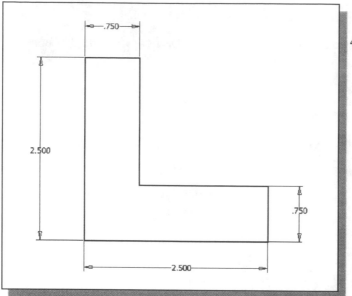

4. On your own, repeat the above steps and adjust the dimensions so that the sketch appears as shown.

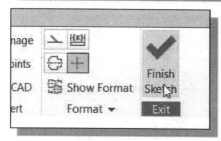

5. In the *Ribbon* toolbar, click once with the **left-mouse-button** to select **Finish Sketch** in the *Ribbon* area to end the Sketch option.

Step 3: Completing the Base Solid Feature

Now that the 2D sketch is completed, we will proceed to the next step: create a 3D part from the 2D profile. Extruding a 2D profile is one of the common methods that can be used to create 3D parts. We can extrude planar faces along a path. We can also specify a height value and a tapered angle. In Autodesk Inventor, each face has a positive side and a negative side; the current face we are working on is set as the default positive side. This positive side identifies the positive extrusion direction and it is referred to as the face's *normal*.

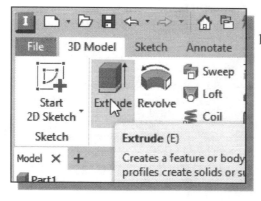

1. In the 3D Model tab, select the **Extrude** command by clicking the left-mouse-button on the icon as shown.

2. In the *Extrude* edit box, enter **2.5** as the extrusion distance. Notice that the sketch region is automatically selected as the extrusion profile.

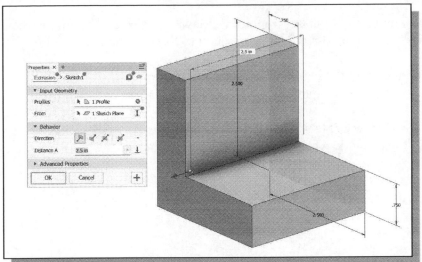

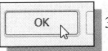 3. Click on the **OK** button to proceed with creating the 3D part.

➢ Note that all dimensions disappeared from the screen. All parametric definitions are stored in the **Autodesk Inventor database** and any of the parametric definitions can be re-displayed and edited at any time.

Isometric View

Autodesk Inventor provides many ways to display views of the three-dimensional design. Several options are available that allow us to quickly view the design to track the overall effect of any changes being made to the model. We will first orient the model to display in the *isometric view* by using the pull-down menu.

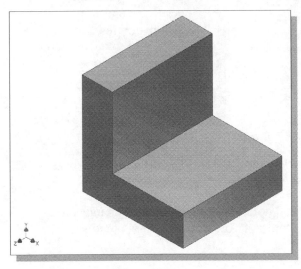

1. Hit the function key **[F6]** once to automatically adjust the display and also reset the display to the *isometric* view.

❖ Note that most of the view-related commands can be accessed in the **ViewCube** and/or the *Navigation* bar located to the right side of the graphics window.

❖ **Dynamic Rotation of the 3D Block – Free Orbit**

The Free Orbit command allows us to:
- Orbit a part or assembly in the graphics window. Rotation can be around the center mark, free in all directions, or around the X/Y-axes in the *3D-Orbit* display.
- Reposition the part or assembly in the graphics window.
- Display isometric or standard orthographic views of a part or assembly.

The Free Orbit tool is accessible while other tools are active. Autodesk Inventor remembers the last used mode when you exit the Orbit command.

1. Click on the **Orbit** icon in the *Navigation* bar.

➢ The *3D Orbit* display is a circular rim with four handles and a center mark. *3D Orbit* enables us to manipulate the view of 3D objects by clicking and dragging with the left-mouse-button:

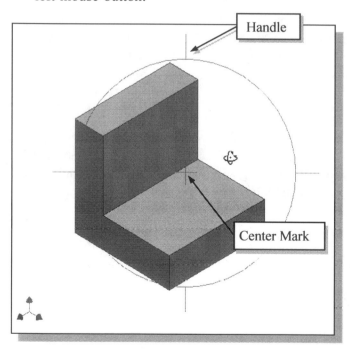

Handle

Center Mark

- Drag with the left-mouse-button near the center for free rotation.

- Drag on the handles to orbit around the horizontal or vertical axes.

- Drag on the rim to orbit about an axis that is perpendicular to the displayed view.

- Single left-click to align the center mark of the view.

2. Inside the *circular rim*, press down the left-mouse-button and drag in an arbitrary direction; the 3D Orbit command allows us to freely orbit the solid model.

3. Move the cursor near the circular rim and notice the cursor symbol changes to a single circle. Drag with the left-mouse-button to orbit about an axis that is perpendicular to the displayed view.

4. Single left-click near the top-handle to align the selected location to the center mark in the graphics window.

5. Activate the **Constrained Orbit** option by clicking on the associated icon as shown.

❖ *The Constrained Orbit can be used to rotate the model about axes in Model Space, equivalent to moving the eye position about the model in latitude and longitude.*

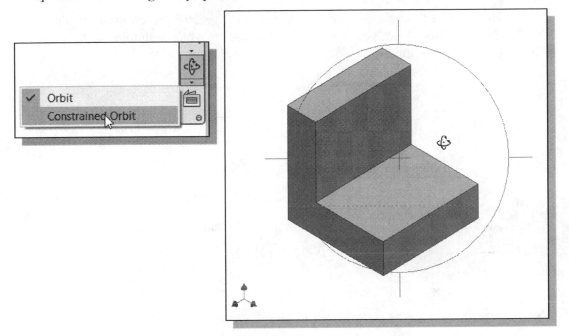

6. On your own, use the different options described in the above steps and familiarize yourself with both of the 3D Orbit commands. Reset the display to the *Isometric* view as shown in the above figure before continuing to the next section.

❖ Note that while in the 3D Orbit mode, a horizontal marker will be displayed next to the cursor if the cursor is away from the circular rim. This is the **exit marker**. Left-clicking once will allow you to exit the 3D Orbit command.

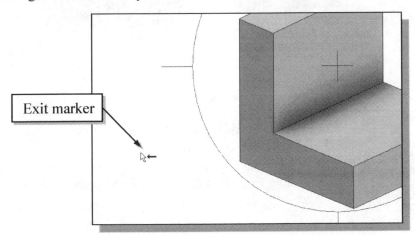

Dynamic Viewing – Quick Keys

We can also use the function keys on the keyboard and the mouse to access the *Dynamic Viewing* functions.

❖ Panning – (1) F2 and drag with the left-mouse-button

Hold the **F2** function key down and drag with the left-mouse-button to pan the display. This allows you to reposition the display while maintaining the same scale factor of the display.

Pan

(2) Press and drag with the mouse wheel

Pressing and dragging with the mouse wheel can also reposition the display.

❖ Zooming – (1) F3 and drag with the left-mouse-button

Hold the **F3** function key down and drag with the left-mouse-button vertically on the screen to adjust the scale of the display. Moving upward will reduce the scale of the display, making the entities display smaller on the screen. Moving downward will magnify the scale of the display.

Zoom

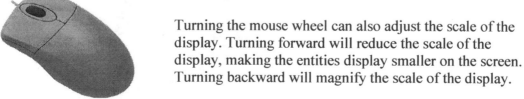

(2) Turning the mouse wheel

Turning the mouse wheel can also adjust the scale of the display. Turning forward will reduce the scale of the display, making the entities display smaller on the screen. Turning backward will magnify the scale of the display.

❖ **3D Dynamic Rotation – Shift and drag with the middle-mouse-button**

Hold the **Shift** key down and drag with the middle-mouse-button to orbit the display. Note that the 3D dynamic rotation can also be activated using the **F4** function key and the left-mouse-button.

Dynamic Rotation   Shift   +   ⇦ ◻MOUSE◻ ⇨

Viewing Tools – Standard Toolbar

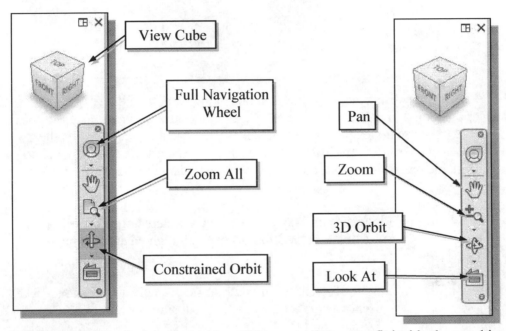

Zoom All – Adjusts the view so that all items on the screen fit inside the graphics window.

Zoom Window – Use the cursor to define a region for the view; the defined region is zoomed to fill the graphics window.

Zoom – Moving upward will reduce the scale of the display, making the entities display smaller on the screen. Moving downward will magnify the scale of the display.

Pan – This allows you to reposition the display while maintaining the same scale factor of the display.

Zoom Selected – In a part or assembly, zooms the selected edge, feature, line, or other element to fill the graphics window. You can select the element either before or after clicking the Zoom button. (Not used in drawings.)

Orbit – In a part or assembly, adds an orbit symbol and cursor to the view. You can orbit the view planar to the screen around the center mark, around a horizontal or vertical axis, or around the X and Y axes. (Not used in drawings.)

Look At – In a part or assembly, zooms and orbits the model to display the selected element planar to the screen or a selected edge or line horizontal to the screen. (Not used in drawings.)

View Cube – The ViewCube is a 3D navigation tool that appears, by default, when you enter Inventor. The ViewCube is a clickable interface which allows you to switch between standard and isometric views.

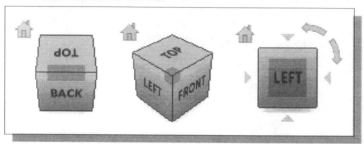

Once the ViewCube is displayed, it is shown in one of the corners of the graphics window over the model in an inactive state. The ViewCube also provides visual feedback about the current viewpoint of the model as view changes occur. When the cursor is positioned over the ViewCube, it becomes active and allows you to switch to one of the available preset views, roll the current view, or change to the Home view of the model.

1. Move the cursor over the ViewCube and notice the different sides of the ViewCube become highlighted and can be activated.

2. Single left-click when the front side is activated as shown. The current view is set to view the front side.

3. Move the cursor over the counterclockwise arrow of the ViewCube and notice the orbit option becomes highlighted.

4. Single left-click to activate the counterclockwise option as shown. The current view is orbited 90 degrees; we are still viewing the front side.

5. Move the cursor over the left arrow of the ViewCube and notice the orbit option becomes highlighted.

6. Single left-click to activate the left arrow option as shown. The current view is now set to view the top side.

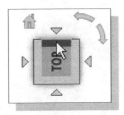

7. Move the cursor over the top edge of the ViewCube and notice the roll option becomes highlighted.

8. Single left-click to activate the roll option as shown. The view will be adjusted to roll 45 degrees.

9. Move the cursor over the ViewCube and drag with the left-mouse-button to activate the **Free Rotation** option.

10. Move the cursor over the home icon of the ViewCube and notice the Home View option becomes highlighted.

11. Single left-click to activate the **Home View** option as shown. The view will be adjusted back to the default *isometric view*.

Full Navigation Wheel – The Navigation Wheel contains tracking menus that are divided into different sections known as wedges. Each wedge on a wheel represents a single navigation tool. You can pan, zoom, or manipulate the current view of a model in different ways. The 3D Navigation Wheel and 2D Navigation Wheel (mostly used in the 2D drawing mode) have some or all of the following options:

Zoom – Adjusts the magnification of the view.
Center – Centers the view based on the position of the cursor over the wheel.
Rewind – Restores the previous view.
Forward – Increases the magnification of the view.
Orbit – Allows 3D free rotation with the left-mouse-button.
Pan – Allows panning by dragging with the left-mouse-button.
Up/Down – Allows panning with the use of a scroll control.
Walk – Allows *walking*, with linear motion perpendicular to the screen, through the model space.
Look – Allows rotation of the current view vertically and horizontally.

3D Full Navigation Wheel

2D Full Navigation Wheel

1. Activate the **Full Navigation Wheel** by clicking on the icon as shown.

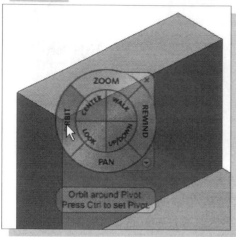

2. Move the cursor in the graphics window and notice the **Full Navigation Wheel** menu follows the cursor on the screen.

3. Move the cursor on the **Orbit** option to highlight the option.

4. Click and drag with the left-mouse-button to activate the **Free Rotation** option.

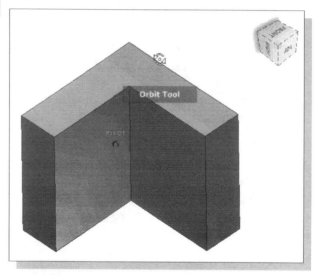

5. Drag with the left-mouse-button and notice the **ViewCube** also reflects the model orientation.

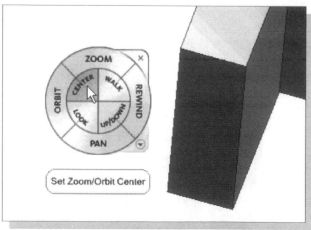

6. Move the cursor to the left side of the model and click the **Center** option as shown. The display is adjusted so the selected point is the new **Zoom/Orbit** center.

7. On your own, experiment with the other available options.

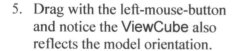

Display Modes

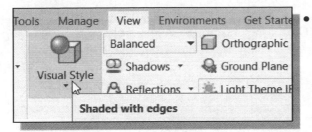

- The **Visual Style** in the *View* tab has eleven display-modes ranging from very realistic renderings of the model to very artistic representations of the model. The more commonly used modes are as follows:

❖ **Realistic Shaded Solid:**

The *Realistic Shaded Solid* display mode generates a high-quality shaded image of the 3D object.

❖ **Standard Shaded Solid:**

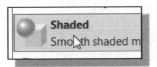

The *Standard Shaded Solid* display option generates a shaded image of the 3D object that requires fewer computer resources compared to the realistic rendering.

❖ **Wireframe Image:**

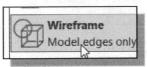

The *Wireframe Image* display option allows the display of the 3D objects using the basic wireframe representation scheme.

❖ **Wireframe with Hidden-Edges:**

The *Wireframe with Hidden-Edges* option can be used to generate an image of the 3D object with all the back lines shown as hidden-lines.

Orthographic vs. Perspective

Besides the above basic display modes, we can also choose orthographic view or perspective view of the display. Click on the triangle icon next to the *display mode button* on the *View* toolbar.

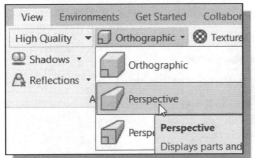

❖ **Orthographic**
The first icon allows the display of the 3D object using the parallel edges representation scheme.

❖ **Perspective**
The second icon allows the display of the 3D object using the perspective, nonparallel edges, and representation scheme.

Disable the Heads-Up Display Option

The **Heads-Up Display** option in Inventor provides mainly the **Dynamic Input** function, which can be quite useful for 2D drafting activities. For example, in the use of a 2D drafting CAD system, most of the dimensions of the design would have been determined by the documentation stage. However, in *parametric modeling*, the usage of the *Dynamic Input* option is quite limited, as this approach does not conform to the "**shape before size**" design philosophy.

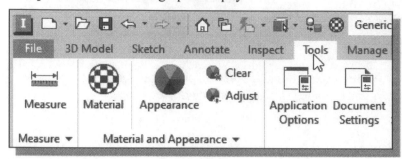

1. Select the **Tools** tab in the *Ribbon* as shown.

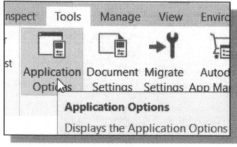

2. Select **Application Options** in the options toolbar as shown.

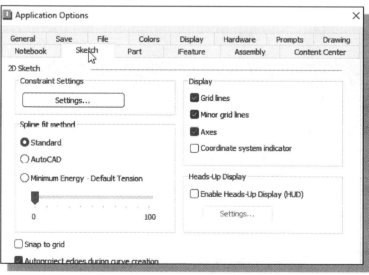

3. Select the **Sketch** tab to display the sketch related settings.

4. In the *Heads-Up Display* section, turn **OFF** the *Enable Heads-Up Display*, *Snap to Grid* options and switch **On** the **Grid lines**, **Minor grid lines** and **Axes** options in the *Display* section as shown.

5. On your own, examine the other sketch settings that are available.

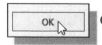

6. Click **OK** to accept the settings.

Step 4-1: Adding an Extruded Feature

1. Activate the *3D Model* tab and select the **Start 2D Sketch** command by left-clicking once on the icon.

2. In the *Status Bar* area, the message "*Select plane to create sketch or an existing sketch to edit*" is displayed. Autodesk Inventor expects us to identify a planar surface where the 2D sketch of the next feature is to be created. Notice that Autodesk Inventor will automatically highlight feasible planes and surfaces as the cursor is on top of the different surfaces. Pick the top horizontal face of the 3D solid object.

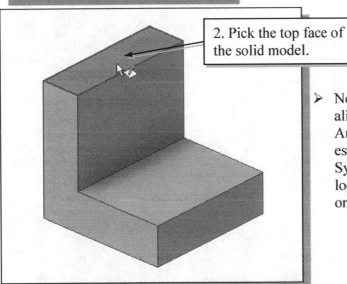

2. Pick the top face of the solid model.

➢ Note that the sketch plane is aligned to the selected face. Autodesk Inventor automatically establishes a User-Coordinate-System (UCS) and records its location with respect to the part on which it was created.

• Next, we will create and profile another sketch, a rectangle, which will be used to create another extrusion feature that will be added to the existing solid object.

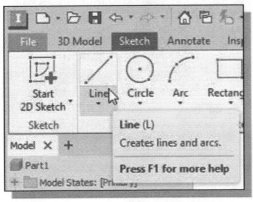

3. Select the **Line** command by clicking once with the **left-mouse-button** on the icon in the *Sketch* tab on the Ribbon.

4. Create a sketch with segments perpendicular/parallel to the existing edges of the solid model as shown below.

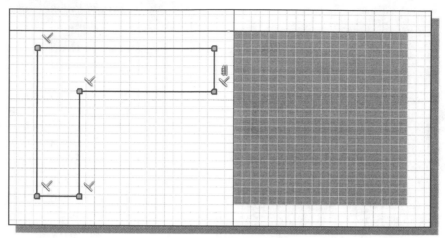

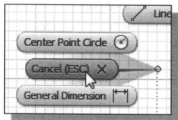

5. Inside the graphics window, click once with the **right-mouse-button** to display the option menu. Select **Cancel (ESC)** in the pop-up menu to end the Line command.

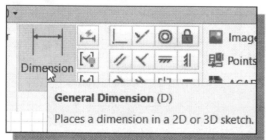

6. Select the **General Dimension** command in the *Sketch* toolbar. The General Dimension command allows us to quickly create and modify dimensions. Left-click once on the icon to activate the General Dimension command.

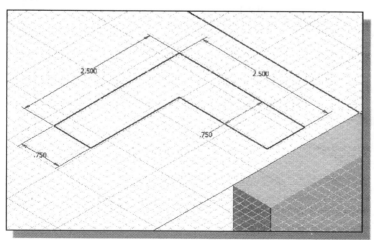

7. Hit the function key **[F6]** once to switch the display to the isometric display as shown.

8. Create the **four dimensions** to describe the size of the sketch as shown in the figure.

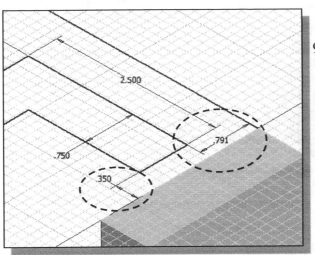

9. Create the two location dimensions to describe the position of the sketch relative to the top corner of the solid model as shown.

10. On your own, modify the two location dimensions to **0.0** and adjust the size dimensions as shown in the figure below.

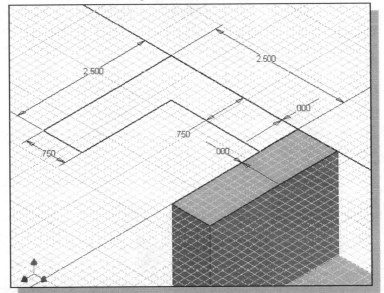

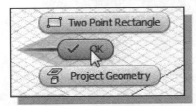

11. Inside the graphics window, click once with the **right-mouse-button** to display the option menu. Select **OK** in the pop-up menu to end the General Dimension command.

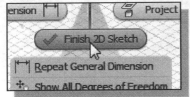

12. Inside the graphics window, click once with the **right-mouse-button** to display the option menu. Select **Finish 2D Sketch** in the pop-up menu to end the Sketch option.

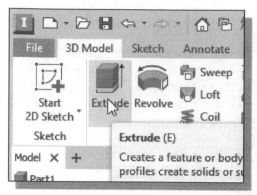

13. In the *3D Model* tab select the **Extrude** command by left-clicking on the icon.

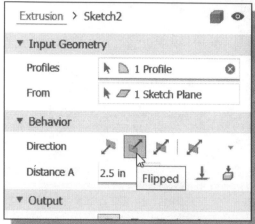

14. In the *Extrude* dialog box, enter **2.5** as the extrude distance as shown.

15. Click on the **Flipped icon** to reverse the *Extrusion Direction* as shown.

16. Confirm the *Boolean* option is set to **Join** and click on the **OK** button to proceed with creating the extruded feature.

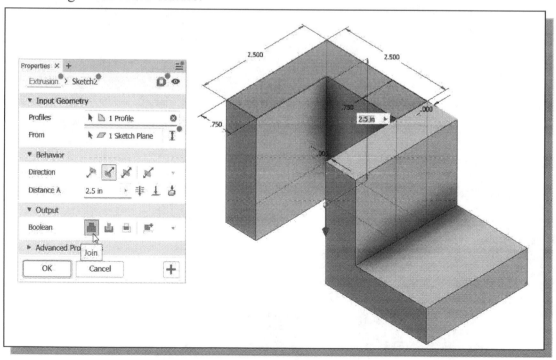

Step 4-2: Adding a Cut Feature

Next, we will create and profile a circle, which will be used to create a cut feature that will be added to the existing solid object.

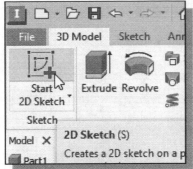

1. In the *3D Model* tab select the **Start 2D Sketch** command by left-clicking once on the icon.

2. In the *Status Bar* area, the message "*Select plane to create sketch or an existing sketch to edit.*" is displayed. Autodesk Inventor expects us to identify a planar surface where the 2D sketch of the next feature is to be created. Pick the top horizontal face of the 3D solid model as shown.

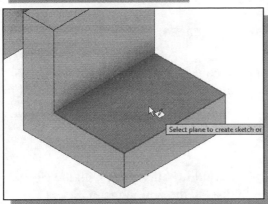

➢ Note that the sketch plane is aligned to the selected face. Autodesk Inventor automatically establishes a User-Coordinate-System (UCS) and records its location with respect to the part on which it was created.

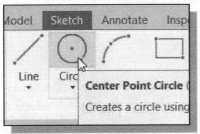

3. Select the **Center point circle** command by clicking once with the **left-mouse-button** on the icon in the *Sketch* tab on the Ribbon.

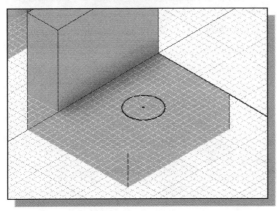

4. Create a circle of arbitrary size on the top face of the solid model as shown.

5. On your own, create and modify the dimensions of the sketch as shown in the figure.

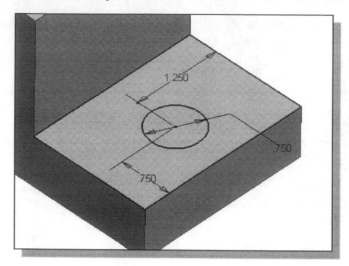

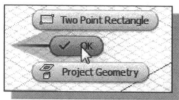

6. Inside the graphics window, click once with the **right-mouse-button** to display the option menu. Select **OK** in the pop-up menu to end the General Dimension command.

7. Inside the graphics window, click once with the **right-mouse-button** to display the option menu. Select **Finish 2D Sketch** in the pop-up menu to end the Sketch option.

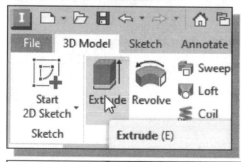

8. In the *3D Model* tab select the **Extrude** command by clicking the left-mouse-button on the icon.

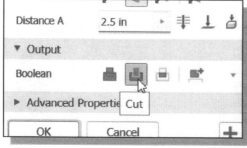

9. Select the **CUT** option in the *Output - Boolean option* list to set the extrusion operation to *Cut*.

10. Set the *Extents* option to **Through All** as shown. The *All* option instructs the software to calculate the extrusion distance and assures the created feature will always cut through the full length of the model.

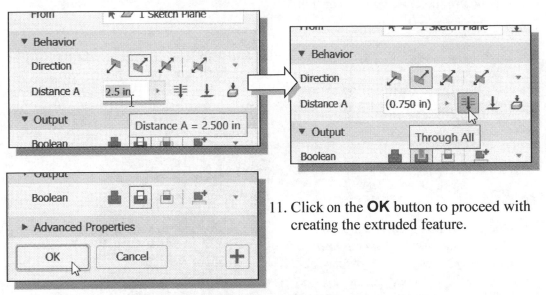

11. Click on the **OK** button to proceed with creating the extruded feature.

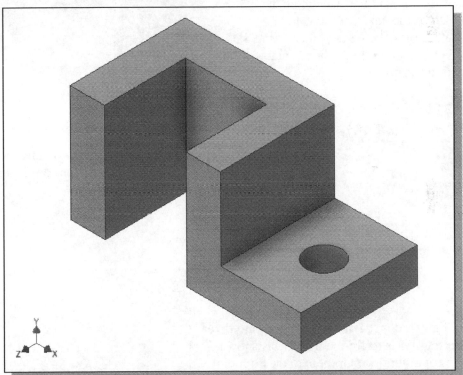

- In *Autodesk Inventor*, the **Extrude** command can be used to create solid features by either adding or subtracting extruded features.

Step 4-3: Adding another Cut Feature

Next, we will create and profile a triangle, which will be used to create a cut feature that will be added to the existing solid object.

1. In the *3D Model* tab select the **Start 2D Sketch** command by left-clicking once on the icon.

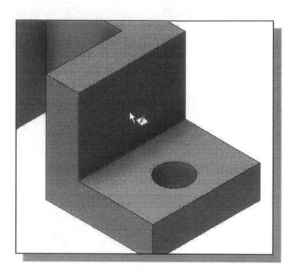

2. In the *Status Bar* area, the message "*Select plane to create sketch or an existing sketch to edit.*" is displayed. Autodesk Inventor expects us to identify a planar surface where the 2D sketch of the next feature is to be created. Pick the vertical face of the 3D solid model next to the horizontal section as shown.

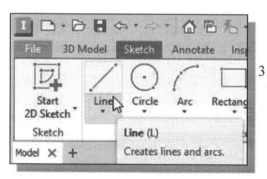

3. Select the **Line** command by clicking once with the **left-mouse-button** on the icon in the *Sketch* ribbon.

4. Start at the upper left corner and create **three line segments** to form a triangle as shown. (Do not align the endpoints to the midpoint of the existing edges.)

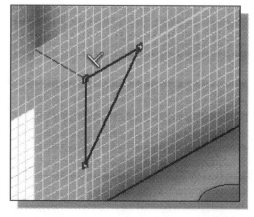

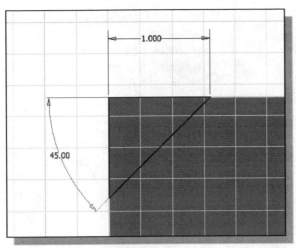

5. On your own, create and modify the two dimensions of the sketch as shown in the figure. (Hint: create the angle dimension by selecting the two adjacent lines and place the angular dimension inside the desired quadrant.)

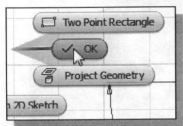

6. Inside the graphics window, click once with the **right-mouse-button** to display the option menu. Select **OK** in the pop-up menu to end the General Dimension command.

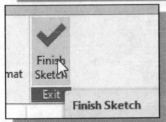

7. Select **Finish 2D Sketch** in the *Ribbon toolbar* to end the Sketch option.

8. In the *3D Model* tab, select the **Extrude** command by left-clicking the icon.

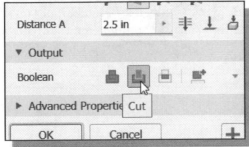

9. Select the **CUT** option in the *Boolean option* list to set the extrusion operation to *Cut*.

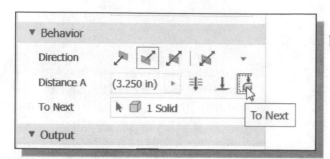

10. Set the *Extents* option to **To Next** as shown. The *To Next* option instructs the software to calculate the extrusion distance and assures the created feature will always cut through the proper length of the model.

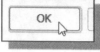

11. Click on the **OK** button to proceed with creating the extruded feature.

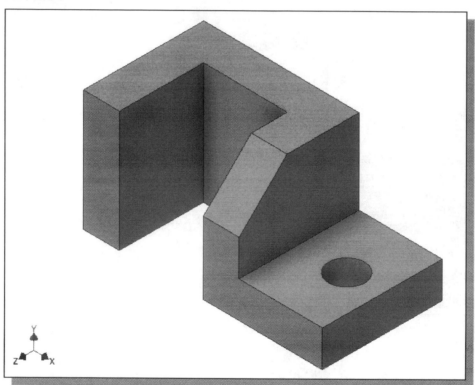

Save the Model

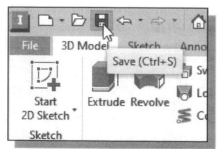

1. Select **Save** in the *Quick Access* toolbar, or you can also use the "**Ctrl-S**" combination (hold down the "Ctrl" key and hit the "S" key once) to save the part.

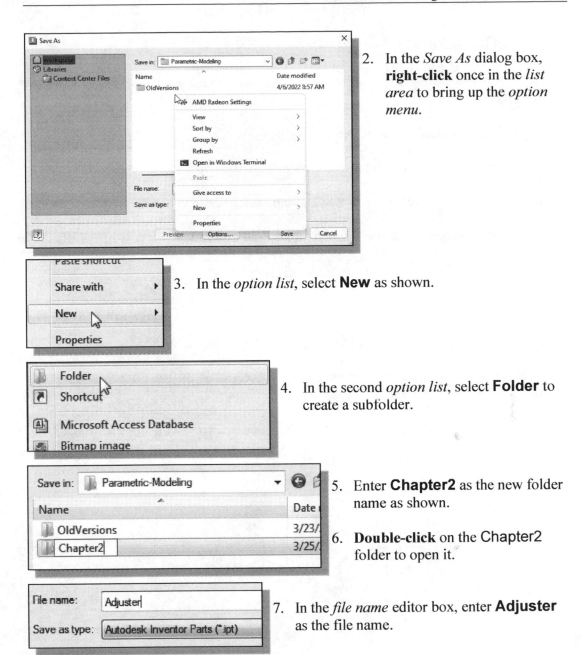

2. In the *Save As* dialog box, **right-click** once in the *list area* to bring up the *option menu*.

3. In the *option list*, select **New** as shown.

4. In the second *option list*, select **Folder** to create a subfolder.

5. Enter **Chapter2** as the new folder name as shown.

6. **Double-click** on the Chapter2 folder to open it.

7. In the *file name* editor box, enter **Adjuster** as the file name.

8. Click on the **Save** button to save the file.

❖ You should form a habit of saving your work periodically, just in case something goes wrong while you are working on it. In general, you should save your work at an interval of every 15 to 20 minutes. You should also save before making any major modifications to the model.

Review Questions: (Time: 20 minutes)

1. What is the first thing we should set up in Autodesk Inventor when creating a new model?

2. Describe the general *parametric modeling* procedure.

3. Describe the general guidelines in creating *Rough Sketches*.

4. What is the main difference between a rough sketch and a *profile*?

5. List two of the geometric constraint symbols used by Autodesk Inventor.

6. What was the first feature we created in this lesson?

7. How many solid features were created in the tutorial?

8. How do we control the size of a feature in parametric modeling?

9. Which command was used to create the last cut feature in the tutorial? How many dimensions do we need to fully describe the cut feature?

10. List and describe three differences between parametric modeling and traditional 2D Computer Aided Drafting techniques.

Exercises:

(Time: 150 minutes. All dimensions are in inches.)

1. **Inclined Support** (Thickness: **.5**)

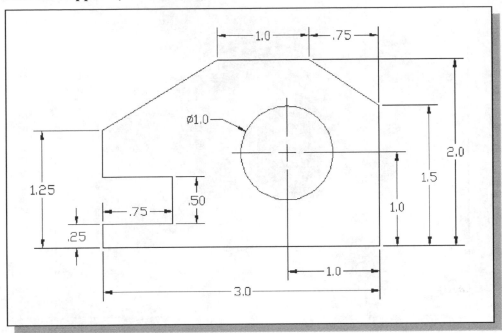

2. **Spacer Plate** (Thickness: **.125**)

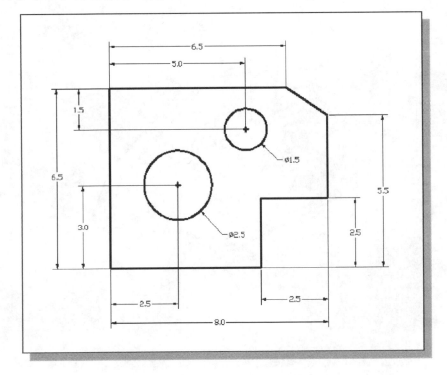

3. **Positioning Stop**

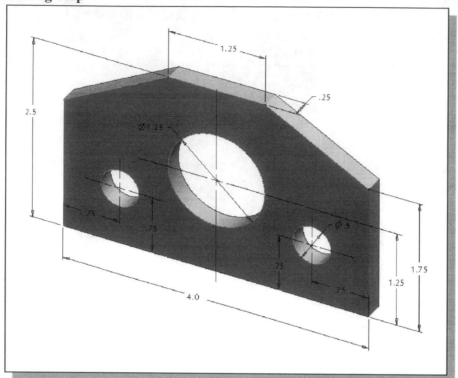

4. **Guide Block**

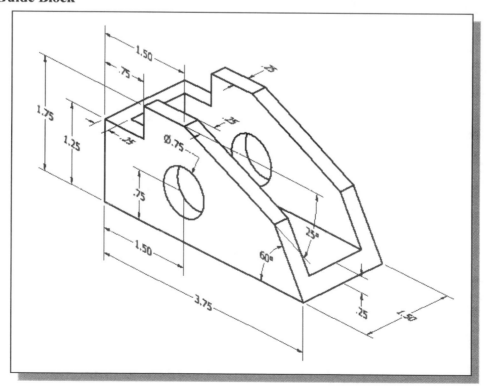

5. **Slider Block**

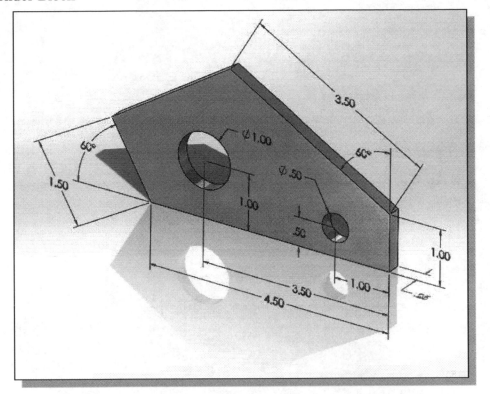

6. **Angle Lock**

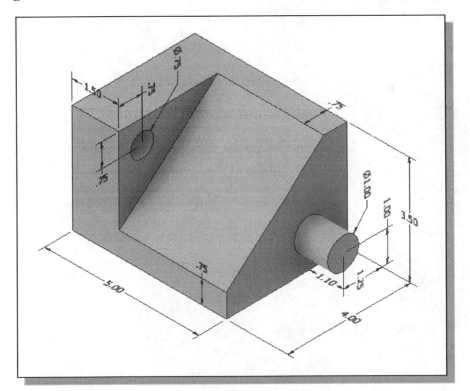

Notes:

Chapter 3
Constructive Solid Geometry Concepts

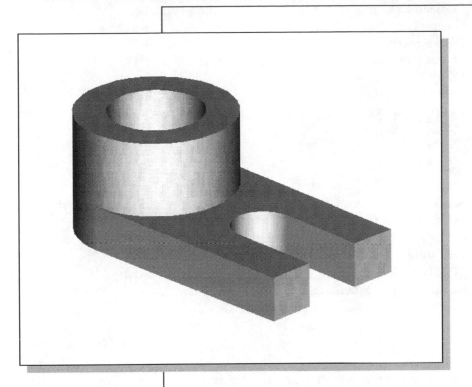

Learning Objectives

- ♦ **Understand Constructive Solid Geometry Concepts**
- ♦ **Create a Binary Tree**
- ♦ **Understand the Basic Boolean Operations**
- ♦ **Set up Grid and Snap Intervals**
- ♦ **Understand the Importance of Order of Features**
- ♦ **Create Placed Features**
- ♦ **Use the Different Extrusion Options**

Autodesk Inventor Certified User Exam Objectives Coverage

Parametric Modeling Basics

Section 3: Sketches

Objectives: Creating 2D Sketches, Draw Tools, Sketch Constraints, Pattern Sketches, Modify Sketches, Format Sketches, Sketch Doctor, Shared Sketches, Sketch Parameters

Section 4: Parts

Objectives: Creating parts, Work Features, Pattern Features, Part Properties

Introduction

In the 1980s, one of the main advancements in **solid modeling** was the development of the **Constructive Solid Geometry** (CSG) method. CSG describes the solid model as combinations of basic three-dimensional shapes (**primitive solids**). The basic primitive solid set typically includes Rectangular-prism (Block), Cylinder, Cone, Sphere, and Torus (Tube). Two solid objects can be combined into one object in various ways using operations known as **Boolean operations**. There are three basic Boolean operations: **JOIN (Union)**, **CUT (Difference)**, and **INTERSECT**. The **JOIN** operation combines the two volumes included in the different solids into a single solid. The **CUT** operation subtracts the volume of one solid object from the other solid object. The **INTERSECT** operation keeps only the volume common to both solid objects. The CSG method is also known as the **Machinist's Approach**, as the method is parallel to machine shop practices.

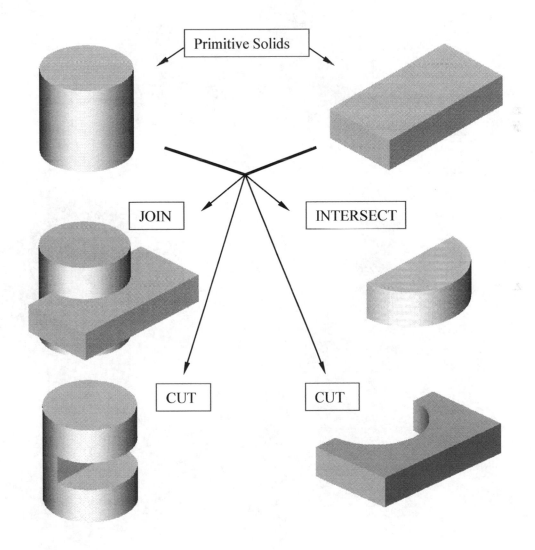

Binary Tree

The CSG is also referred to as the method used to store a solid model in the database. The resulting solid can be easily represented by what is called a **binary tree**. In a binary tree, the terminal branches (leaves) are the various primitives that are linked together to make the final solid object (the root). The binary tree is an effective way to keep track of the *history* of the resulting solid. By keeping track of the history, the solid model can be re-built by re-linking through the binary tree. This provides a convenient way to modify the model. We can make modifications at the appropriate links in the binary tree and re-link the rest of the history tree without building a new model.

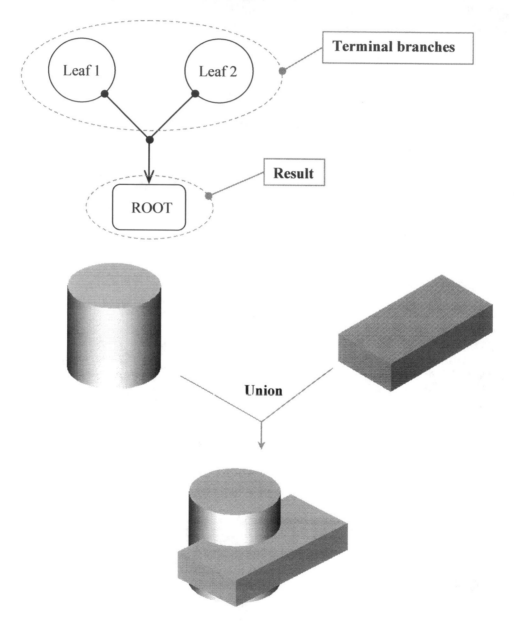

The Locator Design

The CSG concept is one of the important building blocks for feature-based modeling. In Autodesk Inventor, the CSG concept can be used as a planning tool to determine the number of features that are needed to construct the model. It is also a good practice to create features that are parallel to the manufacturing process required for the design. With parametric modeling, we are no longer limited to using only the predefined basic solid shapes. In fact, any solid features we create in Autodesk Inventor are used as primitive solids; parametric modeling allows us to maintain full control of the design variables that are used to describe the features. In this lesson, a more in-depth look at the parametric modeling procedure is presented. The equivalent CSG operation for each feature is also illustrated.

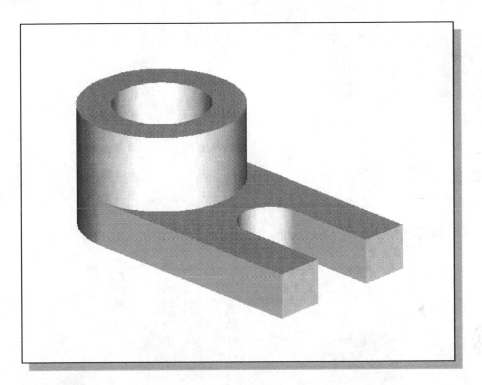

> ➤ Before going through the tutorial, on your own make a sketch of a CSG binary tree of the *Locator* design using only two basic types of primitive solids: cylinder and rectangular prism. In your sketch, how many *Boolean operations* will be required to create the model? What is your choice of the first primitive solid to use, and why? Take a few minutes to consider these questions and do the preliminary planning by sketching on a piece of paper. Compare the sketch you make to the CSG binary tree steps shown on the next page. Note that there are many different possibilities in combining the basic primitive solids to form the solid model. Even for the simplest design, it is possible to take several different approaches to creating the same solid model.

Modeling Strategy – CSG Binary Tree

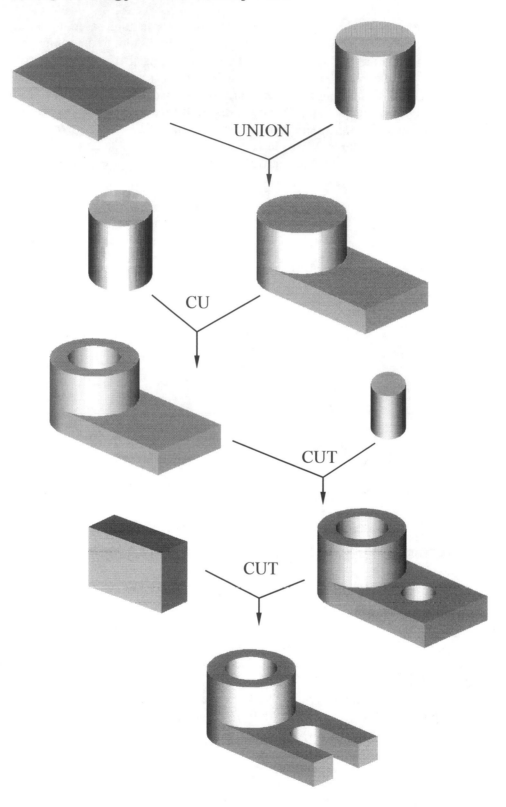

Starting Autodesk Inventor

1. Select the **Autodesk Inventor** option on the *Start* menu or select the **Autodesk Inventor** icon on the desktop to start Autodesk Inventor. The Autodesk Inventor main window will appear on the screen.

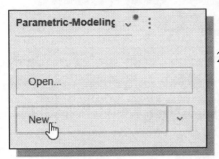

2. Select the **New File** icon with a single click of the left-mouse-button as shown.

❖ Every object we construct in a CAD system is measured in units. We should determine the value of the units within the CAD system before creating the first geometric entities. For example, in one model, a unit might equal one millimeter of the real-world object; in another model, a unit might equal an inch. In Autodesk Inventor, the *Choose Template* option is used to control how Autodesk Inventor interprets the coordinate and angle entries.

3. Select the **Metric** tab as shown below. We will use the millimeter (mm) setting for this example.

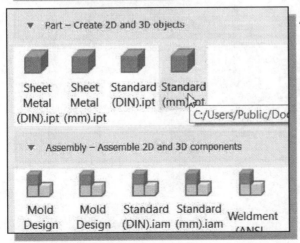

4. In the *New File – Part Template* area, select the **Standard(mm).ipt** icon as shown.

5. Confirm the *Parametric-Modeling* project is activated; note the **Projects** button is available to view/modify the active project.

6. Pick **Create** in the *Startup* dialog box to accept the selected settings.

Base Feature

In *parametric modeling*, the first solid feature is called the **base feature**, which usually is the primary shape of the model. Depending upon the design intent, additional features are added to the base feature.

Some of the considerations involved in selecting the base feature are:

- **Design intent** – Determine the functionality of the design; identify the feature that is central to the design.

- **Order of features** – Choose the feature that is the logical base in terms of the order of features in the design.

- **Ease of making modifications** – Select a base feature that is more stable and is less likely to be changed.

1. Activate the **Start 2D Sketch** icon with a single click of the left-mouse-button.

2. Move the cursor over the edge of the *XZ Plane* in the graphics area. When the *XZ Plane* is highlighted, click once with the **left-mouse-button** to select the *Plane* as the sketch plane for the new sketch.

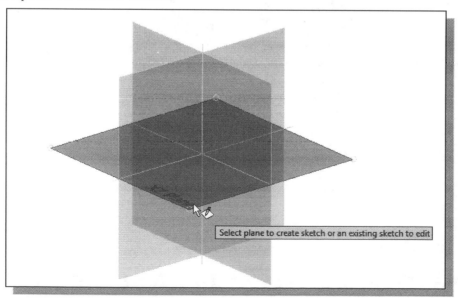

GRID Display Setup

1. In the *Ribbon* toolbar panel, select
   [Tools] → [Document Settings].

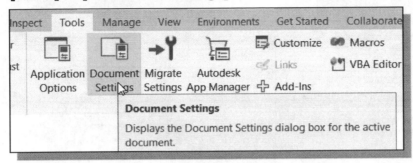

2. In the *Document Settings* dialog box, click on the **Sketch** tab as shown in the below figure.

3. Set the *X* and *Y Snap Spacing* to **5 mm**.

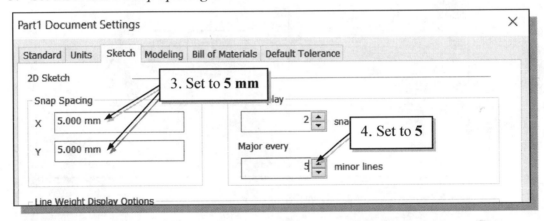

4. Change *Grid Display* to display one *major line* every **5** *minor lines*.

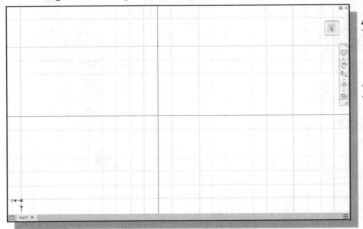

5. Pick **OK** to exit the *Sketch Settings* dialog box.

➢ Note that although the **Snap to grid** option is also available, its usage in parametric modeling is not recommended.

➢ On your own, use the dynamic **Zoom** function to view the grid setup. Refer to Page 2-26 on how to switch on the *grid lines display* options if necessary.

➢ A rectangular block will be first created as the base feature of the *Locator* design.

6. Click on the **rotate-left arrow** on the view cube to rotate the display.

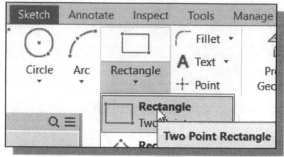

7. Switch back to the *Sketch* tab and select the **Two point rectangle** command by clicking once with the **left-mouse-button**.

8. Create a rectangle of arbitrary size by selecting two locations on the screen as shown below.

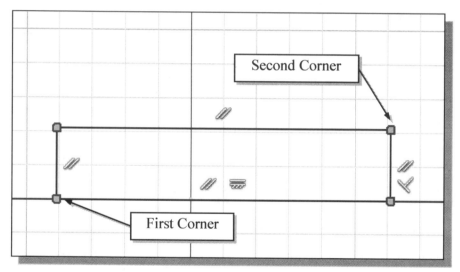

Second Corner

First Corner

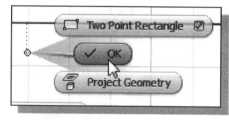

9. Inside the graphics window, click once with the **right-mouse-button** to bring up the option menu.

10. Select **OK** to end the Rectangle command.

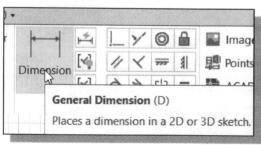

11. Activate the **General Dimension** command by clicking once with the left-mouse-button. The General Dimension command allows us to quickly create and modify dimensions.

12. Inside the graphics window, click once with the right-mouse-button to bring up the option menu and click **Edit Dimension** to turn **OFF** the editing option while creating dimensions.

13. The message "*Select Geometry to Dimension*" is displayed in the *Status Bar* area at the bottom of the Autodesk Inventor window. Select the bottom horizontal line by left-clicking once on the line.

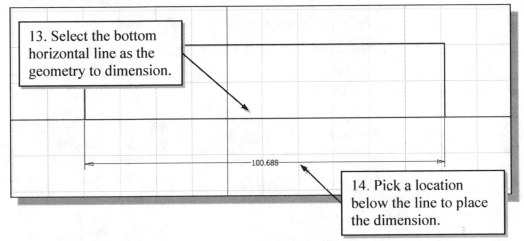

13. Select the bottom horizontal line as the geometry to dimension.

100.688

14. Pick a location below the line to place the dimension.

14. Move the graphics cursor below the selected line and left-click to place the dimension. (Note that the value displayed on your screen might be different than what is shown in the above figure.)

15. On your own, create a vertical size dimension of the sketched rectangle as shown.

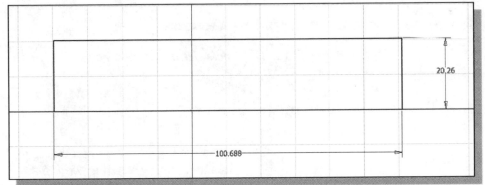

20.26

100.688

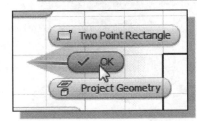

16. Inside the graphics window, click once with the right-mouse-button to bring up the option menu and click **OK** to end the *Dimension* command.

Model Dimensions Format

1. In the *Ribbon* tabs, select
    [Tools] → [Document Settings].

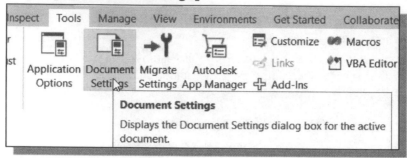

2. In the *Document Settings* dialog box, set the *Modeling Dimension Display* to
    Display as value as shown in the figure.

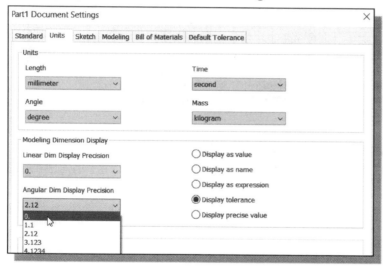

3. Also set the precision
    to **no digits** after the
    decimal point for both
    the *linear dimension*
    and *angular dimension*
    displays as shown in
    the above figure.

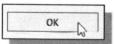

4. Pick **OK** to exit the *Document Settings* dialog box.

Modifying the Dimensions of the Sketch

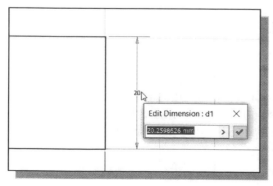

1. Select the height dimension that is to the
    right side of the sketch by ***double-
    clicking*** with the left-mouse-button on the
    dimension text.

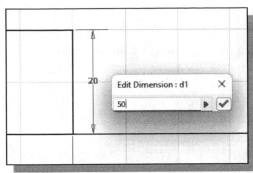

2. In the *Edit Dimension* window, the current length of the line is displayed. Enter **50** to set the selected length of the sketch to 50 millimeters.

3. Click on the **Accept** icon to accept the entered value.

➢ Autodesk Inventor will now update the profile with the new dimension value.

4. On your own, repeat the above steps and adjust the dimensions so that the sketch appears as shown below. Also **exit** the Dimension command.

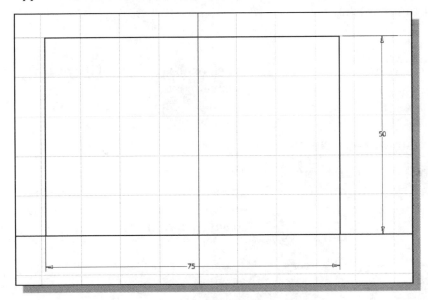

Repositioning Dimensions

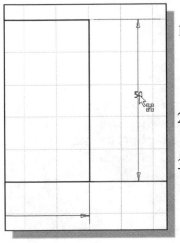

1. Move the cursor near the vertical dimension; note that the dimension is highlighted. Move the cursor slowly until a small arrows marker appears next to the cursor, as shown in the figure.

2. Drag with the **left-mouse-button** to reposition the selected dimension.

3. Repeat the above steps to reposition the horizontal dimension.

Using the Measure Tools

Autodesk Inventor also provides several measuring tools that allow us to measure area, perimeter and additional information of the constructed 2D sketches.

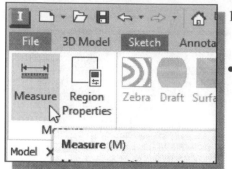

1. In the *Inspect Ribbon tab*, left-click once on the **Measure** option as shown.

    • Note that **other** measurement options are also available in the toolbar.

2. Click on the top edge of the rectangle as shown.

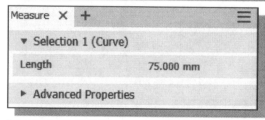

3. The associated length measurement of the selected geometry is displayed in the *Length* dialog box as shown.

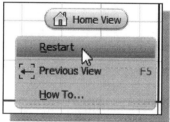

4. Inside the *graphics window*, right-click once to bring up the **Option menu** and select **Restart** as shown.

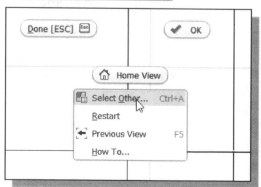

5. Move the cursor on top of any of the edges and click once with the right-mouse-button to bring up the **Option menu** and choose **Select Other...** as shown.

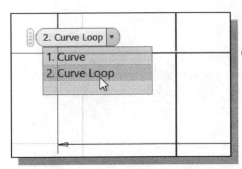

6. In the *Selection list*, left-click once to pick the **Curve Loop** option as shown.

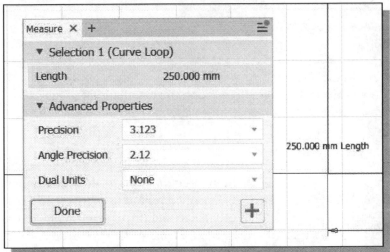

- The *perimeter* of the rectangle is displayed as shown.

250.000 mm Length

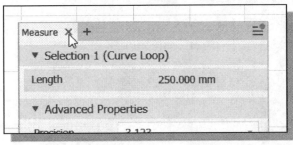

7. Click on the [**X**] icon to end the Measure command as shown.

8. In the *Inspect Ribbon tab*, left-click once on the **Region Properties** option as shown.

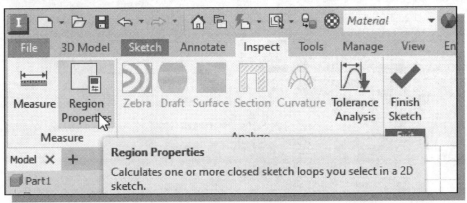

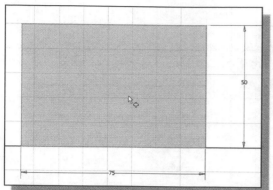

9. Click on the inside of the rectangle; notice the region is highlighted as the cursor is moved inside the rectangle, as shown.

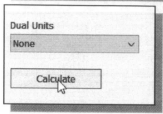

10. In the *Region Properties* dialog box, click on the **Calculate** button to perform the calculations of the associated geometry information.

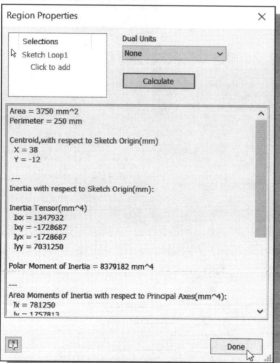

❖ In the *Region Properties* dialog box, the detailed region properties are calculated and displayed, including the *Area Moments of Inertia*, *Area* and *Perimeter*.

11. Click **Done** to exit the **Region Properties** command.

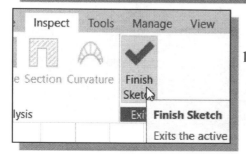

12. Select **Finish Sketch** in the *Ribbon* to end the Sketch option.

Completing the Base Solid Feature

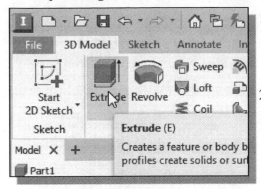

1. In the *3D Model tab* select the **Extrude** command by clicking the left-mouse-button on the icon.

2. In the *Extrude* pop-up window, enter **15** as the extrusion distance. Notice that the sketch region is automatically selected as the extrusion profile.

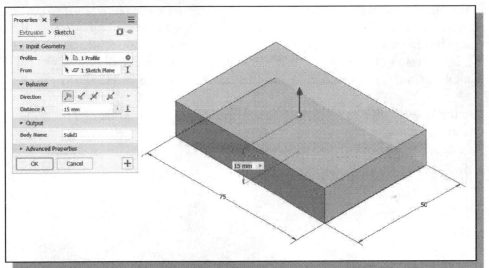

3. Click on the **OK** button to proceed with creating the 3D part. Use the *Dynamic Viewing* options to view the created part. Press **F6** to change the display to the isometric view as shown before going to the next section.

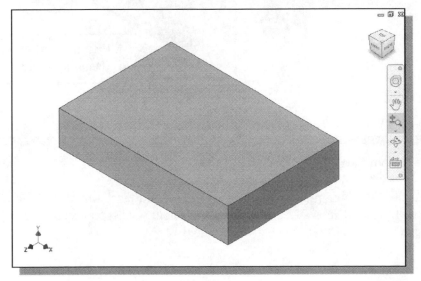

Creating the Next Solid Feature

1. In the *3D Model tab* select the **Start 2D Sketch** command by left-clicking once on the icon.

2. In the *Status Bar* area, the message "*Select plane to create sketch or an existing sketch to edit*" is displayed. Autodesk Inventor expects us to identify a planar surface where the 2D sketch of the next feature is to be created. Move the graphics cursor on the 3D part and notice that Autodesk Inventor will automatically highlight feasible planes and surfaces as the cursor is on top of the different surfaces.

3. Use the **ViewCube** to adjust the display viewing the bottom face of the solid model as shown below.

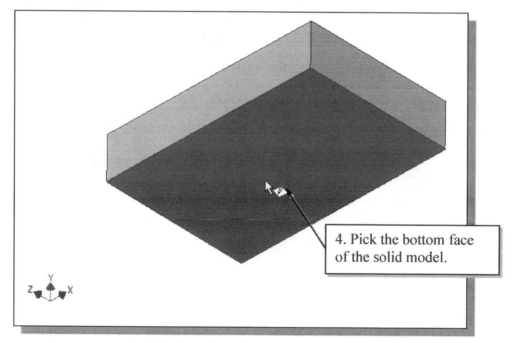

4. Pick the bottom face of the solid model.

4. Pick the bottom face of the 3D model as the sketching plane.

➤ Note that the sketching plane is aligned to the selected face. Autodesk Inventor automatically establishes a User-Coordinate-System (UCS) and records its location with respect to the part on which it was created.

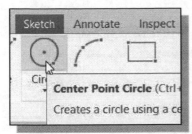

5. Select the **Center Point Circle** command by clicking once with the left-mouse-button on the icon in the *Sketch* tab.

➢ We will align the center of the circle to the midpoint of the base feature.

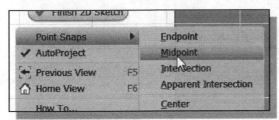

6. Inside the graphics window, click once with the **right-mouse-button** to bring up the option menu and choose the snap to **Midpoint** option.

7. Select the bottom edge of the base feature to align the center point of the new circle.

8. Select the green dot to align the midpoint of the line.

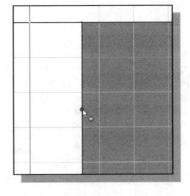

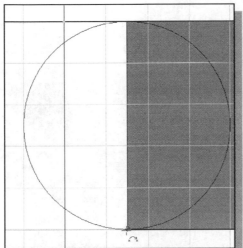

9. Select the bottom corner of the base feature to create a circle as shown in the figure.

10. Inside the *graphics window*, click once with the right-mouse-button to display the option menu. Select **OK** to end the Circle command.

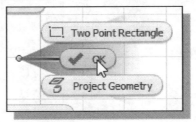

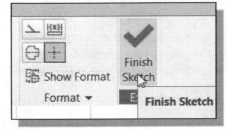

11. In the *Ribbon* toolbar, select **Finish Sketch** to exit the Sketch mode.

12. Press the function key **F6** once or select **Home View** in the **ViewCube** to change the display to the isometric view as shown.

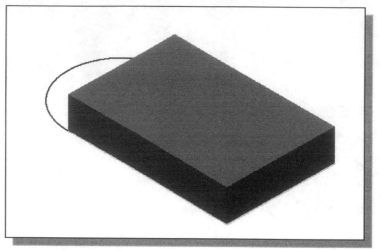

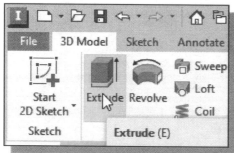

13. In the *3D Model tab*, select the **Extrude** command by left-clicking the icon.

14. Autodesk Inventor next expects us to select the region to be used to create the feature. First select inside the semi-circle region under the solid feature as shown.

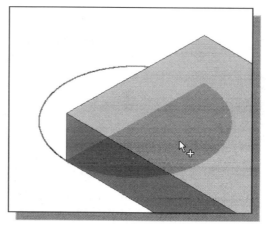

15. Select inside the other semi-circle region outside the solid feature as shown.

- Note that Autodesk Inventor creates the extruded feature downward as shown.

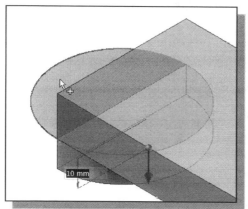
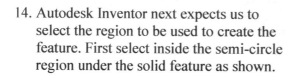

16. In the *Extrude* pop-up control, enter **40** as the blind extrusion distance as shown below. Set the solid operation to **Join** and click on the **Flip direction** button to reverse the direction of extrusion (upward) as shown below.

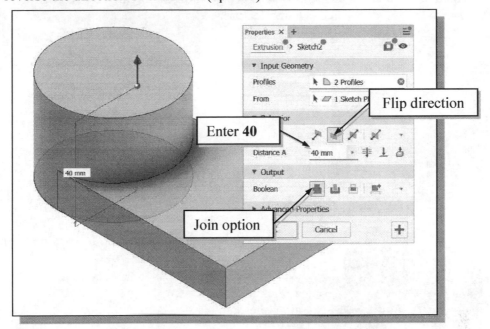

➢ Note that most of the settings can also be set through the icons displayed on the screen.

17. Click on the **OK** button to proceed with the *Join* operation.

• The two features are joined together into one solid part; the *CSG-Union* operation was performed.

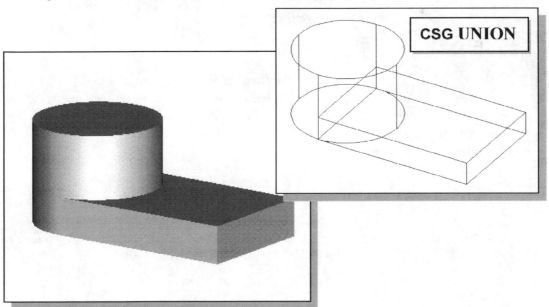

Creating a Cut Feature

We will create a circular cut as the next solid feature of the design. We will align the sketch plane to the top of the last cylinder feature.

1. In the *3D Model tab* select the **Start 2D Sketch** command by left-clicking once on the icon.

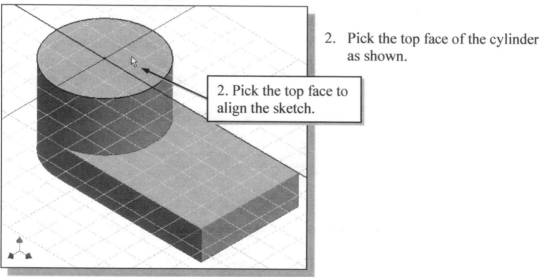

2. Pick the top face of the cylinder as shown.

2. Pick the top face to align the sketch.

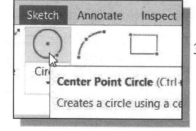

3. Select the **Center Point Circle** command by clicking once with the **left-mouse-button** on the icon in the *Sketch tab* of the Ribbon toolbar.

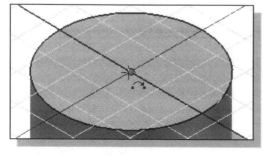

4. Select the **Center** point of the top face of the 3D model as shown.

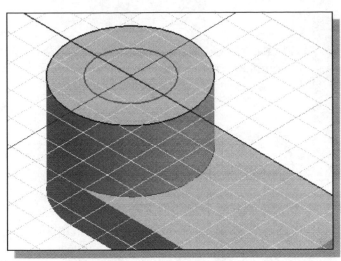

5. Sketch a circle of arbitrary size inside the top face of the cylinder as shown to the left.

6. Use the right-mouse-button to display the option menu and select **OK** in the pop-up menu to end the Circle command.

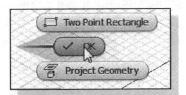

7. Inside the graphics window, click once with the **right-mouse-button** to display the option menu. Select the **General Dimension** option in the pop-up menu.

8. Create a dimension to describe the size of the circle and set it to **30mm**.

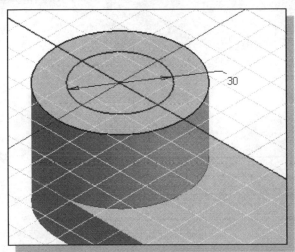

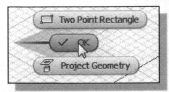

9. Inside the graphics window, click once with the right-mouse-button to display the option menu. Select **OK** in the pop-up menu to end the Dimension command.

10. Inside the graphics window, click once with the right-mouse-button to display the option menu. Select **Finish 2D Sketch** in the pop-up menu to end the Sketch option.

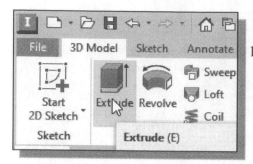

11. In the *3D Model tab*, select the **Extrude** command by left-clicking on the icon.

12. Click on the inside of the sketched circle as the profile to be extruded.

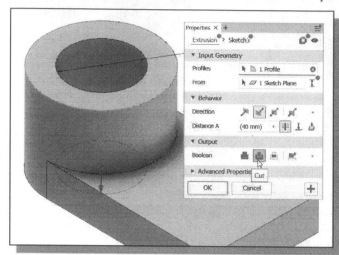

13. In the *Extrusion* pop-up window, set the operation option to **Cut**. Select **Through All** as the *Extents* option, as shown below. Confirm the arrowhead points downward.

14. Click on the **OK** button to proceed with the *Cut* operation.

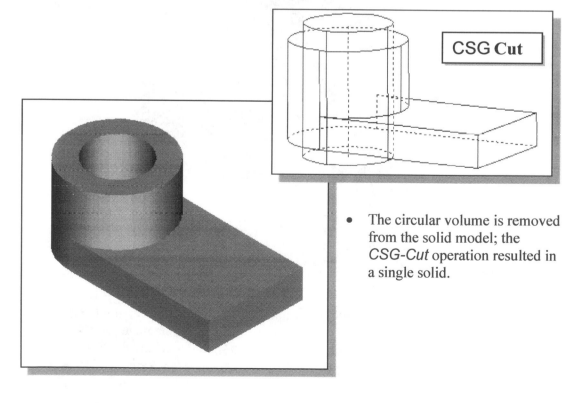

CSG Cut

• The circular volume is removed from the solid model; the *CSG-Cut* operation resulted in a single solid.

Creating a Placed Feature

In Autodesk Inventor, there are two types of geometric features: **placed features** and **sketched features**. The last cut feature we created is a *sketched feature*, where we created a rough sketch and performed an extrusion operation. We can also create a hole feature, which is a placed feature. A *placed feature* is a feature that does not need a sketch and can be created automatically. Holes, fillets, chamfers, and shells are all placed features.

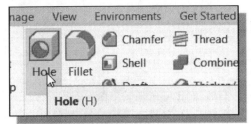

1. In the *Create* toolbar, select the **Hole** command by left-clicking on the icon.

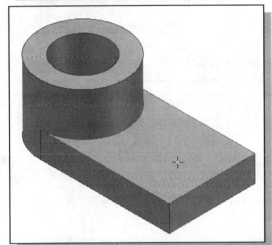

2. Pick a location inside the top horizontal surface of the base feature as shown.

3. Enter **20 mm** as the diameter of the hole as shown. **Do Not** click the OK button yet.

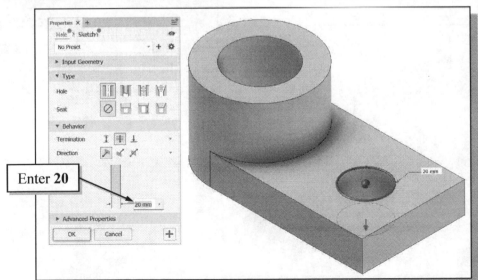

Enter **20**

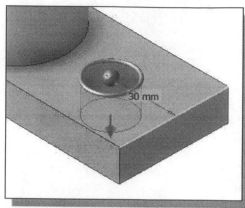

4. Pick the **right-edge** of the top face of the base feature as shown. This will be used as the first reference for placing the hole on the plane.

5. Enter **30** mm as the distance as shown. **Do Not** click the OK button yet.

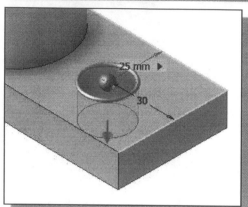

6. Pick the **adjacent edge** of the top face as shown. This will be used as the second reference for placing the hole on the plane.

7. Enter **25** mm as the distance as shown.

8. In *Holes* dialog box, set the *Termination* option to **Through All**.

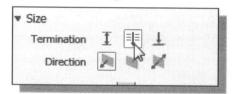

9. Click on the **OK** button to proceed with the *Hole* feature.

• The circular volume is removed from the solid model; the *CSG-Cut* operation resulted in a single solid.

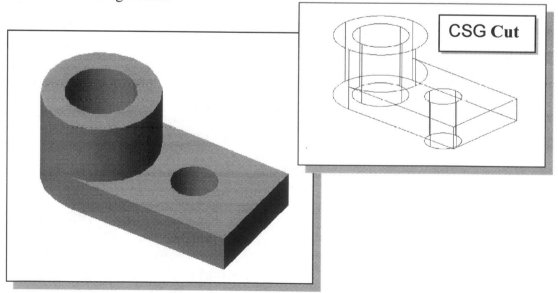

Creating a Rectangular Cut Feature

Next create a rectangular cut as the last solid feature of the *Locator*.

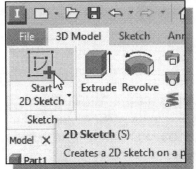

1. In the *3D Model* tab select the **Start 2D Sketch** command by left-clicking once on the icon.

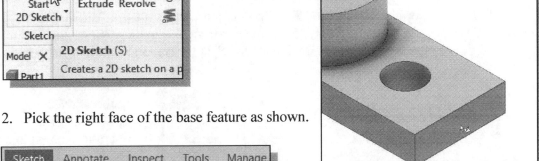

2. Pick the right face of the base feature as shown.

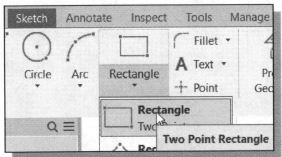

3. Select the **Two point rectangle** command by clicking once with the left-mouse-button on the icon in the *Sketch tab* of the Ribbon toolbar.

4. Create a rectangle that is aligned to the top and bottom edges of the base feature as shown. (Hit **[F6]** to set the display orientation if necessary.)

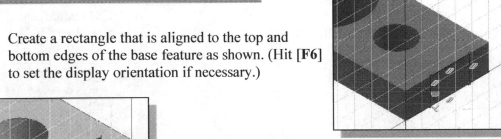

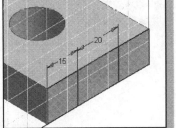

5. On your own, create and modify the two dimensions as shown.

6. Select **Finish Sketch** in the *Ribbon* toolbar to end the Sketch option.

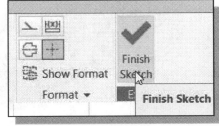
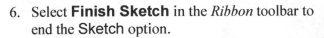

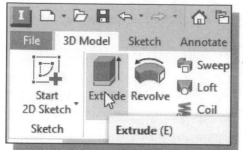

7. In the *3D Model* tab, select the **Extrude** command by left-clicking on the icon.

8. In the *Extrude* pop-up window, the **Profile** button is pressed down; Autodesk Inventor expects us to identify the profile to be extruded. Move the cursor inside the rectangle we just created and left-click once to select the region as the profile to be extruded.

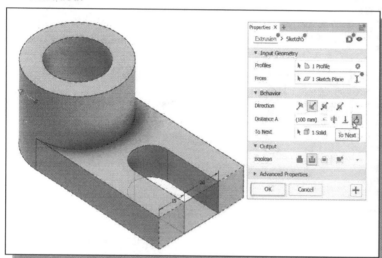

9. In the *Extrude* pop-up window, set the operation option to **Cut**. Select **To Next** as the *Distance* option as shown. Set the arrowhead points toward the center of the solid model.

10. Click on the **OK** button to create the *Cut* feature and complete the design.

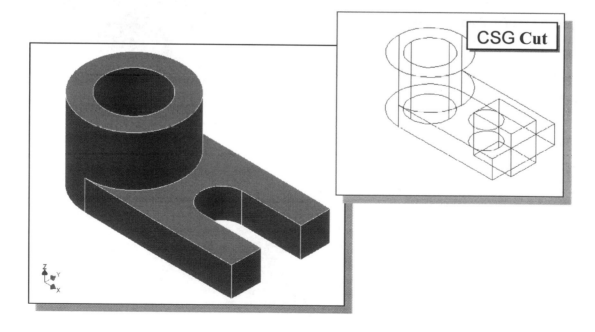

CSG Cut

Save the Model

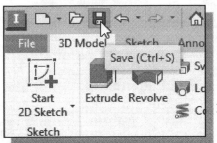

1. Select **Save** in the *Quick Access Toolbar*, or you can also use the "**Ctrl-S**" combination (hold down the "Ctrl" key and hit the "S" key once) to save the part.

2. Switch to the **Parametric Modeling** *folder* if it is not the current folder.

3. In the *Save As* dialog box, **right-click** once in the *list area* to bring up the *option menu*.

4. In the *option list*, select **New** as shown.

5. In the second *option list*, select **Folder** to create a subfolder.

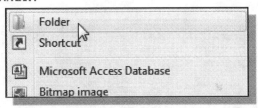

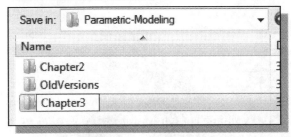

6. Enter **Chapter3** as the new folder name as shown.

7. **Double-click** on the Chapter3 folder to open it.

8. In the *file name* editor box, enter **Locator** as the file name.

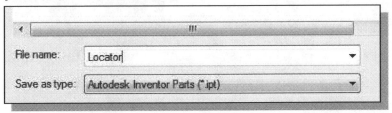

9. Click on the **Save** button to save the file.

Review Questions: (Time: 20 minutes.)

1. List and describe three basic *Boolean operations* commonly used in computer geometric modeling software.

2. What is a *primitive solid*?

3. What does *CSG* stand for?

4. Which *Boolean operation* keeps only the volume common to the two solid objects?

5. What is the main difference between a *CUT feature* and a *HOLE feature* in Autodesk Inventor?

6. Create the following 2D Sketch and measure the associated area and perimeter.

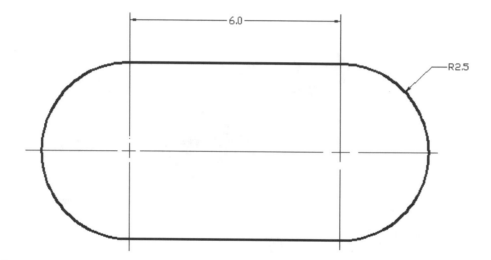

7. Using the CSG concepts, create *Binary Tree* sketches showing the steps you plan to use to create the two models shown on the next page:

Exercises: Create and save the exercises in the Chapter3 folder.
(Time: 180 minutes.)

1. **Latch Clip** (Dimensions are in inches. Thickness: **0.25** inches.)

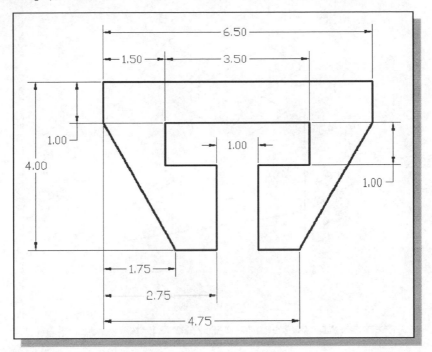

2. **Guide Plate** (Dimensions are in inches. Thickness: **0.25** inches. Boss height **0.125** inches.)

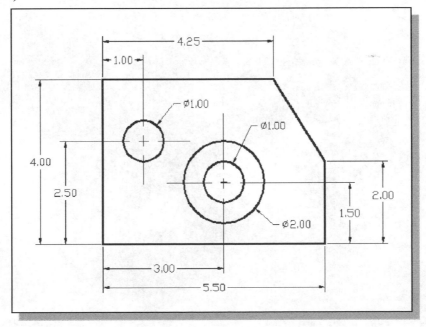

3. **Angle Slider** (Dimensions are in Millimeters.)

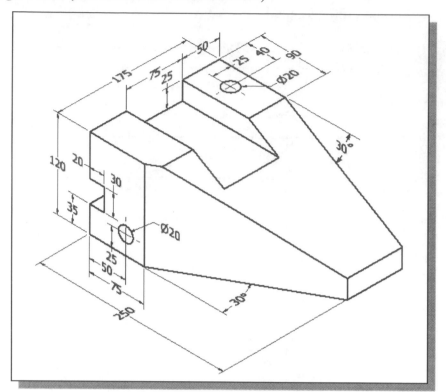

4. **Coupling Base** (Dimensions are in inches.)

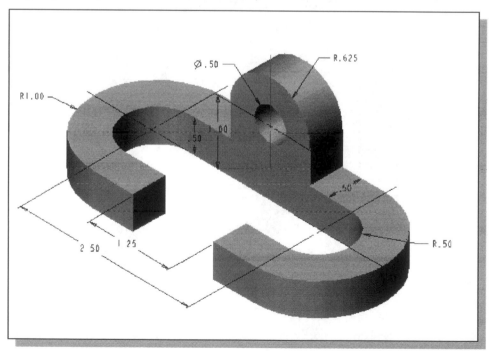

5. **Indexing Guide** (Dimensions are in inches.)

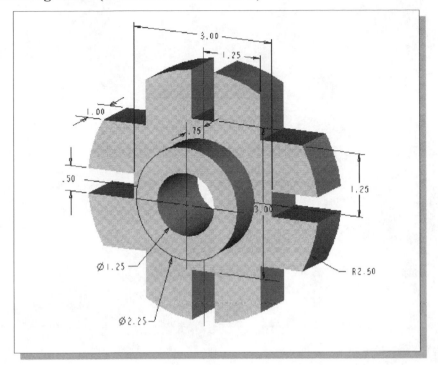

6. **L-Bracket** (Dimensions are in inches.)

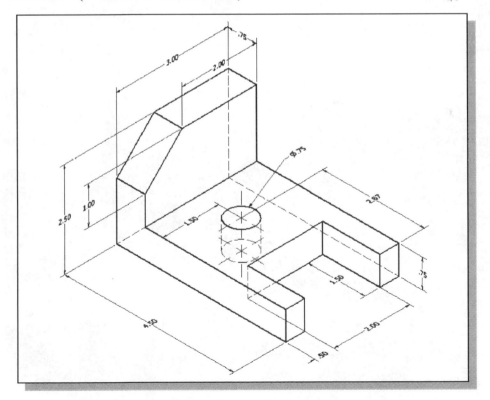

Notes:

Chapter 4
Geometric Constructions

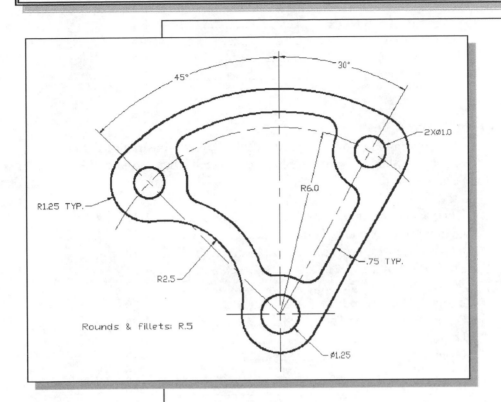

45° 30°

2XØ1.0

R6.0

R1.25 TYP.

.75 TYP.

R2.5

Rounds & fillets: R.5

Ø1.25

Learning Objectives

♦ **Understand the Classic Geometric Construction Tools and Methods**
♦ **Create Geometric Relations**
♦ **Use Dimensional Variables**
♦ **Display, Add, and Delete Geometric Relations**
♦ **Understand and Apply Different Geometric Relations**
♦ **Display and Modify Parametric Relations**
♦ **Create Fully Defined Sketches**

Certified Autodesk Inventor User Exam Objectives Coverage

Parametric Modeling Basics

Section 3: Sketches

Objectives: Creating 2D Sketches, Draw Tools, Sketch Constraints, Pattern Sketches, Modify Sketches, Format Sketches, Sketch Doctor, Shared Sketches, Sketch Parameters

Geometric Constructions

The creation of designs usually involves the manipulation of geometric shapes. Traditionally, manual graphical construction uses simple hand tools like the T-square, straightedge, scales, triangles, compass, dividers, pencils, and paper. The manual drafting tools are designed specifically to assist the construction of geometric shapes. For example, a T-square and drafting machine can be used to construct parallel and perpendicular lines very easily and quickly. Today, modern CAD systems provide designers much better control and accuracy in the construction of geometric shapes.

In technical drawings, many of the geometric shapes are constructed with specific geometric properties, such as perpendicularity, parallelism and tangency. For example, in the drawing below, quite a few **implied** geometric properties are present.

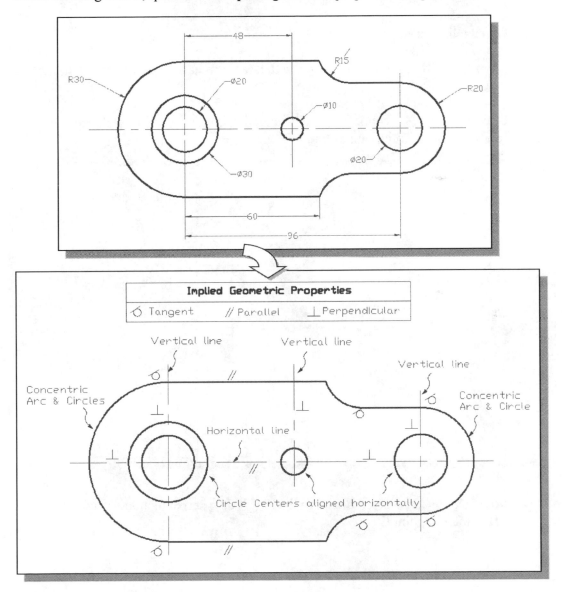

Geometric Constructions – Classical Methods

Geometric constructions are done by applying geometric rules to graphics entities. Knowledge of the principles of geometric construction and its applications are essential to Designers, Engineers and CAD users.

For 2D drawings, it is crucial to be able to construct geometric entities at specified angles to each other, various plane figures, and other graphic representations. In this section, we will examine both the traditional graphical methods and the CAD methods of the basic geometric constructions commonly used in engineering graphics. This chapter provides information that will aid you in drawing different types of geometric constructions.

- ## Bisection of a Line or Arc

    1. Given a line or an arc AB.

    2. From A and B draw two equal arcs with a radius that is greater than one half of line AB.

    3. Construct a line by connecting the intersection points, D and E, with a straight line to locate the midpoint of line AB.

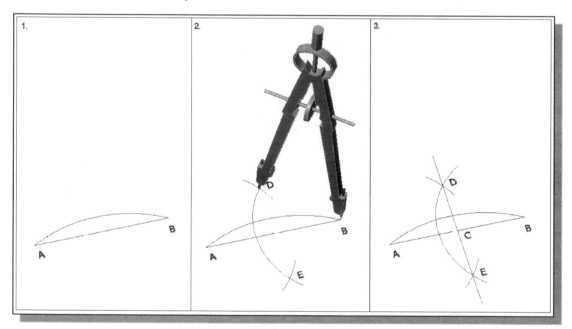

❖ Note that the constructed bisecting line DE is also perpendicular to the given line AB at the midpoint C.

• Bisection of an Angle

1. Given an angle ABC.

2. From A draw an arc with an arbitrary radius.

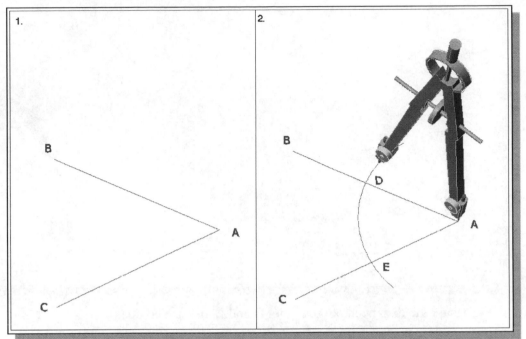

3. Construct two equal radius arcs at D and E.

4. Construct a straight line by connecting point A to the intersection of the two arcs.

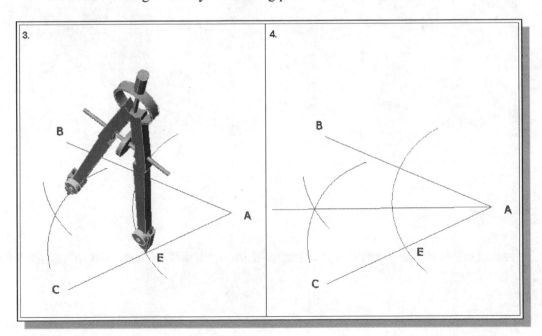

- ## Transfer of an Angle

1. Given an angle ABC, transfer the angle to line XY.

2. Create two arcs, at A and X, with an arbitrary radius R.

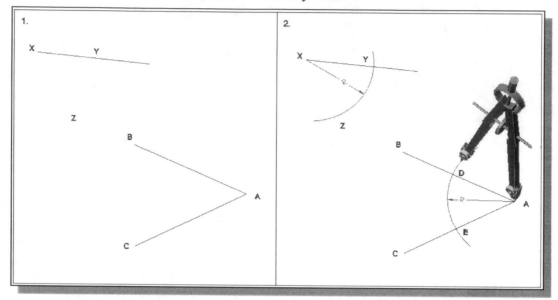

3. Measure the distance between point D and E, using a compass.

4. Construct an arc at Y, using the distance measured in the previous step.

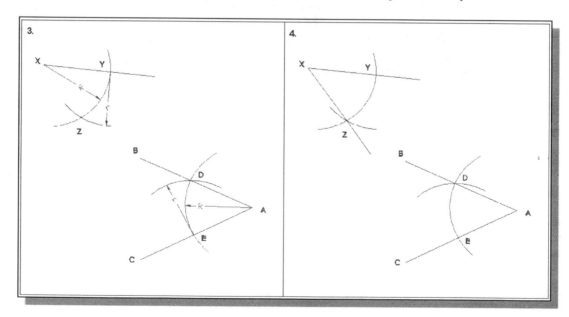

• Dividing a Given Line into a Number of Equal Parts

1. Given a line AB, the line is to be divided into five equal parts.

2. Construct another line at an arbitrary angle. Measure and mark five units along the line.

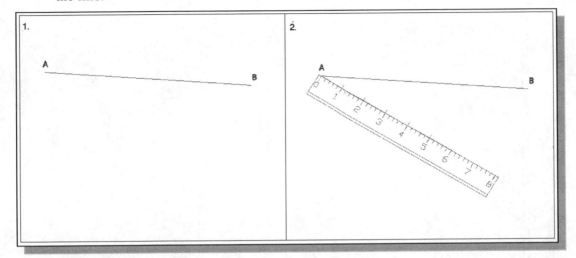

3. Construct a line connecting the fifth mark to point B.

4. Create four lines parallel to the constructed line through the marks.

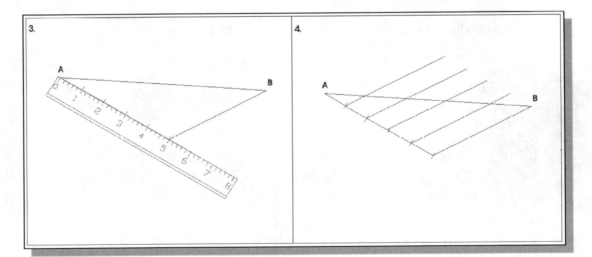

• Circle through Three Points

1. Given three points A, B and C.

2. Construct a bisecting line through line AB.

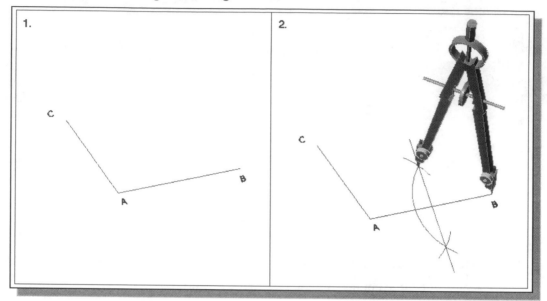

3. Construct a second bisecting line through line AC. The two bisecting lines intersect at point D.

4. Create the circle at point D using DA, DB or DC as radius.

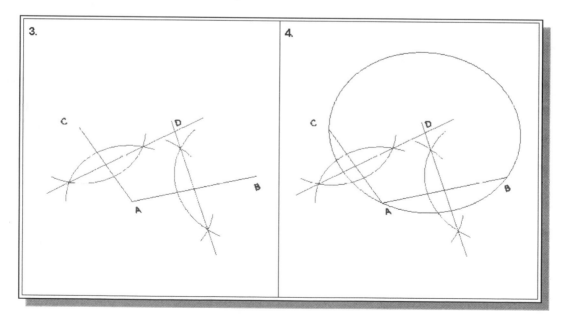

- ## A Line Tangent to a Circle

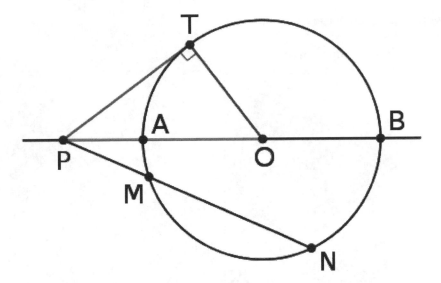

A line tangent to a circle intersects the circle at a single point. For comparison, secant lines intersect a circle at two points, whereas another line may not intersect a circle at all. In machine design, a smooth transition from surface to surface is typically desired, both for aesthetic and functionality considerations. Tangency is therefore a common implied geometric property in Mechanical and Manufacturing Engineering practices.

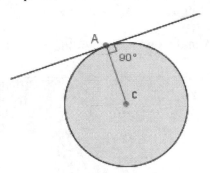

One unique property exists for the tangency between a line and a circle: *The radius of a circle is perpendicular to the tangent line through its endpoint on the circle's circumference*. Conversely, the perpendicular to a radius through the same endpoint is a tangent line.

Note that tangency can also be established in between curves.

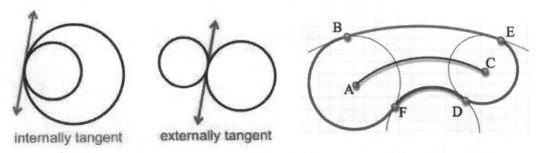

internally tangent   externally tangent

- ## Line Tangent to a Circle from a Given Point

  1.   Given a circle, center point C and a point A.

  2.   Create a bisecting line through line AC.

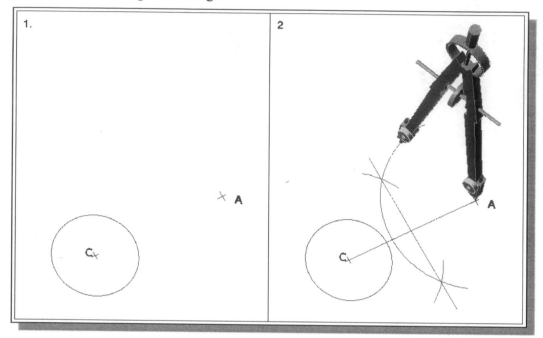

  3.   Create an arc at point B (midpoint on line AC), with AB or BC as the radius.

  4.   Construct the tangent line by connecting point A at the intersection of the arc and the circle.

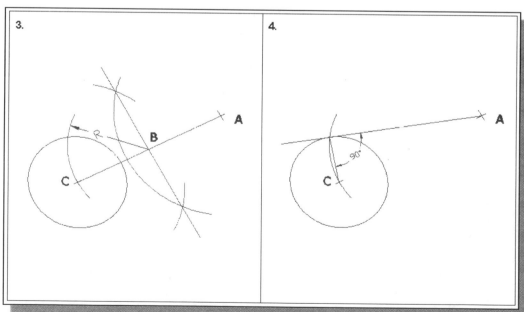

• Circle of a Given Radius Tangent to Two Given Lines

1. Given two lines and a given radius R.

2. Create a parallel line by creating two arcs of radius R, and draw a line tangent to the two arcs.

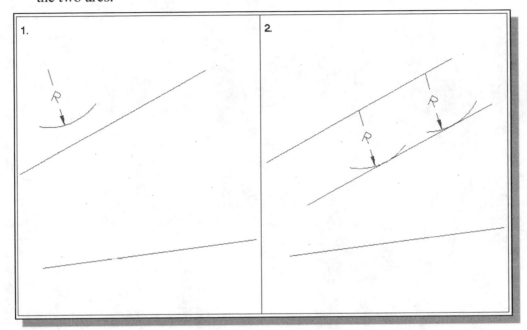

3. Create another line parallel to the bottom edge by first drawing two arcs.

4. Construct the required circle at the intersection of the two lines using the given radius R.

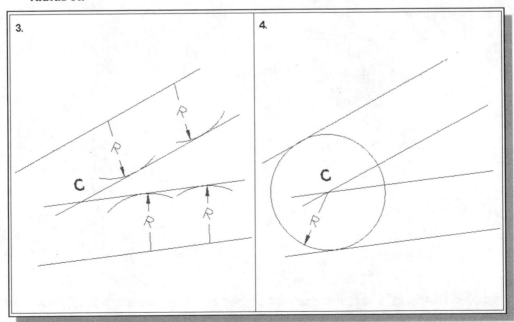

- ## Circle of a Given Radius Tangent to an Arc and a Line

    1. Given a radius R, a line and an arc.

    2. Create a parallel line by creating two arcs of radius R, and draw a line tangent to the two arcs.

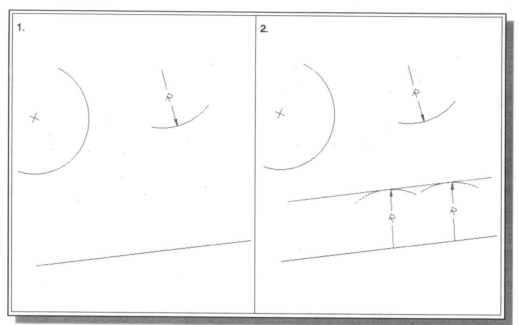

    3. Create a concentric arc at the center of the given arc using a radius that is r+R.

    4. Create the desired circle at the intersection using the given radius R.

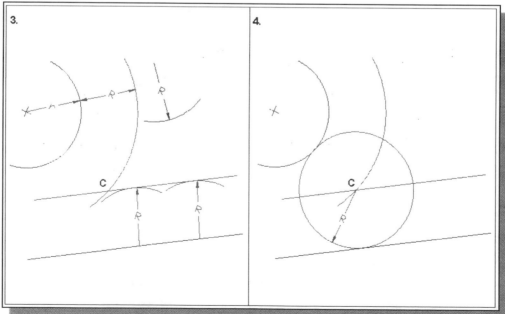

• Circle of a Given Radius Tangent to Two Arcs

1. Given a radius R and two arcs.

2. Create a concentric arc at the center of the small arc, using a radius that is R distance more than the original radius.

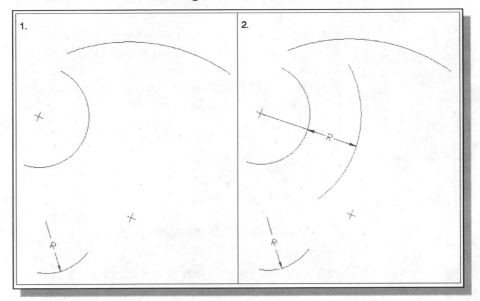

3. Create another concentric arc at the center of the large arc using a radius that is R distance smaller than the original radius.

4. Create the desired circle at the intersection using the given radius R.

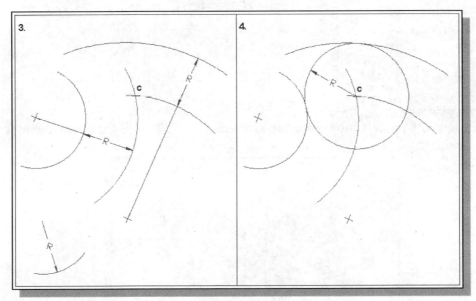

Starting Autodesk Inventor

1. Select the **Autodesk Inventor** option on the *Start* menu or select the **Autodesk Inventor** icon on the desktop to start Autodesk Inventor. The Autodesk Inventor main window will appear on the screen. Once the program is loaded into memory, the *Startup* dialog box appears at the center of the screen.

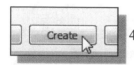

2. Select the **New** icon with a single click of the left-mouse-button in the *Launch* toolbar as shown.

3. Select the **English** tab, and in the *Part template* area select **Standard(in).ipt**.

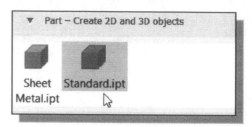

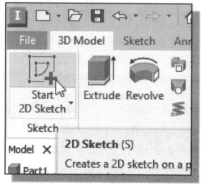

4. Click **Create** in the *New File* dialog box to accept the selected settings to start a new model.

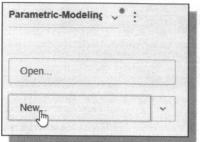

5. Click once with the left-mouse-button to select the **Create 2D Sketch** command.

6. Click once with the **left-mouse-button** to select the *XY Plane* as the sketch plane for the new sketch.

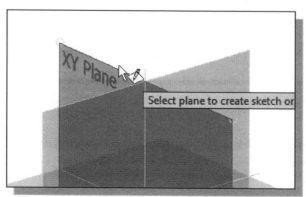

Geometric Construction – CAD Method

The main characteristic of any CAD system is its ability to create and modify 2D/3D geometric entities quickly and accurately. Most CAD systems provide a variety of object construction and editing tools to relieve the designer of the tedious drudgery of this task, so that the designer can concentrate more on design content. A good understanding of the computer geometric construction techniques will enable the CAD users to fully utilize the capability of the CAD systems.

Note that with CAD systems, besides following the classic geometric construction methods, quite a few options are also feasible. In the following sections, additional Autodesk Inventor sketching/editing tools will be used to illustrate some of the classical geometric construction methods.

- ### Bisection of a Line or Arc

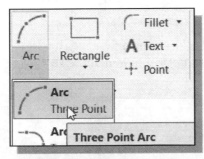

1. Use the **3 Point Arc** command and create an arbitrary arc at any angle on the screen.

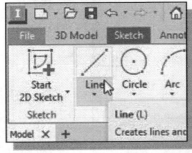

2. Using the **Line** command, create a line connecting the two endpoints of the arc.

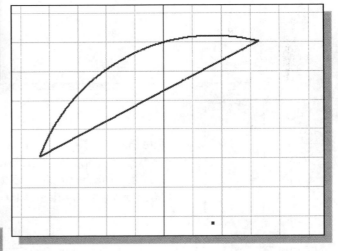

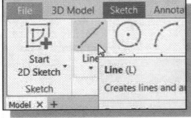

3. Activate the **Line** command again by clicking on the icon as shown.

4. Move the cursor along the arc and notice a **Green marker** appeared on the arc indicating the midpoint. Click once with the **left mouse button** to attach the first endpoint of the line at this location.

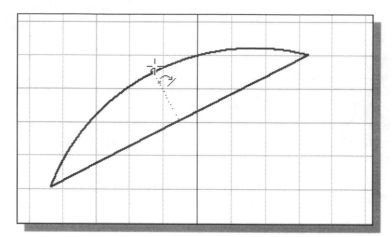

5. Move the cursor along the line and notice a **Green marker** appeared on the arc indicating the midpoint. Click once with the **left mouse button** to attach the second endpoint of the line at this location.

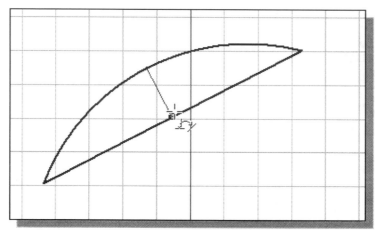

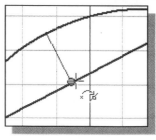

➢ Note the small symbol next to the cursor shows the geometric property at the current location. In this case, the new line will be aligned to the midpoint of the existing line. (The Green dot indicates the alignment to the midpoint.)

➢ The constructed bisecting line is perpendicular to the line and passes through the midpoints of both the line and arc.

Dimensions and Relations

A primary and essential difference between parametric modeling and previous generation computer modeling is that parametric modeling captures the *design intent*. In the previous lessons, we have seen that the design philosophy of *"shape before size"* is implemented through the use of Autodesk Inventor's Smart Dimension commands. In performing geometric constructions, dimensional values are necessary to describe the **SIZE** and **LOCATION** of constructed geometric entities. Besides using dimensions to define the geometry, we can also apply geometric rules to control geometric entities. More importantly, Autodesk Inventor can capture design intent through the use of **geometric relations**, **dimensional constraints** and **parametric relations**.

Geometric relations are geometric restrictions that can be applied to geometric entities; for example, *horizontal*, *parallel*, *perpendicular*, and *tangent* are commonly used *geometric relations* in parametric modeling. For part modeling in Autodesk Inventor relations are applied to *2D sketches*. They can be added automatically as the sketch is created or by using the **Constrain Toolbar**.

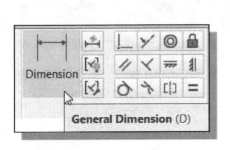
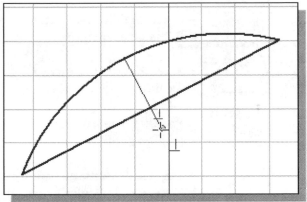

Dimensional constraints are used to describe the SIZE and LOCATION of individual geometric shapes. They are added using the Autodesk Inventor **Smart Dimension** command. You should also realize that depending upon the way the geometric relations and dimensional constraints are applied, the same results can be accomplished by applying different constraints to the geometric entities.

In Autodesk Inventor, **parametric relations** can be applied using **Global Variables** and **Equations**. Global variables are used when multiple dimensions have the same value. The dimension value is applied through the use of a named variable. Autodesk Inventor Equations are user-defined mathematical relations between model dimensions, using dimension names as variables. In parametric modeling, features are made of geometric entities with dimensional, geometric, and parametric constraints describing individual design intent. In this lesson, we will discuss the fundamentals of geometric relations and equations.

Applying Geometric/Dimensional Constraints

In Autodesk Inventor, twelve types of constraints are available for 2D sketches.

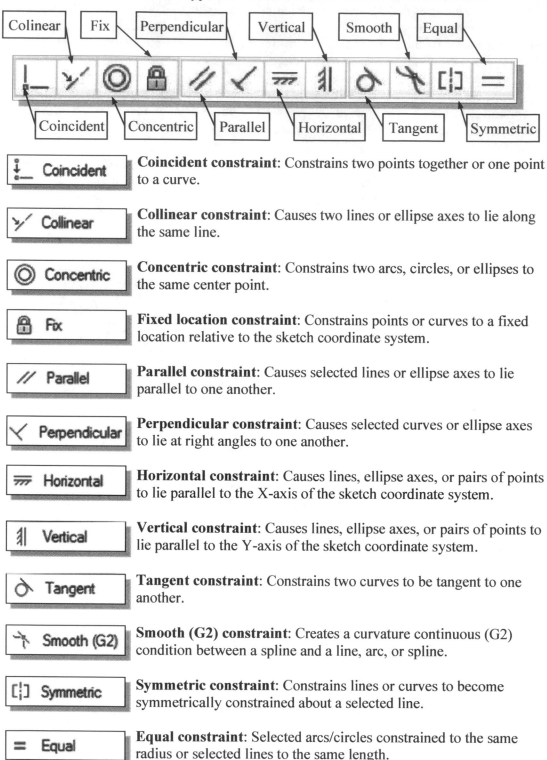

Coincident constraint: Constrains two points together or one point to a curve.

Collinear constraint: Causes two lines or ellipse axes to lie along the same line.

Concentric constraint: Constrains two arcs, circles, or ellipses to the same center point.

Fixed location constraint: Constrains points or curves to a fixed location relative to the sketch coordinate system.

Parallel constraint: Causes selected lines or ellipse axes to lie parallel to one another.

Perpendicular constraint: Causes selected curves or ellipse axes to lie at right angles to one another.

Horizontal constraint: Causes lines, ellipse axes, or pairs of points to lie parallel to the X-axis of the sketch coordinate system.

Vertical constraint: Causes lines, ellipse axes, or pairs of points to lie parallel to the Y-axis of the sketch coordinate system.

Tangent constraint: Constrains two curves to be tangent to one another.

Smooth (G2) constraint: Creates a curvature continuous (G2) condition between a spline and a line, arc, or spline.

Symmetric constraint: Constrains lines or curves to become symmetrically constrained about a selected line.

Equal constraint: Selected arcs/circles constrained to the same radius or selected lines to the same length.

• Bisection of an Angle

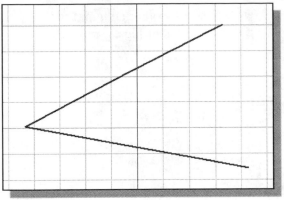

1. Using the **Line** command, create an arbitrary angle as shown in the figure.

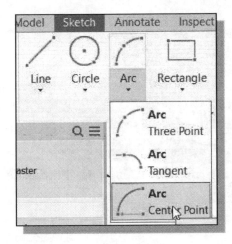

2. Activate the **Centerpoint Arc** command by clicking on the icon as shown.

3. Create the arc by first selecting the vertex point of the angle as the center of the new arc.

4. Move the cursor along the bottom leg of the angle and place the first endpoint on the line as shown. Confirm the other end of the arc intersects with the other leg as shown.

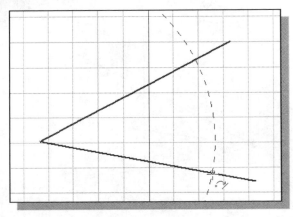

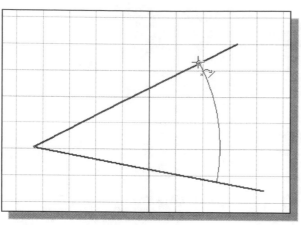

5. Place the other endpoint of the arc on the other leg as shown.

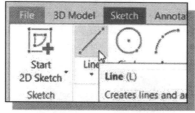

6. Activate the **Line** command by clicking on the icon as shown.

7. Create a bisecting line of the angle as shown.

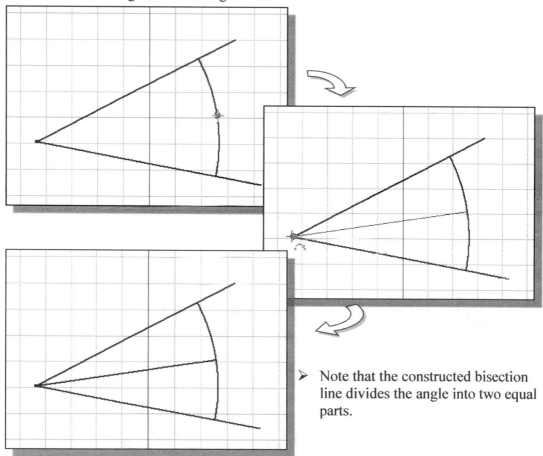

➢ Note that the constructed bisection line divides the angle into two equal parts.

• Dividing a Given Line into a Number of Equal Parts

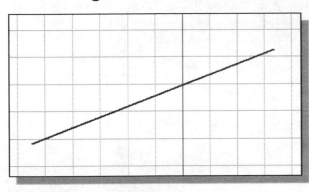

1. On your own, create a line at an arbitrary angle; the line is to be divided into five equal parts.

2. In the *Constrain Toolbar* select the **Fix constraint.**

3. Select the line to apply the *Fix* constraint and lock the line.

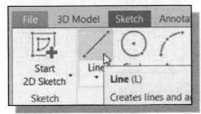

4. Activate the **Line** command by clicking on the icon as shown.

5. Click on the lower endpoint of the line to attach the new line and create a perpendicular line as shown.

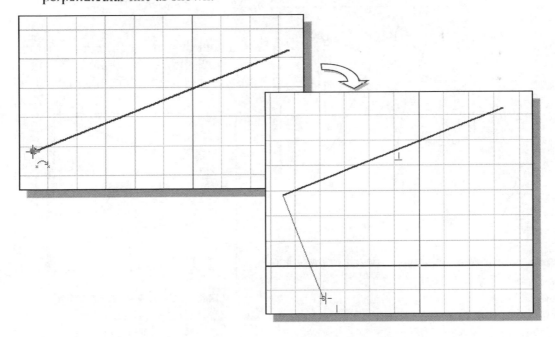

6. On your own, create four additional short line segments at arbitrary angles toward the right side of the previous line as shown. Note the last line segment on the right side is a vertical line as the *vertical symbol* is displayed next to the line.

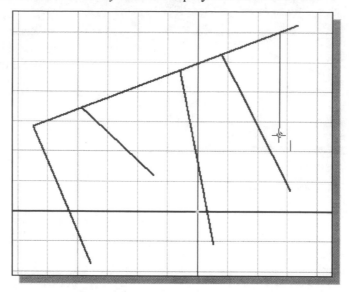

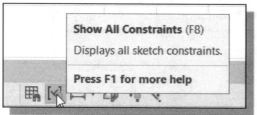

7. Select **Show All Constraints** in the *Status* toolbar as shown.

➢ Note the icons in the status toolbar act as a toggle switch; each click turns on or off the selected option.

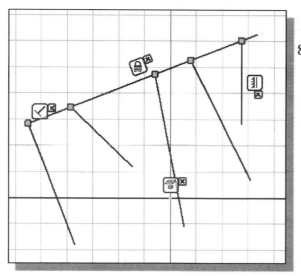

8. On your own, move the cursor on top of the displayed constraint icon. The affected objects will be highlighted.

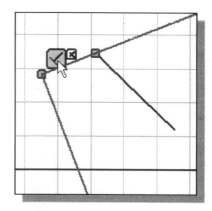

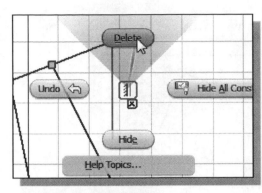

9. Move the cursor on top of the Vertical constraint.

10. **Right-click** once to bring up the option list and select **Delete** as shown.

➤ Note that any of the applied geometric relations can be deleted.

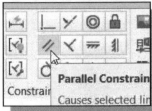

11. Click on the **Parallel Constraint** icon to activate the constraint.

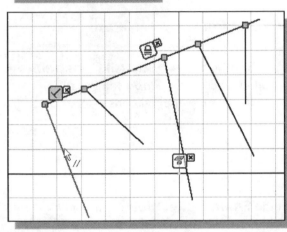

12. Select the short line segment to the left, the one with the perpendicular constraint, as shown.

13. Select the line next to the previously selected line to make both lines parallel toward each other.

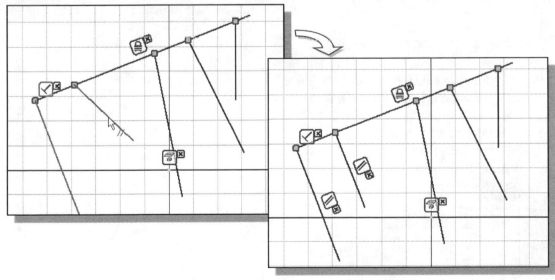

14. On your own, repeat the above steps and make all of the short line segments parallel to each other as shown.

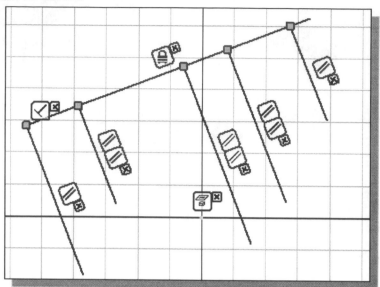

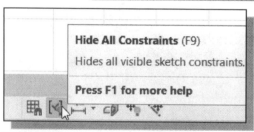

15. Click the **Show/Hide All Constraints** icon twice in the *Status* toolbar to redisplay the applied constraints.

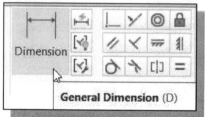

16. Select the **Smart Dimension** command by clicking once with the **left-mouse-button** on the icon in the *Constrain* toolbar.

17. On your own, use the **Aligned** option and create the dimension above the line as shown. Note the dimension value may be different than what is shown here.

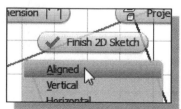

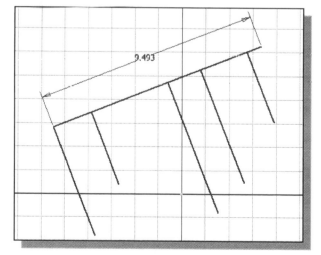

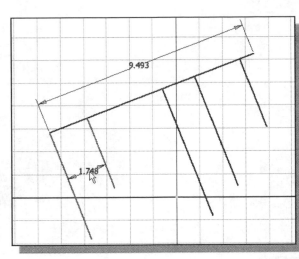

18. On your own, create a dimension between the first two parallel lines toward the left side.

19. Place the dimension in between the two lines as shown.

20. Click on the dimension if the Edit dimension dialog box does not appear on the screen.

21. Click on the top dimension and notice the name of the selected dimension appears in the Edit dimension dialog box as shown.

22. Enter **/5** (divided by 5) to create a parametric equation as shown.

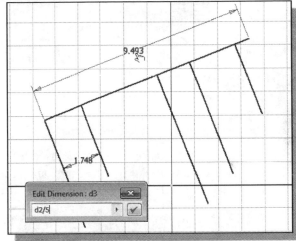

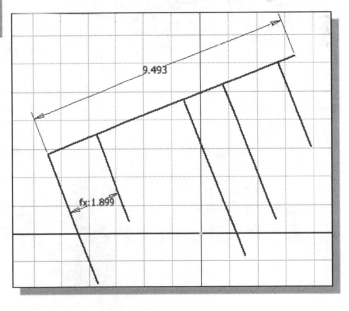

23. Click **OK** to accept the entered equation as shown.

- Notice the derived dimension value is displayed with **fx** in front of the numbers.

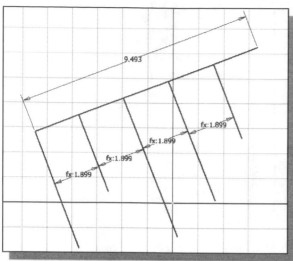

24. On your own, repeat the above steps and create the other three parametric equations as shown.

25. Click once with the left-mouse-button on the **Manage** tab and select **Parameters** in the toolbar as shown.

- The *Parameters* dialog box can be used to perform all tasks related to equations, including editing existing equations, and creating additional design variables and equations. We will use it here to show an alternate method to add/edit an equation.

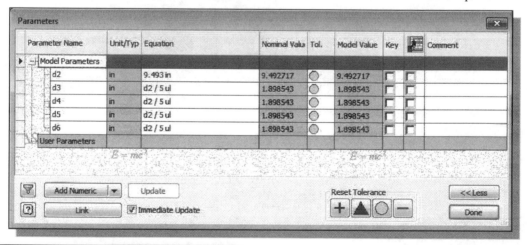

Parameter Name	Unit/Typ	Equation	Nominal Valu	Tol.	Model Value	Key		Comment
Model Parameters								
d2	in	9.493 in	9.492717	○	9.492717	□	□	
d3	in	d2 / 5 ul	1.898543	○	1.898543	□	□	
d4	in	d2 / 5 ul	1.898543	○	1.898543	□	□	
d5	in	d2 / 5 ul	1.898543	○	1.898543	□	□	
d6	in	d2 / 5 ul	1.898543	○	1.898543	□	□	
User Parameters								

Parameter Name	Unit/Typ	Equation
Model Parameters		
d2	in	10 in
d3	in	d2 / 5 ul

26. Click inside the first dimension and change the value to **10** as shown.

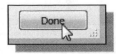 27. Click **Done** to accept the settings and close the *Equations* dialog box.

- Notice all of the dimensions are updated with the new values.

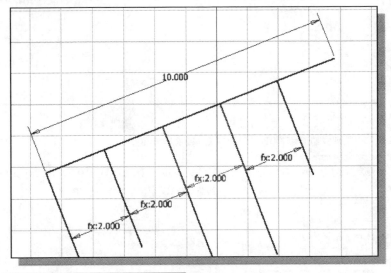

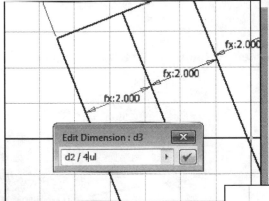

28. On your own, change the dimension between any two parallel to **/4** and see the effect.

- In parametric modeling, dimensions can be linked to each other through the use of equations.

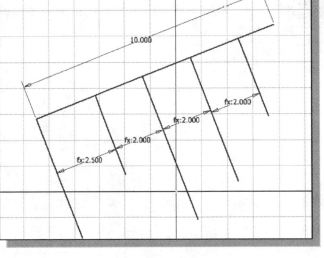

- ## Arc through Three Points

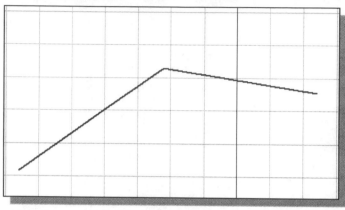

1. Create two arbitrary connected line segments as shown.

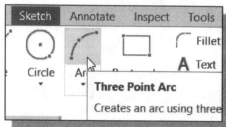

2. Select the **Three Point Arc** command in the *Sketch* toolbar as shown.

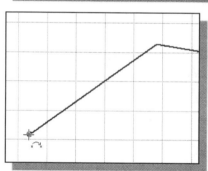

3. Select the **left endpoint** of the left line segment as shown.

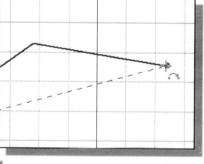

4. Select the **right endpoint** of the right line segment as shown.

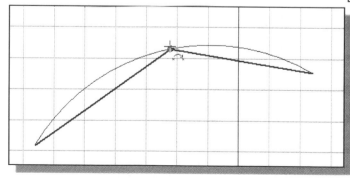

5. Select the point on top to create the arc that passes through all three points.

• Line Tangent to a Circle from a Given Point

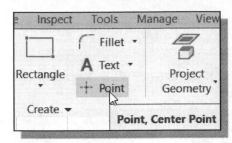

1. Create a circle and a point. (Use the Point command to create the point.)

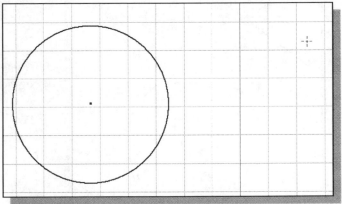

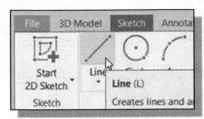

2. Activate the **Line** command in the *Sketch* toolbar as shown.

3. Select the point as the starting point of the new line.

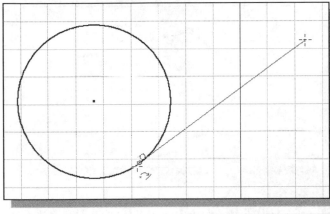

4. Move the cursor on the lower right side of the circle and notice the Tangent symbol is displayed.

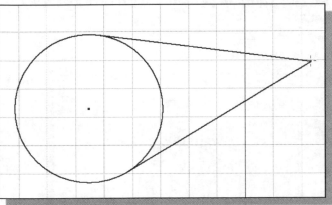

➢ Note that we can create two tangent lines, one to the top and one to the bottom of the circle, from the point.

- ## Circle of a Given Radius Tangent to Two Given Lines

Option I: Using the Add Relations Option

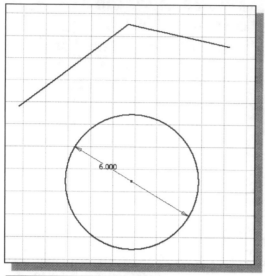

1. Create **two arbitrary line segments** as shown.

2. Create a **circle** below the two line segments as shown.

3. Use the **Smart Dimension** command and adjust the size of the circle to **6.00**.

4. On your own, apply the **Fix** constraint on the two line segments.

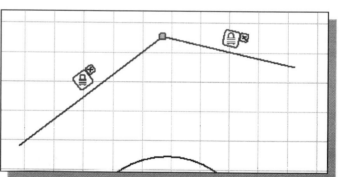

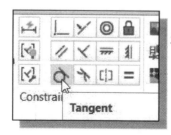

5. Click on the **Tangent Constraint** icon to activate the constraint.

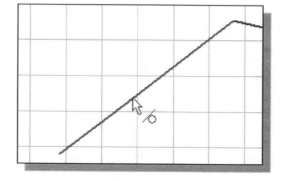

6. Select the left line as the first object.

7. Select the circle as the second object to apply the tangent constraint.

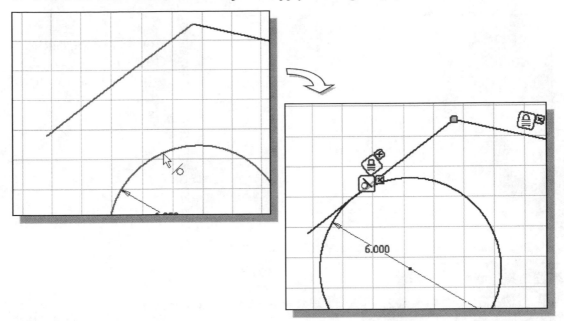

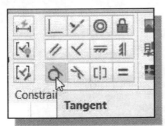

8. On your own, repeat the above steps and add another tangent constraint to complete the construction.

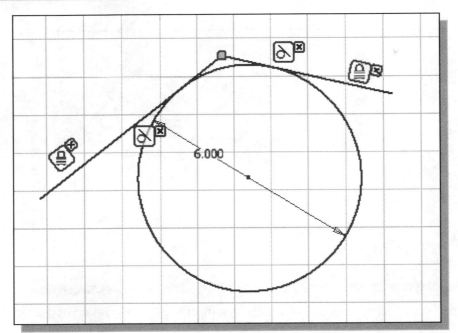

Option II: Fillet Command

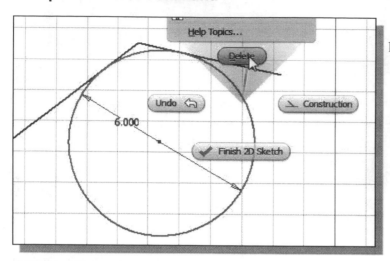

1. On your own, delete the circle by using the **Delete** option with the right-click option list as shown.

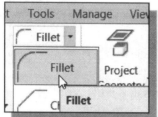

2. Select **Fillet** in the *Sketch* toolbar as shown.

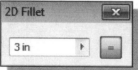

3. Enter *3* as the new radius of the Fillet command.

4. Select the two lines and create the tangent arc as shown.

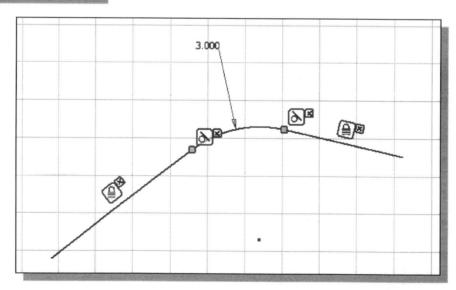

➢ Note that the Fillet command automatically trims the edges as shown in the figure.

Adding Geometric Relations and Fully Defined Geometry

In Autodesk Inventor, as we create 2D sketches, **geometric relations** such as *horizontal* and *parallel* are automatically added to the sketched geometry. In most cases, additional relations and dimensions are needed to fully describe the sketched geometry beyond the geometric relations added by the system. By carefully applying proper **geometric relations**, very intelligent models can be created. Although we can use Autodesk Inventor to build partially constrained or totally unconstrained solid models, the models may behave unpredictably as changes are made. In most cases, it is important to consider the design intent, develop a modeling strategy, and add proper constraints to geometric entities. In the following sections, a simple triangle is used to illustrate the different tools that are available in Autodesk Inventor to create/modify geometric relations and dimensional constraints.

Starting a New Drawing

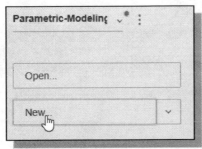

1. Select the **New** icon with a single click of the left-mouse-button on the *Menu Bar* area.

2. Select the **English** tab and in the *Part template* area and select **Standard(in).ipt**.

3. Click **Create** in the *New File* dialog box to accept the selected settings to start a new model.

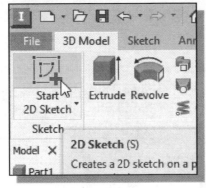

4. Click once with the left-mouse-button to select the **Create 2D Sketch** command.

5. Click once with the **left-mouse-button** to select the *XY Plane* as the sketch plane for the new sketch.

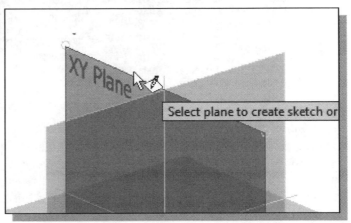

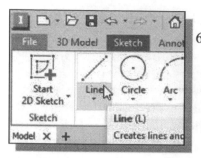

6. Click the **Line** icon in the *Sketch* toolbar to activate the command. A *Help-tip box* appears next to the cursor, and a brief description of the command is displayed at the bottom of the drawing screen: "*Creates Straight line segments and tangent arcs.*"

7. Create a triangle of arbitrary size positioned near the center of the screen as shown below. (Note that the base of the triangle is horizontal.)

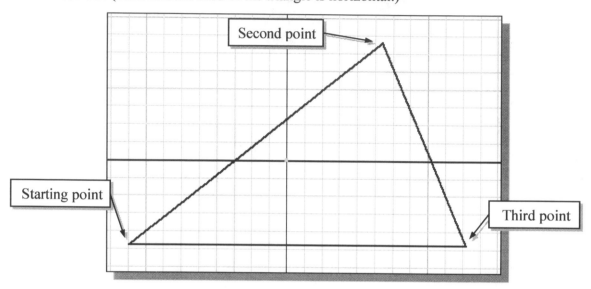

Displaying Existing Constraints

1. Select the **Show Constraints** command in the *Constrain* toolbar. This icon allows us to display constraints that are already applied to the 2D profiles. Left-click once on the icon to activate **Show Constraints**.

> In *parametric modeling*, constraints are typically applied as geometric entities are created. Autodesk Inventor will attempt to add proper constraints to the geometric entities based on the way the entities were created. Constraints are displayed as symbols next to the entities as they are created. The current profile consists of three line entities: three straight lines. The horizontal line has a **Horizontal** *constraint* applied to it.

2. Move the cursor on top of the horizontal line and notice the number of constraints applied is displayed in the message area.

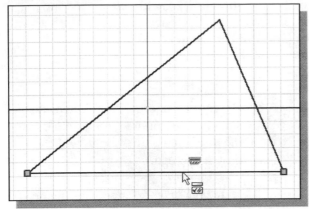

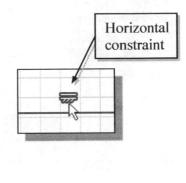

Horizontal constraint

3. On your own, move the cursor on top of the other two lines and notice no additional constraints exist on the other two entities.

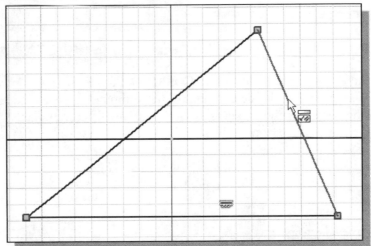

4. Move the cursor on top of the constraints displayed and notice the highlighted endpoint/line indicating the location/entities where the constraints are applied.

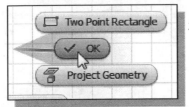

5. Inside the graphics window, right-click to bring up the option menu and select **OK** to end the Show Constraints command.

6. Right-click on the **horizontal** constraint icon to bring up the option list and select the Hide All Constraints option as shown.

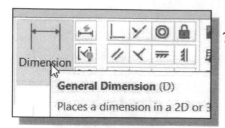

7. Select the **General Dimension** command in the *Sketch* toolbar. The General Dimension command allows us to quickly create and modify dimensions. Left-click once on the icon to activate the General Dimension command.

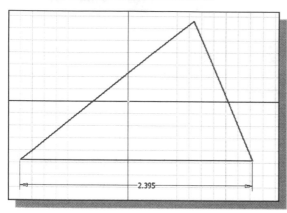

8. On your own, create the dimension as shown in the figure below. (Note that the displayed value might be different on your screen.)

❖ Note that *Inventor* identifies the need for additional dimensions at the bottom of the screen.

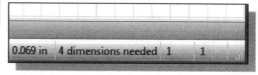

9. Move the cursor on top of the different **Constraint** icons. A *Help-tip box* appears next to the cursor and a brief description of the command is displayed at the bottom of the drawing screen as the cursor is moved over the different icons.

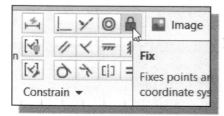

10. Click on the **Fix** constraint icon to activate the command.

11. Pick the **lower right corner** of the triangle to make the corner a fixed point.

12. Inside the graphics window, right-click to bring up the option menu and select **OK** to end the Fix Constraints command.

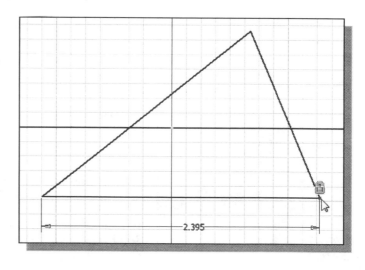

➤ Geometric constraints can be used to control the direction in which changes can occur. For example, in the current design we are adding a horizontal dimension to control the length of the horizontal line. If the length of the line is modified to a greater value, Autodesk Inventor will lengthen the line toward the left side. This is due to the fact that the Fix constraint will restrict any horizontal movement of the horizontal line toward the right side.

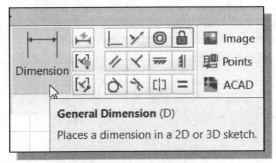

13. Select the **General Dimension** command in the *2D Sketch* toolbar.

14. Click on the dimension text to open the *Edit Dimension* window.

15. Enter a value that is greater than the displayed value to observe the effects of the modification.

16. On your own, use the **Undo** command to reset the dimension value to the previous value.

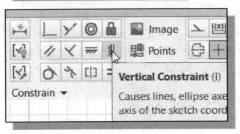

17. Select the **Vertical** constraint icon in the *2D Constraints* toolbar.

18. Pick the inclined line on the right to make the line vertical as shown in the figure below.

19. Hit the [**Esc**] key once to end the Vertical Constraint command.

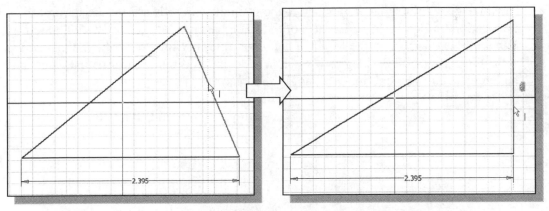

➤ You should think of the constraints and dimensions as defining elements of the geometric entities. How many more constraints or dimensions will be necessary to fully constrain the sketched geometry? Which constraints or dimensions would you use to fully describe the sketched geometry?

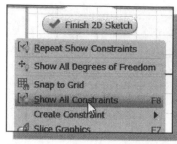

20. Inside the graphics window, click once with the **right-mouse-button** to display the option menu. Select **Show All Constraints** in the pop-up menu to show all the applied constraints. (Note that function key **F8** can also be used to activate this command.)

21. Move the cursor on top of the top corner of the triangle. (Note the display of the two Coincident constraints at the top corner.)

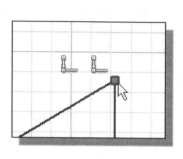

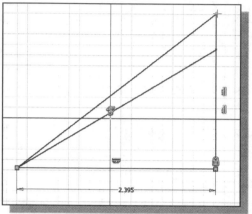

22. Drag the top corner of the triangle and note that the corner can be moved to a new location. Release the mouse button at a new location and notice the corner is adjusted only in an upward or downward direction. Note that the two adjacent lines are automatically adjusted to the new location.

23. On your own, experiment with dragging the other corners to new locations.

• The three constraints that are applied to the geometry provide a full description for the location of the two lower corners of the triangle. The Vertical constraint, along with the Fix constraint at the lower right corner, does not fully describe the location of the top corner of the triangle. We will need to add additional information, such as the length of the vertical line or an angle dimension.

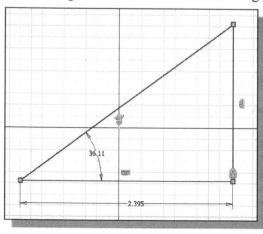

24. On your own, add an angle dimension to the left corner of the triangle.

25. Press the [**Esc**] key once to exit the General Dimension command.

26. On your own, try changing the angle to **45°** and observe the adjustment of the triangle with the adjustment.

• Note the sketched geometry is now fully constrained with the added dimension.

Over-Constraining and Driven Dimensions

We can use Autodesk Inventor to build partially constrained or totally unconstrained solid models. In most cases, these types of models may behave unpredictably as changes are made. However, Autodesk Inventor will not let us over-constrain a sketch; additional dimensions can still be added to the sketch, but they are used as references only. These additional dimensions are called *driven dimensions*. *Driven dimensions* do not constrain the sketch; they only reflect the values of the dimensioned geometry. They are enclosed in parentheses to distinguish them from normal (parametric) dimensions. A *driven dimension* can be converted to a normal dimension only if another dimension or geometric constraint is removed.

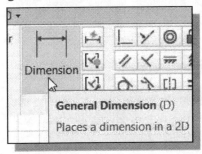

1. Select the **General Dimension** command in the *Sketch* toolbar.

2. Select the **vertical line**.

3. Pick a location that is to the right side of the triangle to place the dimension text.

4. A warning dialog box appears on the screen stating that the dimension we are trying to create will over-constrain the sketch. Click on the **Accept** button to proceed with the creation of a driven dimension.

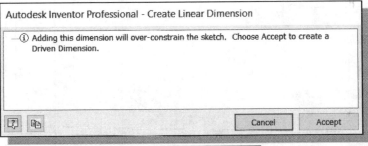

Autodesk Inventor Professional - Create Linear Dimension

ⓘ Adding this dimension will over-constrain the sketch. Choose Accept to create a Driven Dimension.

Cancel   Accept

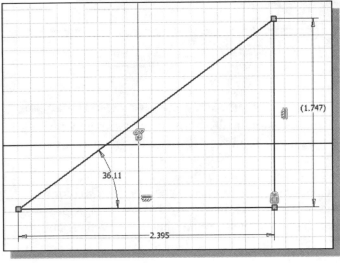

(1.747)

36.11

2.395

5. On your own, modify the angle dimension to **35°** and observe the changes of the 2D sketch and the driven dimension.

❖ Note that Inventor has indicated the sketch is **Fully Constrained**.

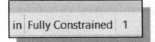

in Fully Constrained 1

Deleting Existing Constraints

1. On your own, display all the active constraints if they are not already displayed. (Hint: Use the Show Constraints command in the *Sketch* toolbar or Show All Constraints in the option menu.)

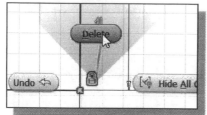

2. Move the cursor on top of the **Fix** constraint icon and **right-click** once to bring up the option menu.

3. Select **Delete** to remove the Fix constraint that is applied to the lower right corner of the triangle.

❖ Note the removal of the **Fix** constraint has caused the need for **two additional dimensions**.

4. On your own, hide all the constraints with the [**F9**] key and drag the top corner of the triangle upward and note that the entire triangle is free to move in all directions. Release the mouse button to move the triangle to the new location.

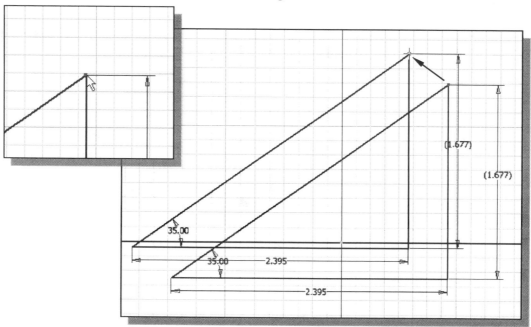

5. On your own, delete the reference dimension on the right and experiment with dragging the corners and the line segments to move the triangle to new locations.

❖ **Dimensional constraints** are used to describe the SIZE and LOCATION of individual geometric shapes. **Geometric constraints** are **geometric restrictions** that can be applied to geometric entities. The constraints applied to the triangle are sufficient to maintain its size and shape, but the geometry can be moved around; its location definition isn't complete.

Using the Auto Dimension Command

In Autodesk Inventor, the Auto Dimension command can be used to assist in creating a fully constrained sketch. **Fully constrained** sketches can be updated more predictably as design changes are implemented. The general procedure for applying dimensions to sketches is to use the General Dimension command to add the more critical dimensions, and then use the Auto Dimension command to add the additional dimensions/constraints to fully constrain the sketch. The Auto Dimension command can also be used to apply the missing dimensions that are needed. It is also important to realize that different sets of dimensions and geometric constraints can be applied to the same sketch to accomplish a fully constrained geometry.

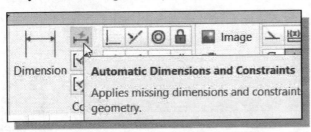

1. Click on the **Auto Dimension** icon in the *Constrain* toolbar.

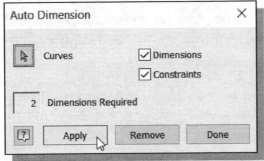

❖ Note that Autodesk Inventor indicates the sketch is missing two locational dimensions.

2. Click **Apply** to the Auto Dimension command.

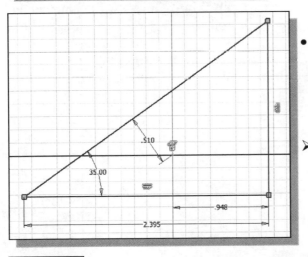

• Note that the Auto Dimension command added two dimensions measuring from the **Origin**, the intersection of the horizontal and vertical axes.

➤ The two newly added dimensions do not change the size of the triangle; they are used to position the triangle.

3. Click **Remove** to undo the Auto Dimension command. Note that only the dimensions created by the Auto Dimension command are affected.

4. Click Done to close the dialog box and exit the Auto Dimension command.

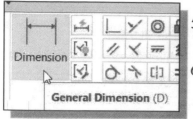

5. Activate the **General Dimension** command in the *Constrain* toolbar.

6. On your own, create the two locational dimensions, measuring the lower right corner to the origin as shown.

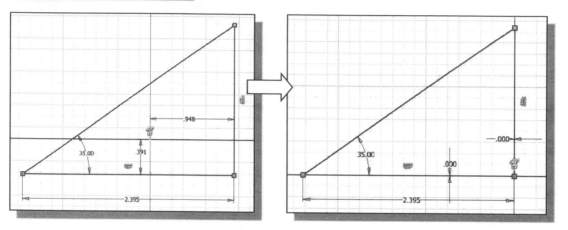

7. On your own, set the two dimensions to **0.0** and observe the alignment of the lower right corner of the triangle to the *Origin*.

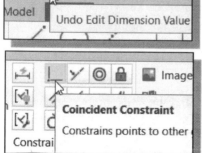

8. On your own, click the **Undo** button several times to return to the point before the two location dimensions were added.

9. Activate the **Coincident Constraint** command in the *Constrain* toolbar.

10. Select the **lower right corner** of the triangle and the **Origin** to fully constrain the sketch as shown.

➤ In parametric modeling, it is desired to create fully constrained sketches.

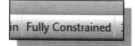

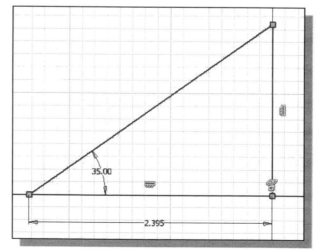

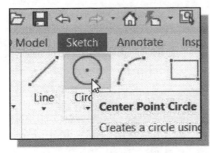

11. Select the **Center Point Circle** command by clicking once with the left-mouse-button on the icon in the *Sketch* toolbar.

12. On your own, create a circle of arbitrary size inside the triangle as shown below.

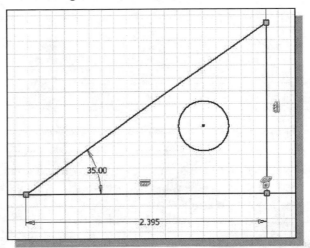

13. Click on the **Auto Dimension** icon in the *Sketch* panel.

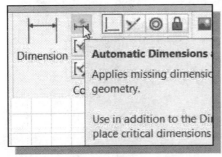

❖ Note that *Autodesk Inventor* confirms that the sketch is not fully constrained and "*3 Dimensions Required*" to fully constrain the circle. What are the dimensions and/or constraints that can be applied to fully constrain the circle?

14. Click **Done** to exit the Auto Dimension command.

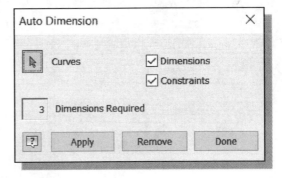

15. Click on the **Tangent** constraint icon in the *Sketch* toolbar.

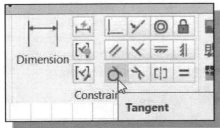

16. Pick the circle by left-clicking once on the geometry.

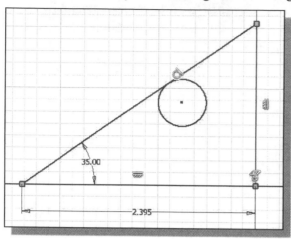

17. Pick the inclined line. The sketched geometry is adjusted as shown.

18. Inside the graphics window, click once with the right-mouse-button to display the option menu. Select **OK** in the pop-up menu to end the **Tangent** command.

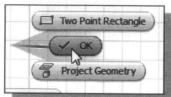

➢ How many more constraints or dimensions do you think will be necessary to fully constrain the circle? Which constraints or dimensions would you use to fully constrain the geometry?

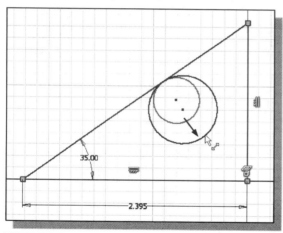

19. Move the cursor on top of the right side of the circle, and then drag the circle toward the right edge of the graphics window. Notice the size of the circle is adjusted while the system maintains the **Tangent** constraint.

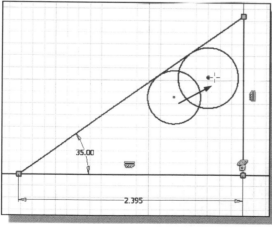

20. Drag the center of the circle toward the upper right direction. Notice the **Tangent** constraint is always maintained by the system.

➢ On your own, experiment with adding additional constraints and/or dimensions to fully constrain the sketched geometry. Use the **Undo** command to undo any changes before proceeding to the next section.

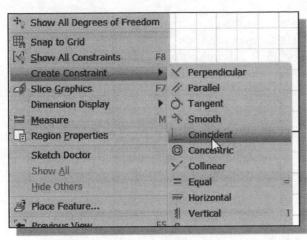

21. Inside the *graphics window*, click once with the **right-mouse-button** to display the option menu. Select **Create Constraint → Coincident** in the pop-up menus.

- The option menu is a quick way to access many of the commonly used commands in Autodesk Inventor.

22. Pick the **vertical line**.

23. Pick the center of the circle to align the **center** of the circle and the vertical line.

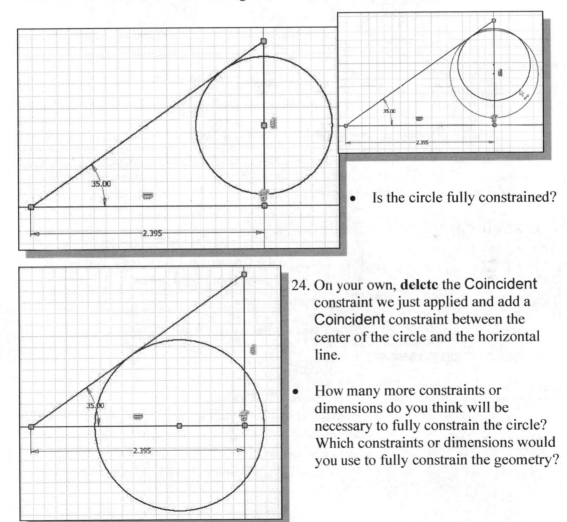

- Is the circle fully constrained?

24. On your own, **delete** the Coincident constraint we just applied and add a Coincident constraint between the center of the circle and the horizontal line.

- How many more constraints or dimensions do you think will be necessary to fully constrain the circle? Which constraints or dimensions would you use to fully constrain the geometry?

❖ The application of different constraints affects the geometry differently. The design intent is maintained in the CAD model's database and thus allows us to create very intelligent CAD models that can be modified and revised fairly easily. On your own, experiment and observe the results of applying different constraints to the triangle. For example: (1) adding another **Fix** constraint to the top corner of the triangle; (2) deleting the horizontal dimension and adding another **Fix** constraint to the left corner of the triangle; and (3) adding another **Tangent** constraint and adding the size dimension to the circle.

25. On your own, modify the 2D sketch as shown below.

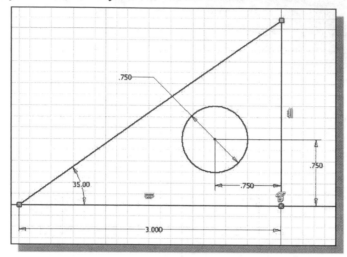

➢ On your own, use the **Extrude** command and create a 3D solid model with a plate thickness of **0.25**. Also experiment with modifying the parametric relations and dimensions through the *part browser*.

Constraint and Sketch Settings

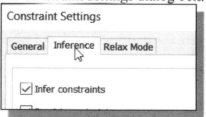

➢ Click on the **Constraint Settings** icon to access the constraint settings dialog box.

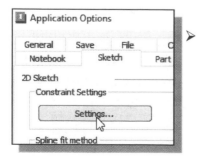

➢ Note the settings can also be accessed through **Application Options** in the **Tools** pull-down menu. Switch to the **Sketch** tab then click on **Constraint Settings** to display and/or modify the constraint settings.

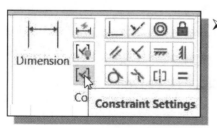

Parametric Relations

In parametric modeling, dimensions are design parameters that are used to control the sizes and locations of geometric features. Dimensions are more than just values; they can also be used as feature control variables. This concept is illustrated by the following example.

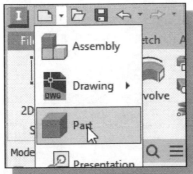

1. Start a new drawing by left-clicking once on the **New → Part** icon in the *Standard* toolbar.

● Another graphics window appears on the screen. We can switch between the two models by clicking on the different graphics windows.

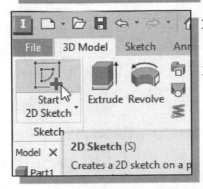

2. Select the **Start 2D Sketch** command with the left-mouse-button.

3. Select the *XY Plane* as the sketch plane for the new sketch with the **left-mouse-button**.

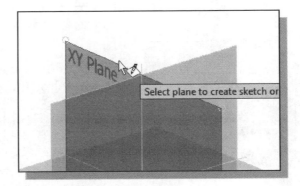

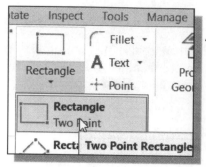

4. Select the **Two point rectangle** command by clicking once with the left-mouse-button on the icon in the *Sketch* toolbar.

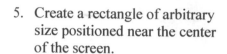

5. Create a rectangle of arbitrary size positioned near the center of the screen.

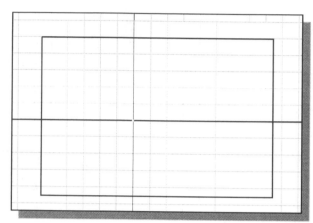

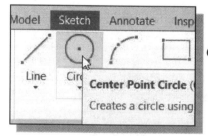

6. Select the **Center Point Circle** command by clicking once with the left-mouse-button on the icon in the *Sketch* toolbar.

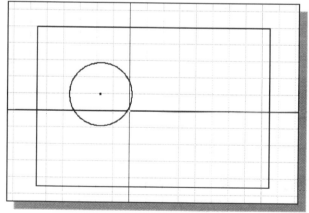

7. Create a **circle** of arbitrary size inside the rectangle as shown.

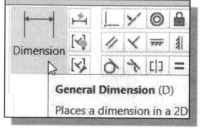

8. Select the **General Dimension** command in the *Sketch* toolbar.

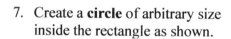

9. On your own, create and adjust the geometry by modifying the dimensions as shown below. (Note: Align the center of the circle to the origin to fully constrain the sketch.)

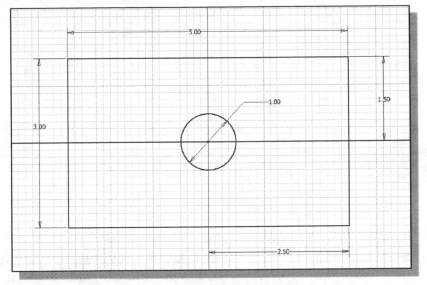

*   On your own, change the overall width of the rectangle to **6.0** and the overall height of the rectangle to **3.6** and observe the location of the circle in relation to the edges of the rectangle. Adjust the dimensions back to **5.0** and **3.0** as shown in the above figure before continuing.

Dimensional Values and Dimensional Variables

Initially in Autodesk Inventor, values are used to create different geometric entities. The text created by the Dimension command also reflects the actual location or size of the entity. Each dimension is also assigned a name that allows the dimension to be used as a control variable. The default format is "dxx," where the "xx" is a number that Autodesk Inventor increments automatically each time a new dimension is added.

Let us look at our current design, which represents a plate with a hole at the center. The dimensional values describe the size and/or location of the plate and the hole. If a modification is required to change the width of the plate, the location of the hole will remain the same as described by the two location dimensional values. This is okay if that is the design intent. On the other hand, the *design intent* may require (1) keeping the hole at the center of the plate and (2) maintaining the size of the hole to be one-third of the height of the plate. We will establish a set of parametric relations using the dimensional variables to capture the design intent described in statements (1) and (2) above.

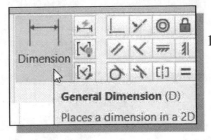

1. Left-click once on the **General Dimension** icon to activate the General Dimension command.

2. Click on the **width dimension** of the rectangle to display the *Edit Dimension* window.

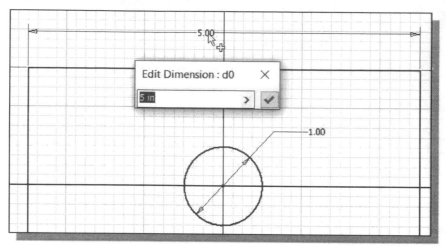

-   **Notice the *variable name* d0 is displayed in the title area of the *Edit Dimension* window and also in the cursor box when the cursor is moved near the text box.**

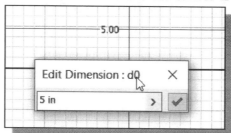

3. **Click on the *check mark* button to close the *Edit Dimension* window.**

Parametric Equations

Each time we add a dimension to a model, that value is established as a parameter for the model. We can use parameters in equations to set the values of other parameters.

1. Click on the **horizontal location dimension** of the circle to display the *Edit Dimension* window.

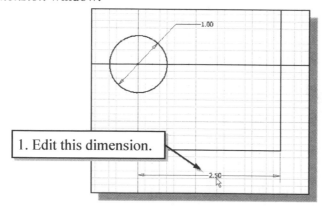

1. Edit this dimension.

2. Click on the width dimension of the rectangle **d0** (value of **5.0**). Notice the selected variable name is automatically entered in the *Edit Dimension* window.

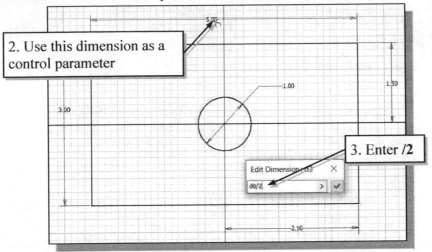

2. Use this dimension as a control parameter

3. Enter **/2**

3. In the *Edit Dimension* window, enter **/2** to set the horizontal location dimension of the circle to be one-half of the width of the rectangle.

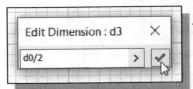

4. Click on the ***check mark*** button to close the *Edit Dimension* window.

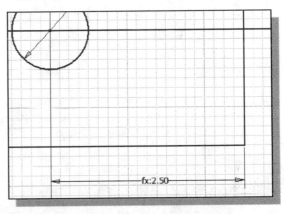

❖   Notice the derived dimension values are displayed with **fx** in front of the numbers. The parametric relations we entered are used to control the location of the circle; the location is based on the height and width of the rectangle.

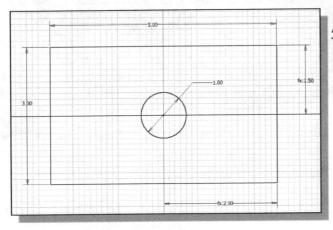

5. On your own, repeat the above steps and set the vertical location dimension to one-half of the height of the rectangle.

Viewing the Established Parameters and Relations

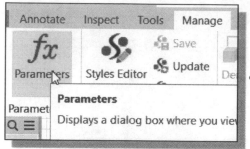

1. In the *Manage* toolbar select the **Parameters** command by left-clicking once on the icon.

- The Parameters command can be used to display all dimensions used to define the model. We can also create additional parameters as design variables, which are called ***user parameters***.

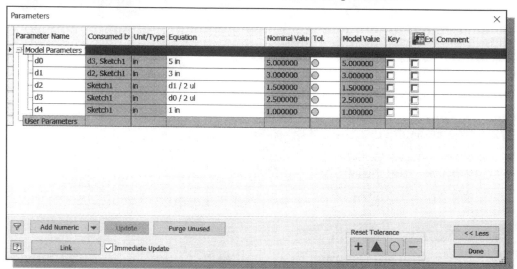

2. Click on the **1.0** value in the equation section of the *Parameters* window. Enter **d1/3** as the parametric relation to set the size of the circle to be one-third of the height of the rectangle as shown below.

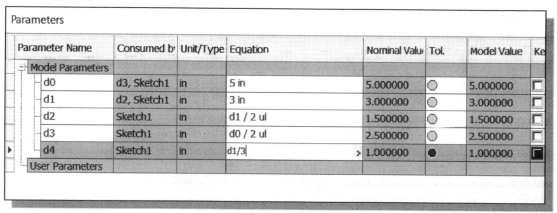

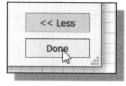

3. Click on the **Done** button to accept the settings.

4. On your own, change the dimensions of the rectangle to **6.0 x 3.6** and observe the changes to the location and size of the circle. (Hint: Double-click the dimension text to bring up the *Edit Dimension* window.)

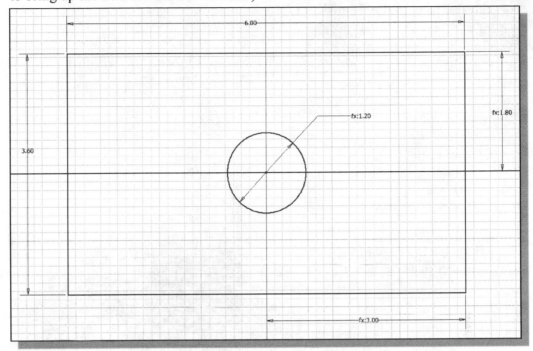

❖ *Autodesk Inventor* automatically adjusts the dimensions of the design, and the parametric relations we entered are also applied and maintained. The dimensional constraints are used to control the size and location of the hole. The design intent, previously expressed by statements (1) and (2) at the beginning of this section, is now embedded into the model.

➢ On your own, use the **Extrude** command and create a 3D solid model with a plate thickness of **0.25**. Also, experiment with modifying the parametric relations and dimensions through the *part browser*.

Saving the Model File

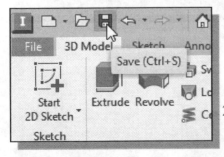

1. Select **Save** in the *Standard* toolbar. We can also use the "**Ctrl-S**" combination (press down the "Ctrl" key and hit the "S" key once) to save the part.

2. In the pop-up window, enter **Plate** as the name of the file.

3. Click on the **SAVE** button to save the file.

Using the Measure Tools

Besides using the measure tools to get geometric information at the 2D level, the measure tools can also be used on 3D models.

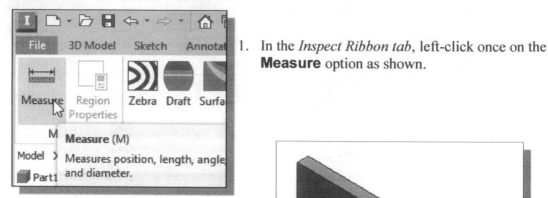

1. In the *Inspect Ribbon tab*, left-click once on the **Measure** option as shown.

2. Click on the top edge of the rectangular plate as shown.

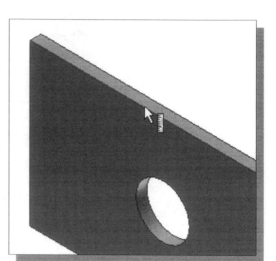

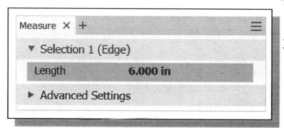

3. The associated length measurement of the selected geometry is displayed in the *Length* dialog box as shown.

4. Click on the **triangular icon** of the **Advanced Settings** option list to display the available options.

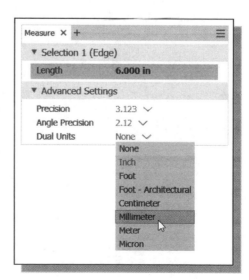

5. Choose [**Dual Units**] → [**Millimeter**] to also display the measurement in mm.

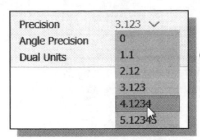

6. On your own, set the precision to show 4 decimal places as shown.

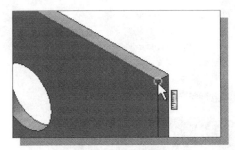

7. Click the **top-right corner** of the 3D model as shown.

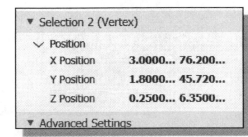

❖ Notice the information regarding the selected point is displayed in the *Measure* dialog box. The absolute position of the point is displayed. (Note the displayed numbers may be different on your screen.)

8. Select the front left bottom corner as the second location for the **Measure Length** command. The distance in between the two selected objects is calculated and displayed as shown.

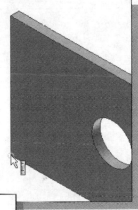

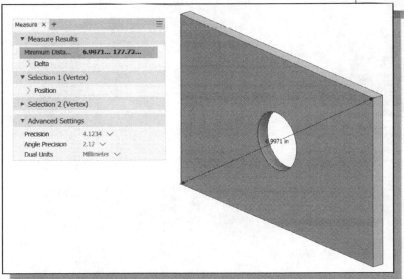

9. Set the display option to **Wireframe Display** by clicking the associated icon in the *Display* toolbar as shown.

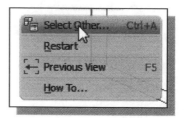

10. Select the left vertical plane by clicking on the selection arrows. (Hint: Use the **Select Other** option if you are having difficulty selecting it.)

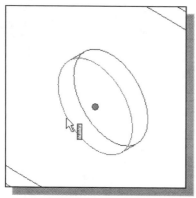

11. Select the circular hole in the front face; notice the center point is highlighted (the actual point used for the calculation) as shown.

❖ Different types of entities can be selected and the measurements are calculated accordingly.

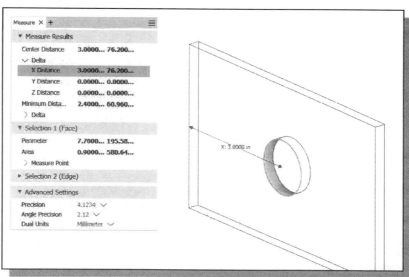

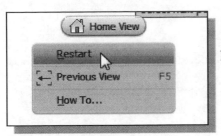

12. On your own, reset the **Measure** option by selecting **Restart** in the option list as shown.

13. Click on the front face of the plate model to display the area of the selected surface.

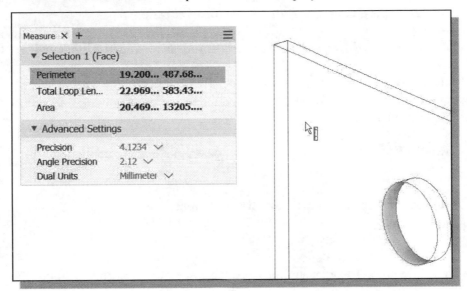

14. Note that we can also measure cylindrical surfaces by selecting the cylindrical surface as shown.

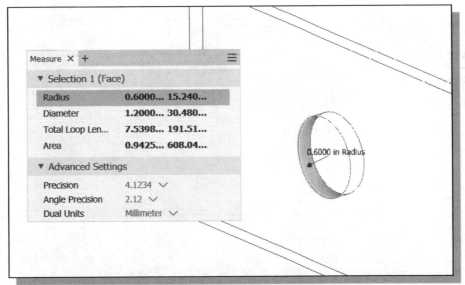

15. On your own, experiment with the different **measure tools** available.

Review Questions: (Time: 30 minutes)

1. What is the difference between *dimensional* constraints and *geometric* constraints?

2. How can we confirm that a sketch is fully constrained?

3. How do we distinguish between derived dimensions and regular dimensions on the screen?

4. Describe the procedure to Display/Edit user-defined equations.

5. List and describe three different geometric constraints available in Autodesk Inventor.

6. Does Autodesk Inventor allow us to build partially constrained or totally unconstrained solid models? What are the advantages and disadvantages of building these types of models?

7. How do we display and examine the existing constraints that are applied to the sketched entities?

8. Describe the advantages of using parametric equations.

9. Can we delete an applied constraint? How?

10. Create the following 2D Sketch and measure the associated area and perimeter.

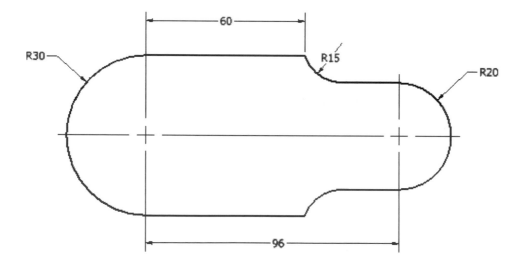

11. Describe the purpose and usage of the Auto Dimension command.

Exercises: (Time: 180 minutes.)

(Create and establish three parametric relations for each of the following designs.)

1. **Swivel Base** (Dimensions are in millimeters. Base thickness: **10 mm.** Boss: **5 mm.**)

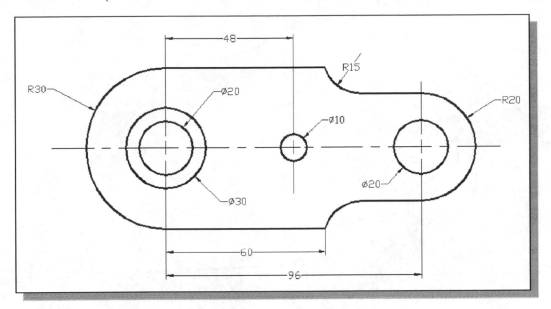

2. **Anchor Base** (Dimensions are in inches.)

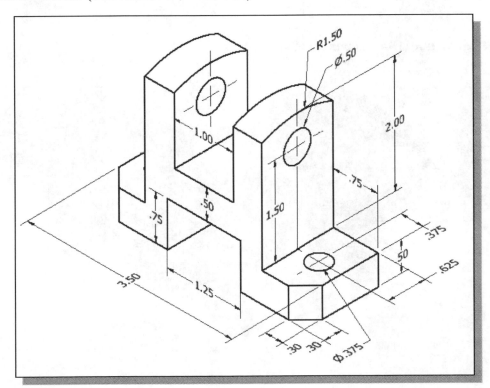

3. **Wedge Block** (Dimensions are in inches.)

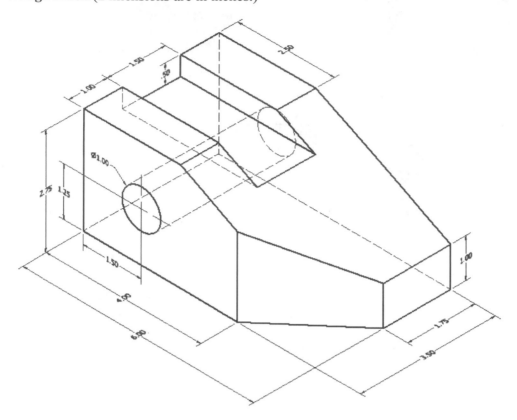

4. **Hinge Guide** (Dimensions are in inches.)

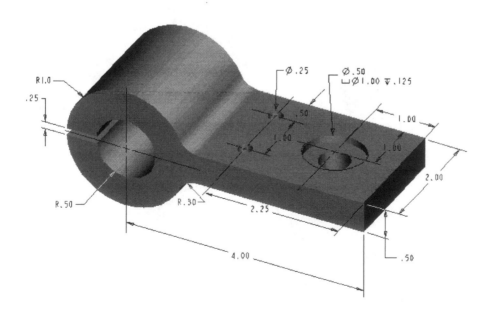

5. **Pivot Holder** (Dimensions are in inches.)

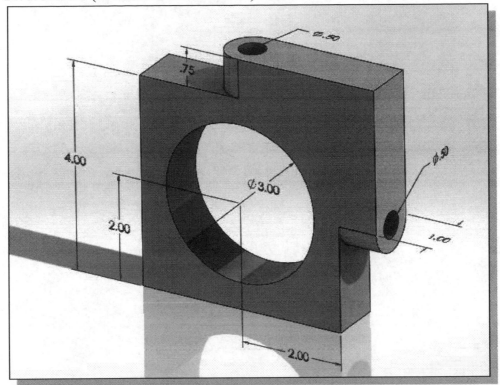

6. **Support Fixture** (Dimensions are in inches.)

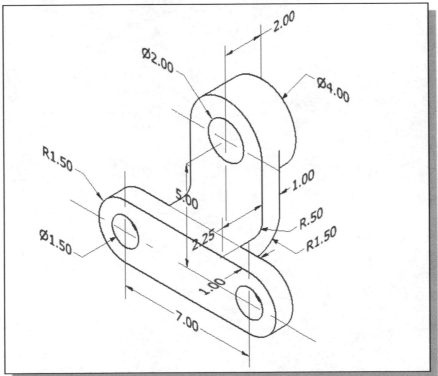

Notes:

Chapter 5
Model History Tree

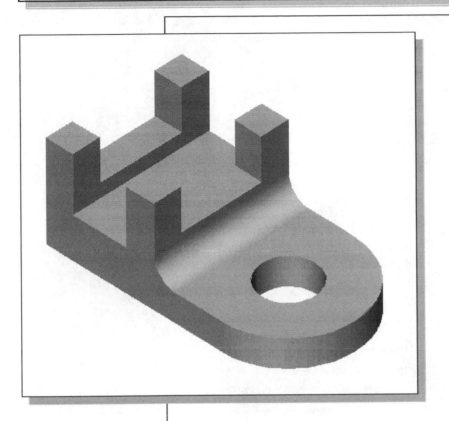

Learning Objectives

- ◆ **Understand Feature Interactions**
- ◆ **Use the Part Browser**
- ◆ **Modify and Update Feature Dimensions**
- ◆ **Perform History-Based Part Modifications**
- ◆ **Change the Names of Created Features**
- ◆ **Implement Basic Design Changes**

Autodesk Inventor Certified User Exam Objectives Coverage

Parametric Modeling Basics

Section 3: Sketches

Objectives: Creating 2D Sketches, Draw Tools, Sketch Constraints, Pattern Sketches, Modify Sketches, Format Sketches, Sketch Doctor, Shared Sketches, Sketch Parameters

Section 4: Parts

Objectives: Creating parts, Work Features, Pattern Features, Part Properties

Autodesk Inventor Certified User Reference Guide

Introduction

In Autodesk Inventor, the **design intents** are embedded into features in the **history tree**. The structure of the model history tree resembles that of a **CSG binary tree**. A CSG binary tree contains only *Boolean relations*, while the **Autodesk Inventor history tree** contains all features, including *Boolean relations*. A history tree is a sequential record of the features used to create the part. This history tree contains the construction steps, plus the rules defining the design intent of each construction operation. In a history tree, each time a new modeling event is created previously defined features can be used to define information such as size, location, and orientation. It is therefore important to think about your modeling strategy before you start creating anything. It is important, but also difficult, to plan ahead for all possible design changes that might occur. This approach in modeling is a major difference in **FEATURE-BASED CAD SOFTWARE**, such as Autodesk Inventor, from previous generation CAD systems.

Sequential record of the construction steps

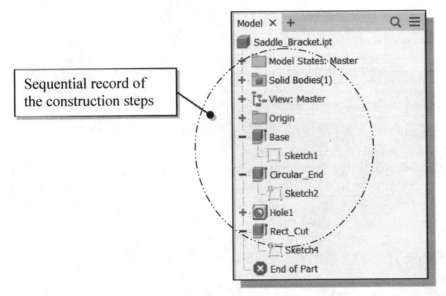

Feature-based parametric modeling is a cumulative process. Every time a new feature is added, a new result is created, and the feature is also added to the history tree. The database also includes parameters of features that were used to define them. All of this happens automatically as features are created and manipulated. At this point, it is important to understand that all of this information is retained, and modifications are done based on the same input information.

In Autodesk Inventor, the history tree gives information about modeling order and other information about the feature. Part modifications can be accomplished by accessing the features in the history tree. It is therefore important to understand and utilize the feature history tree to modify designs. Autodesk Inventor remembers the history of a part, including all the rules that were used to create it, so that changes can be made to any operation that was performed to create the part. In Autodesk Inventor, to modify a feature we access the feature by selecting the feature in the *browser* window.

The Saddle Bracket Design

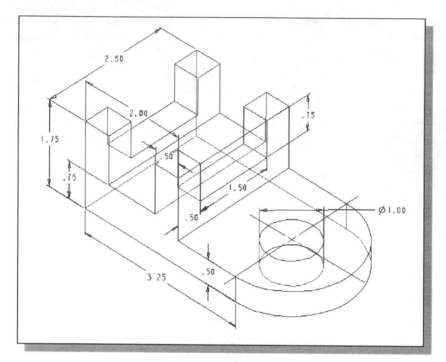

❖ Based on your knowledge of Autodesk Inventor so far, how many features would you use to create the design? Which feature would you choose as the **BASE FEATURE**, the first solid feature, of the model? What is your choice in arranging the order of the features? Would you organize the features differently if additional fillets were to be added in the design? Take a few minutes to consider these questions and do preliminary planning by sketching on a piece of paper. You are also encouraged to create the model on your own prior to following through the tutorial.

Starting Autodesk Inventor

1. Select the **Autodesk Inventor** option on the *Start* menu or select the **Autodesk Inventor** icon on the desktop to start Autodesk Inventor. The Autodesk Inventor main window will appear on the screen.

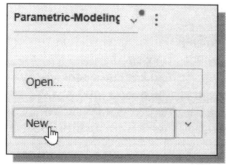

2. Once the program is loaded into memory, select the **New File** icon with a single click of the left-mouse-button in the *Launch* toolbar.

3. Select the **en-US->English** tab, and in the *Part template* area select **Standard(in).ipt**.

4. Click **Create** in the *New File* dialog box to accept the selected settings to start a new model.

Modeling Strategy

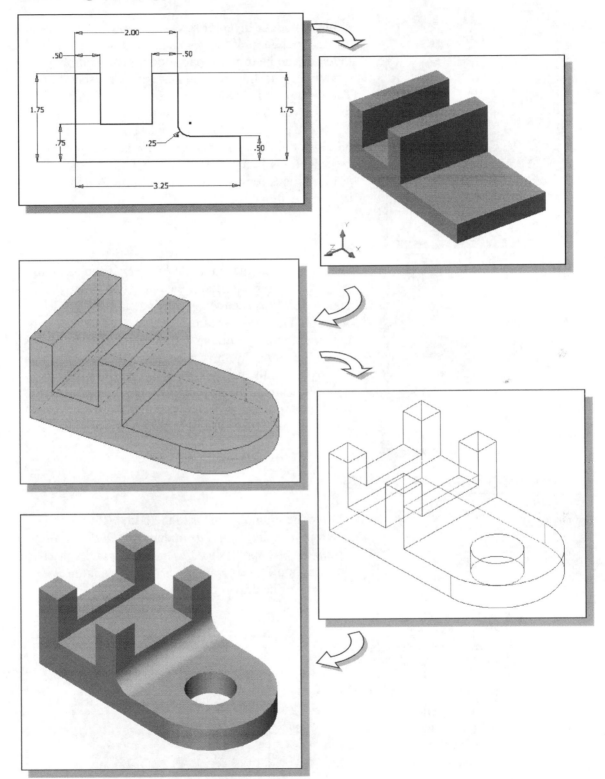

The Autodesk Inventor Browser

- In the Autodesk Inventor screen layout, the *browser* is located to the left of the graphics window. Autodesk Inventor can be used for part modeling, assembly modeling, part drawings, and assembly presentation. The *browser* window provides a visual structure of the features, constraints, and attributes that are used to create the part, assembly, or scene. The *browser* also provides right-click menu access for tasks associated specifically with the part or feature, and it is the primary focus for executing many of the Autodesk Inventor commands.

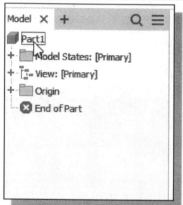

- The first item displayed in the *browser* is the name of the part, which is also the file name. By default, the name "Part1" is used when we first started Autodesk Inventor. The *browser* can also be used to modify parts and assemblies by moving, deleting, or renaming items within the hierarchy. Any changes made in the *browser* directly affect the part or assembly and the results of the modifications are displayed on the screen instantly. The *browser* also reports any problems and conflicts during the modification and updating procedure.

Creating the Base Feature

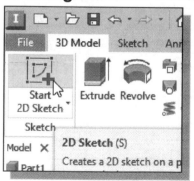

1. Move the graphics cursor to the **Start 2D Sketch** icon in the *Sketch toolbar* under the *3D Model tab*. A *Help-tip box* appears next to the cursor and a brief description of the command is displayed at the bottom of the drawing screen.

2. Move the cursor over the edge of the *XY Plane* in the graphics area. When the *XY Plane* is highlighted, click once with the **left-mouse-button** to select the *Plane* as the sketch plane for the new sketch.

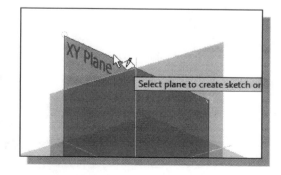

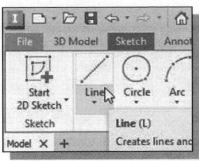

3. Select the **Line** icon by clicking once with the left-mouse-button; this will activate the Line command.

4. On your own, create and adjust the geometry by adding and modifying dimensions as shown below.

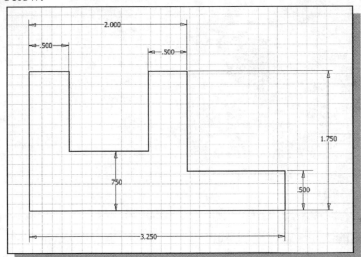

5. Inside the graphics window, click once with the right-mouse-button to display the option menu. Select **Finish 2D Sketch** in the pop-up menu to end the Sketch option.

6. On your own, use the dynamic viewing functions to view the sketch. Click the home view icon to change the display to the *isometric* view before proceeding to the next step.

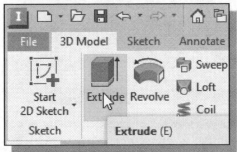

7. In the *Sketch toolbar* under the *3D Model tab*, select the **Extrude** command by left-clicking on the icon.

8. In the *Distance* option box, enter **2.5** as the total extrusion distance.

9. In the *Extrude* pop-up window, left-click once on the **Symmetric** icon. The
    Symmetric option allows us to extrude in both directions of the sketched profile.

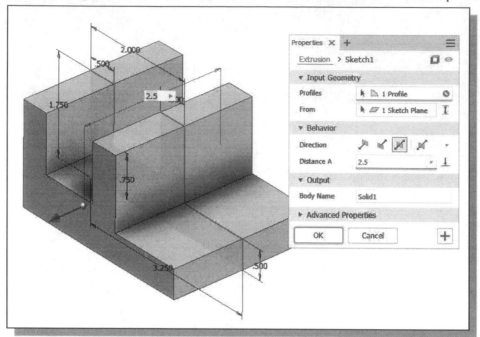

10. Click on the **OK** button to accept the settings and create the base feature.

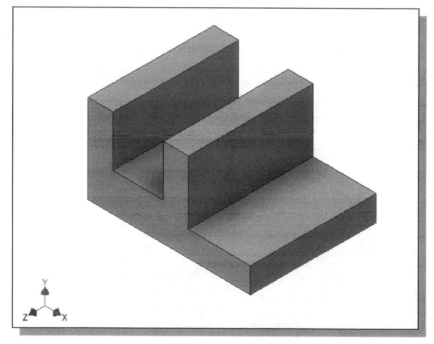

➢ On your own, use the *Dynamic Viewing* functions to view the 3D model. Also notice
    the extrusion feature is added to the *Model Tree* in the *browser* area.

Adding the Second Solid Feature

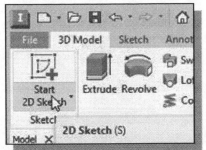

1. In the *Sketch toolbar* under the *3D Model tab* select the **Start 2D Sketch** command by left-clicking once on the icon.

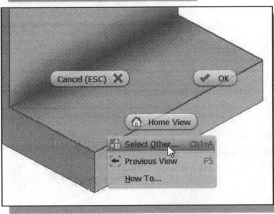

2. In the *Status Bar* area, the message *"Select plane to create sketch or an existing sketch to edit."* is displayed. Move the cursor inside the upper horizontal face of the 3D object as shown below.

3. Click once with the **right-mouse-button** to bring up the option menu and choose **Select Other** to switch to the next feasible choice.

4. On your own, click on the down arrow to examine all possible surface selections.

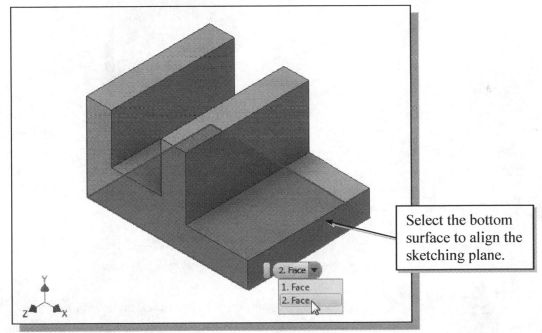

Select the bottom surface to align the sketching plane.

5. Select the **bottom horizontal face** of the solid model when it is highlighted as shown in the above figure.

Creating a 2D Sketch

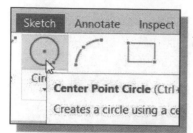

1. Select the **Center Point Circle** command by clicking once with the left-mouse-button on the icon in the *Sketch* tab.

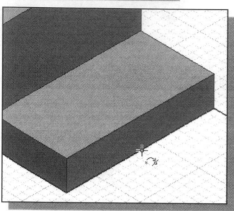

➢ We will align the center of the circle to the midpoint of the base feature.

2. On your own, use the snap to midpoint option to pick the midpoint of the edge when the midpoint is displayed with GREEN color as shown in the figure. (Hit [F6] to set the display orientation if necessary.)

3. Select the front corner of the base feature to create a circle as shown below.

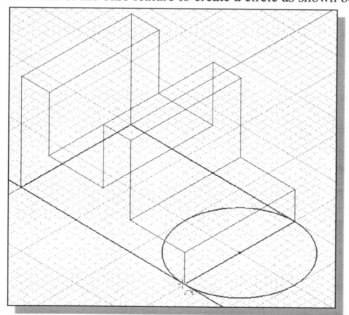

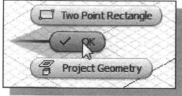

4. Inside the graphics window, click once with the right-mouse-button to display the option menu. Select **OK** in the pop-up menu to end the Circle command.

5. Inside the graphics window, click once with the right-mouse-button to display the option menu. Select **Finish 2D Sketch** in the pop-up menu to end the Sketch option.

6. In the *Features* toolbar (the toolbar that is located to the left side of the graphics window), select the **Extrude** command by clicking the left-mouse-button on the icon.

7. Move the cursor to the outside **semi-circle** we just created and left-click once to select the region as the **profile** to be extruded.

8. In the *Extrude* pop-up control, set the operation option to **Join**.

9. Also set the *Distance* option to **To Selected Face** as shown below.

10. Select the top face of the base feature as the termination surface for the extrusion.

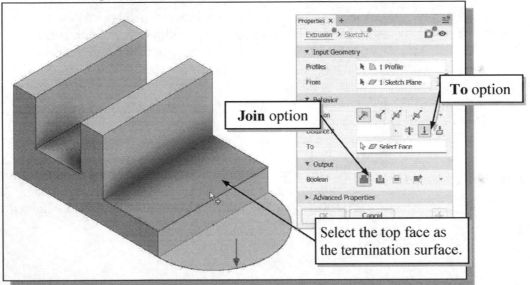

11. Click on the **OK** button to proceed with the *Join* operation.

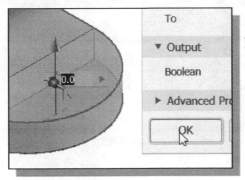

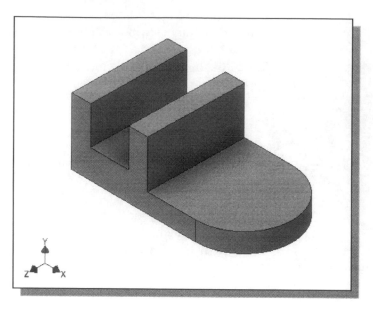

Renaming the Part Features

Currently, our model contains two extruded features. The feature is highlighted in the display area when we select the feature in the *browser* window. Each time a new feature is created, the feature is also displayed in the *Model Tree* window. By default, Autodesk Inventor will use generic names for part features. However, when we begin to deal with parts with a large number of features, it will be much easier to identify the features using more meaningful names. Two methods can be used to rename the features: 1. **Clicking** twice on the name of the feature and 2. Using the **Properties** option. In this example, the use of the first method is illustrated.

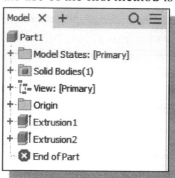

1. Select the first extruded feature in the *model browser* area by left-clicking once on the name of the feature, **Extrusion1**. Notice the selected feature is highlighted in the graphics window.

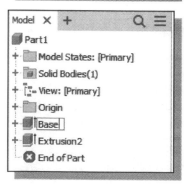

2. Left-click again on the feature name to enter the *Edit* mode as shown.

3. Enter **Base** as the new name for the first extruded feature.

4. On your own, rename the second extruded feature to **Circular_End**.

Adjusting the Width of the Base Feature

One of the main advantages of parametric modeling is the ease of performing part modifications at any time in the design process. Part modifications can be done through accessing the features in the history tree. Autodesk Inventor remembers the history of a part, including all the rules that were used to create it, so that changes can be made to any operation that was performed to create the part. For our *Saddle Bracket* design, we will reduce the size of the base feature from 3.25 inches to 3.0 inches, and the extrusion distance to 2.0 inches.

1. Select the first extruded feature, **Base**, in the *browser* area. Notice the selected feature is highlighted in the graphics window.

2. Inside the *browser* area, **right-mouse-click** on the first extruded feature to bring up the option menu and select the **Show Dimensions** option in the pop-up menu.

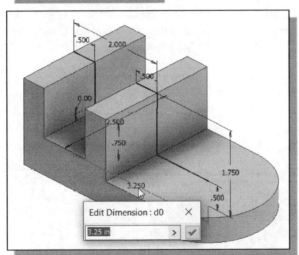

3. All dimensions used to create the **Base** feature are displayed on the screen. Select the overall width of the **Base** feature, the **3.25** dimension value, by double-clicking on the dimension text as shown.

4. Enter **3.0** in the *Edit Dimension* window.

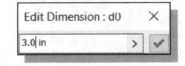

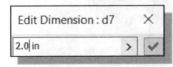

5. On your own, repeat the above steps and modify the extruded distance from **2.5** to **2.0**.

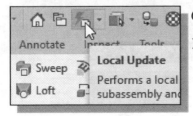

6. Click **Local Update** in the *Quick Access Toolbar*.

➤ Note that Autodesk Inventor updates the model by re-linking all elements used to create the model. Any problems or conflicts that occur will also be displayed during the updating process.

Adding a Placed Feature

1. In the *Sketch tab*, select the **Hole** command by left-clicking on the icon.

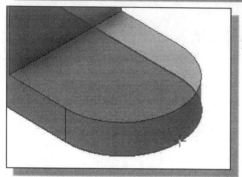

2. Pick the **bottom plane** of the solid model as the placement plane as shown.

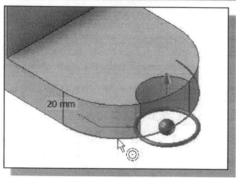

3. Pick the **bottom arc** and notice the *concentric reference symbol* appears indicating the center of the hole will be aligned to the center of the selected arc.

4. Set the hole diameter to **0.75 in** as shown.

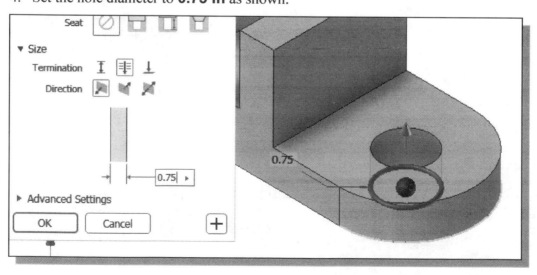

5. Set the termination option to **Through All** as shown.

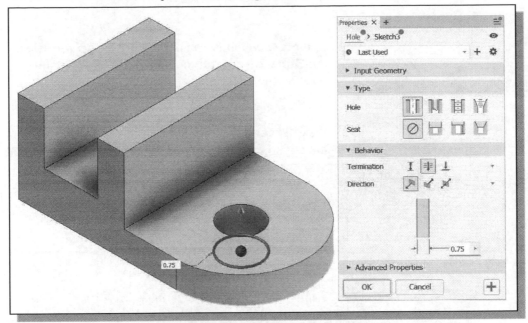

6. Click **OK** to accept the settings and create the *Hole* feature.

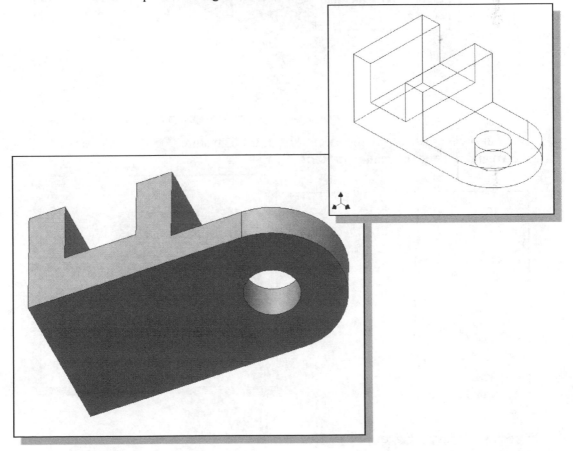

Creating a Rectangular Cut Feature

1. In the *Sketch toolbar* under the *3D Model tab* select the **Start 2D Sketch** command by left-clicking once on the icon.

2. Pick the **vertical face** of the solid as shown. (Note the alignment of the origin of the sketch plane.)

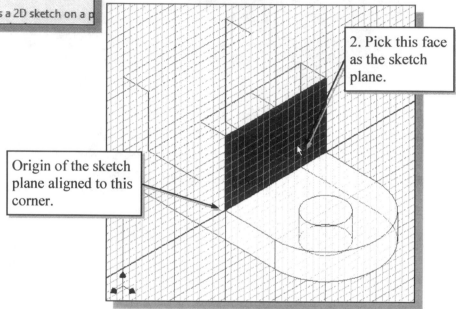

2. Pick this face as the sketch plane.

Origin of the sketch plane aligned to this corner.

➤ On your own, create a rectangular cut (**1.0 x 0.75**) feature (**To Next** option) as shown and rename the feature to **Rect_Cut**.

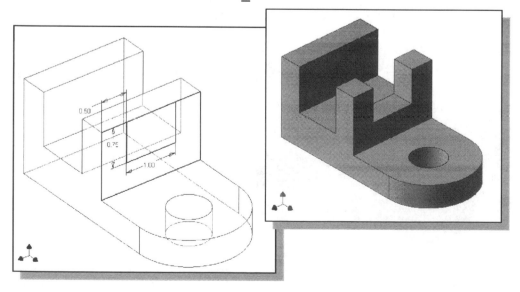

History-Based Part Modifications

Autodesk Inventor uses the *history-based part modification* approach, which enables us to make modifications to the appropriate features and re-link the rest of the history tree without having to reconstruct the model from scratch. We can think of it as going back in time and modifying some aspects of the modeling steps used to create the part. We can modify any feature that we have created. As an example, we will adjust the depth of the rectangular cutout.

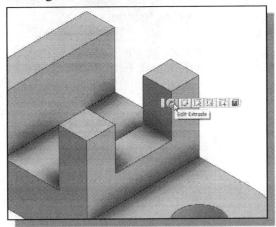

1. In the *graphics* window, select the last cut feature, **Rect_Cut**, by left-clicking on one surface of the feature.

2. Select **Edit Feature** in the pop-up menu. Notice the *Extrude* dialog box appears on the screen.

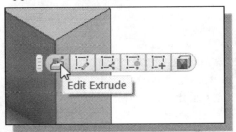

- Note that the feature pop-up options are also available through the Model History Tree, as shown on the next page.

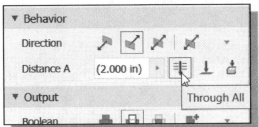

3. In the *Extrude* dialog box, set the termination *Extents* to the **Through All** option.

4. Click on the **OK** button to accept the settings.

- As can be seen, the history-based modification approach is very straight forward, and it only took a few seconds to adjust the cut feature to the **Through All** option.

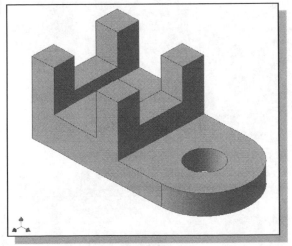

A Design Change

Engineering designs usually go through many revisions and changes. Autodesk Inventor provides an assortment of tools to handle design changes quickly and effectively. We will demonstrate some of the tools available by changing the **Base** feature of the design.

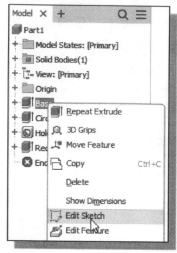

1. In the *browser* window, select the **Base** feature by left-clicking once on the name of the feature.

2. In the *browser*, right-click once on the **Base** feature to bring up the option menu; then pick **Edit Sketch** in the pop-up menu.

3. Click **Home** to reset the display to *Isometric*.

❖ Autodesk Inventor will now display the original 2D sketch of the selected feature in the graphics window. We have literally gone back to the point where we first created the 2D sketch. Notice the feature being modified is also highlighted in the desktop *browser*.

4. Click on the **Look At** icon in the *Standard* toolbar area.

•   The **Look At** command automatically aligns the *sketch plane* of a selected entity to the screen.

5. Select any line segment of the 2D sketch to reset the display to align to the 2D sketch.

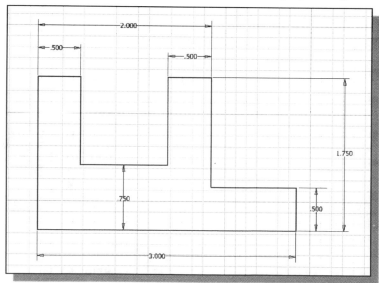

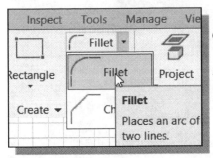

6. Select the **Fillet** command in the *2D Sketch* toolbar.

7. In the graphics window, enter **0.25** as the new radius of the fillet.

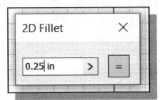

8. Select the two edges as shown to create the fillet.

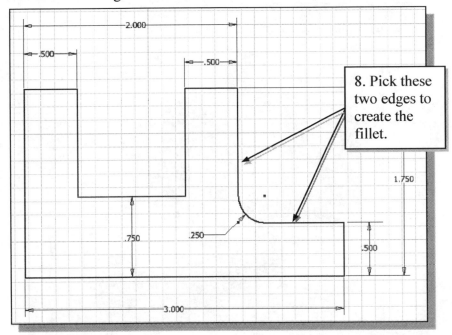

8. Pick these two edges to create the fillet.

- Note that the fillet is created automatically with the dimension attached. The attached dimension can also be modified through the history tree.

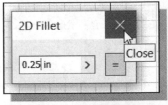

9. Click on the [**X**] icon in the *2D Fillet* window to end the Fillet command.

10. Select **Finish Sketch** in the *Ribbon* toolbar to end the Sketch option.

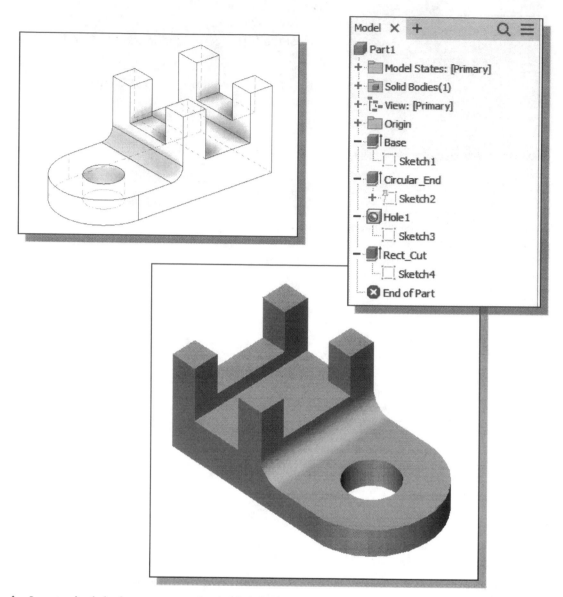

❖ In a typical design process, the initial design will undergo many analyses, testing, and reviews. The *history-based part modification* approach is an extremely powerful tool that enables us to quickly update the design. At the same time, it is quite clear that PLANNING AHEAD is also important in doing feature-based modeling.

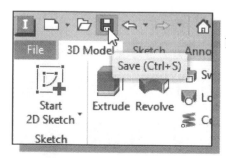

11. On your own, create the **Chapter5** folder and save the model as *Saddle_Bracket.ipt*.

Assigning and Calculating the Associated Physical Properties

Autodesk Inventor models have properties called **iProperties**. The *iProperties* can be used to create reports and update assembly bills of materials, drawing parts lists, and other information. With *iProperties*, we can also set and calculate physical properties for a part or assembly using the material library. This allows us to examine the physical properties of the model, such as weight or center of gravity.

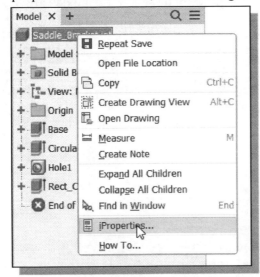

1. In the *browser*, **right-click** once on the *part name* to bring up the option menu; then pick **iProperties** in the *pop-up* menu.

2. On your own, look at the different information listed in the *iProperties* dialog box.

3. Click on the **Physical** tab; this is the page that contains the physical properties of the selected model.

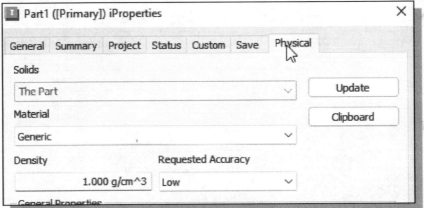

- Note that the *Material* option is not assigned, and none of the physical properties are shown.

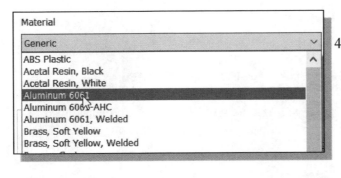

4. Click the down-arrow in the *Material* option to display the material list, and select **Aluminum-6061** as shown.

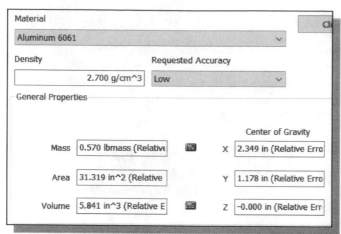

❖ Note the *General Properties* area now has the *Mass*, *Area*, *Volume* and the *Center of Gravity* information of the model based on the density of the selected material.

5. Click on the **Global** button to display the *Mass Moments Inertial* of the design.

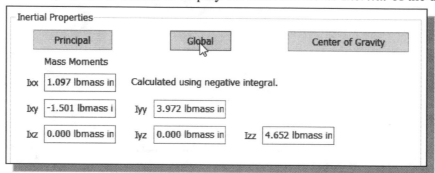

6. On your own, select **Cast Iron** as the *Material* type and compare the differences in using the different materials.

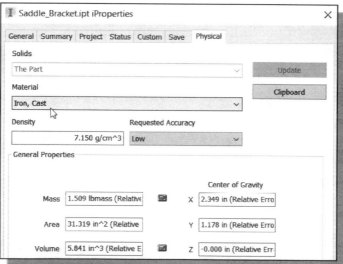

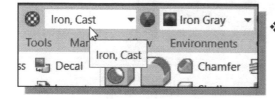

❖ Also note the *Material* can be assigned through the quick access menu as shown.

Review Questions: (Time: 30 minutes.)

1. What are stored in the Autodesk Inventor *History Tree*?

2. When extruding, what is the difference between *Distance* and *Through All*?

3. Describe the *history-based part modification* approach.

4. What determines how a model reacts when other features in the model change?

5. Describe the steps to rename existing features.

6. Describe two methods available in Autodesk Inventor to *modify the dimension values* of parametric sketches.

7. Create *History Tree sketches* showing the steps you plan to use to create the two models shown on the next page:

Ex.1)

Ex.2)

Exercises: Create and save the exercises in the Chapter5 folder.
(Time: 180 minutes.)

1. **C-Clip** (Dimensions are in inches. Plate thickness: **0.25 inches**.)

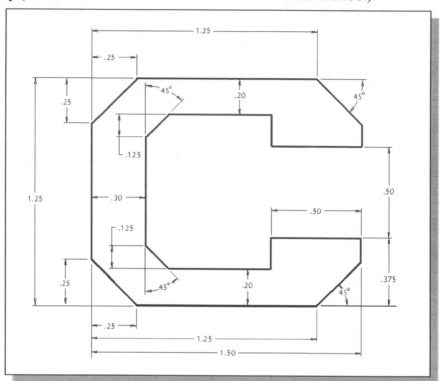

2. **Tube Mount** (Dimensions are in inches.)

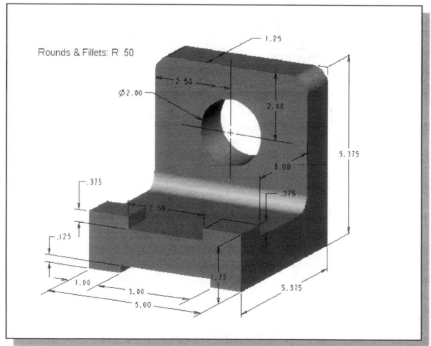

3. **Hanger Jaw** (Dimensions are in inches. Volume =?)

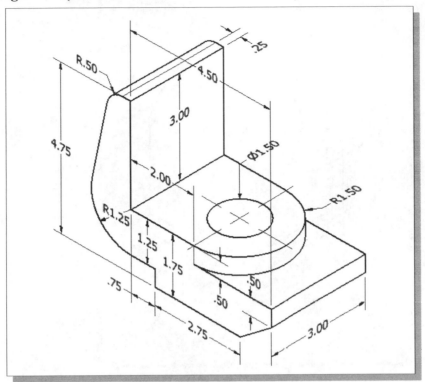

4. **Transfer Fork** (Dimensions are in inches. Material: **Cast Iron.** Volume =?)

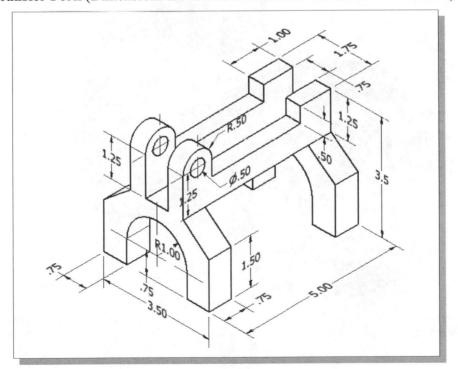

5. **Guide Slider** (Material: **Cast Iron**. Weight and Volume =?)

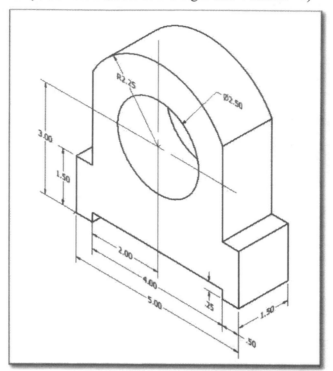

6. **Shaft Guide** (Material: **Aluminum-6061**. Mass and Volume =?)

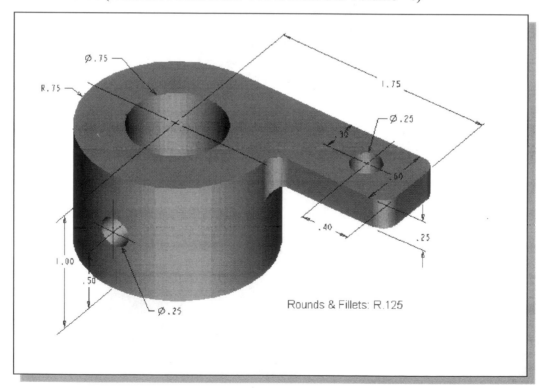

Chapter 6
Geometric Construction Tools

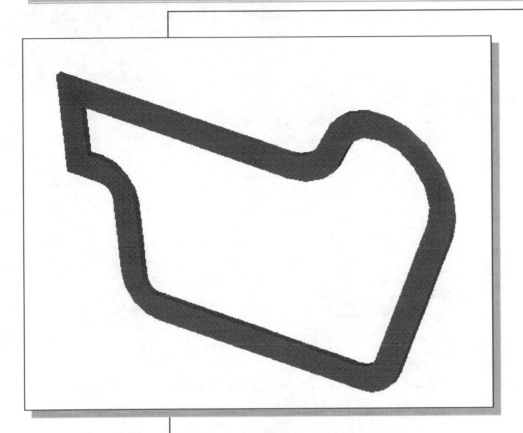

Learning Objectives

- ◆ **Apply Geometry Constraints**
- ◆ **Use the Trim/Extend Command**
- ◆ **Use the Offset Command**
- ◆ **Understand the Profile Sketch Approach**
- ◆ **Create Projected Geometry**
- ◆ **Understand and Use Reference Geometry**
- ◆ **Edit with Click and Drag**
- ◆ **Use the Auto Dimension Command**

Autodesk Inventor Certified User Exam Objectives Coverage

Parametric Modeling Basics

Section 3: Sketches

Objectives: Creating 2D Sketches, Draw Tools, Sketch Constraints, Pattern Sketches, Modify Sketches, Format Sketches, Sketch Doctor, Shared Sketches, Sketch Parameters.

Section 4: Parts

Objectives: Creating parts, Work Features, Pattern Features, Part Properties.

Introduction

The main characteristics of solid modeling are the accuracy and completeness of the geometric database of the three-dimensional objects. However, working in three-dimensional space using input and output devices that are largely two-dimensional in nature is potentially tedious and confusing. Autodesk Inventor provides an assortment of two-dimensional construction tools to make the creation of wireframe geometry easier and more efficient. Autodesk Inventor includes two types of wireframe geometry: ***curves*** and ***profiles***. Curves are basic geometric entities such as lines, arcs, etc. Profiles are a group of curves used to define a boundary. A *profile* is a closed region and can contain other closed regions. Profiles are commonly used to create extruded and revolved features. An *invalid profile* consists of self-intersecting curves or open regions. In this lesson, the basic geometric construction tools, such as Trim and Extend, are used to create profiles. The Autodesk Inventor's ***profile sketch*** approach to creating profiles is also introduced. Mastering the geometric construction tools along with the application of proper constraints and parametric relations is the true essence of *parametric modeling*.

The Gasket Design

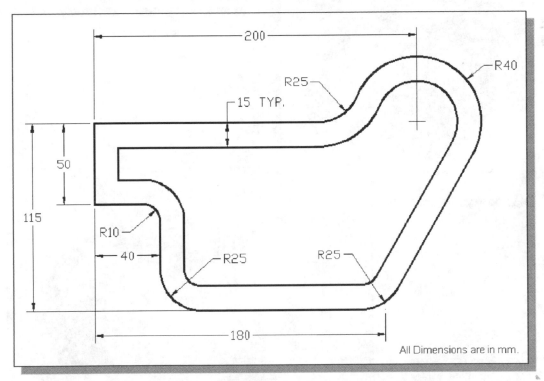

❖ Based on your knowledge of Autodesk Inventor so far, how would you create this design? What is the most difficult geometry involved in the design? Take a few minutes to consider a modeling strategy and do preliminary planning by sketching on a piece of paper. You are also encouraged to create the design on your own prior to following through the tutorial.

Modeling Strategy

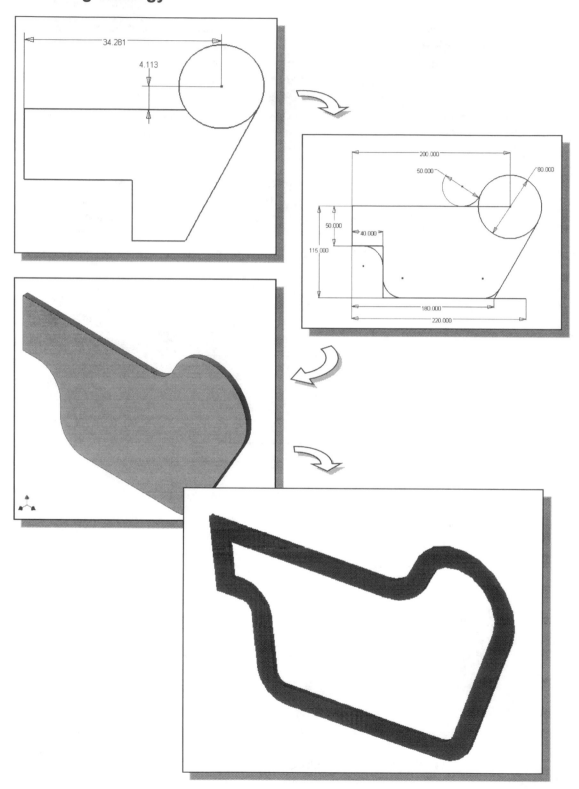

Starting Autodesk Inventor

1. Select the **Autodesk Inventor** option on the *Start* menu or select the **Autodesk Inventor** icon on the desktop to start Autodesk Inventor. The Autodesk Inventor main window will appear on the screen.

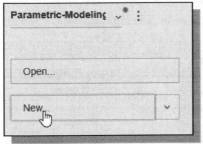

2. Select the **New File** icon with a single click of the left-mouse-button as shown.

3. Select the **Metric** tab as shown. We will use the millimeter (mm) setting for this design.

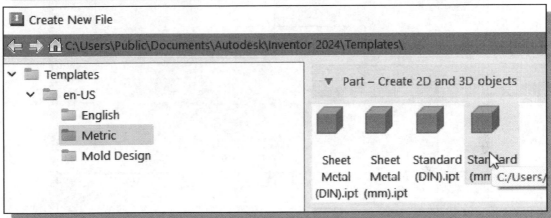

4. In the *New File* dialog box, use the scroll bar on the right and select the **Standard(mm).ipt** icon as shown.

5. Click **Create** in the *New File* dialog box to accept the selected settings to start a new model.

6. Click once with the left-mouse-button to select the **Start 2D Sketch** command. Click once with the **left-mouse-button** to select the *XY Plane* as the sketch plane for the new sketch.

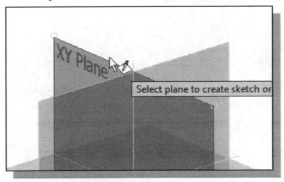

Create a 2D Sketch

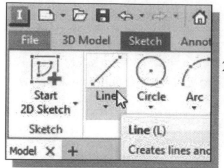

1. Click on the **Line** icon in the *Sketch* tab on the Ribbon.

2. Create a sketch as shown in the figure below. Start the sketch from the top right corner. The line segments are all parallel and/or perpendicular to each other. We will intentionally create arbitrary line segments, as it is quite common during the initial design stage that most of the shapes and forms are undetermined.

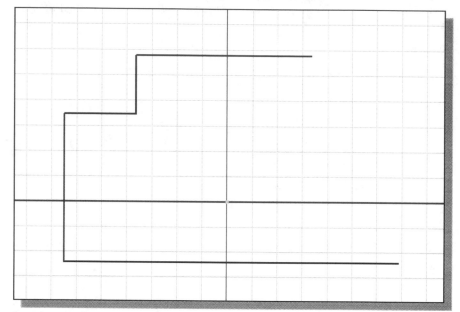

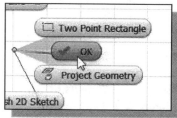

3. Inside the graphics window, right-click to bring up the option menu, and select **OK** to end the Line command.

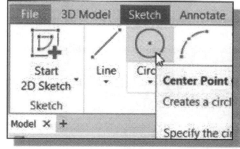

4. Select the **Center Point Circle** command by clicking once with the left-mouse-button on the icon in the *Sketch* tab on the Ribbon.

5. Pick a location that is above the bottom horizontal line as the center location of the circle.

6. Move the cursor toward the right and create a circle of arbitrary size by clicking once with the left-mouse-button.

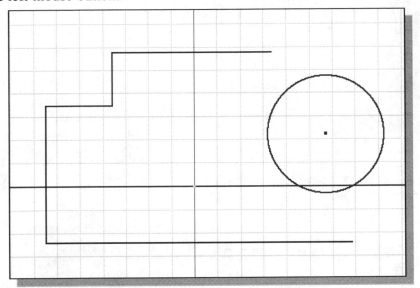

7. Click on the **Line** icon in the *Sketch* tab on the Ribbon.

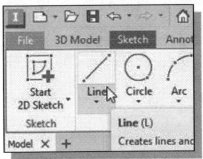

8. Move the cursor near the upper portion of the circle and pick a location on the circle when the **Coincident** constraint symbol is displayed.

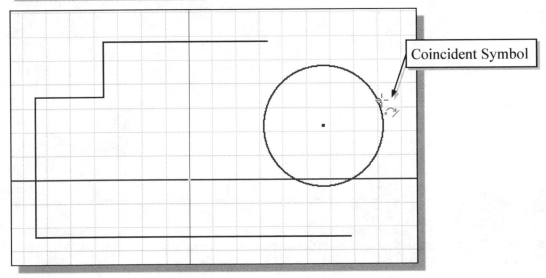

Coincident Symbol

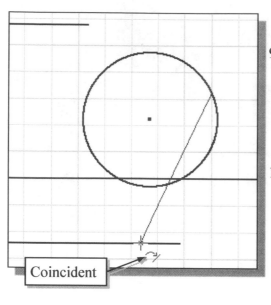

Coincident

9. For the other end of the line, select a location that is on the lower horizontal line and about one-third from the right endpoint. Notice the **Coincident** constraint symbol is displayed when the cursor is on the horizontal line.

10. Inside the graphics window, right-click to bring up the option menu and select **OK** to end the **Line** command.

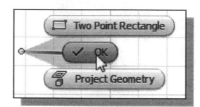

Edit the Sketch by Dragging the Sketched Entities

In Autodesk Inventor, we can click and drag any under-constrained curve or point in the sketch to change the size or shape of the sketched profile. As illustrated in the previous chapter, this option can be used to identify under-constrained entities. This *Editing by dragging* method is also an effective visual approach that allows designers to quickly make changes.

1. Move the cursor on the lower left vertical edge of the sketch. Click and drag the edge to a new location that is toward the right side of the sketch.

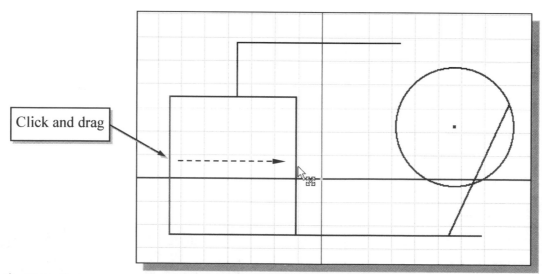

Click and drag

❖ Note that we can only drag the vertical edge horizontally; the connections to the two horizontal lines are maintained while we are moving the geometry.

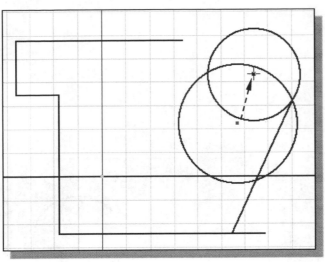

2. Click and drag the center point of the circle to a new location.

❖   Note that as we adjust the size and the location of the circle, the connection to the inclined line is maintained.

3. Click and drag the lower endpoint of the inclined line to a new location.

❖   Note that as we adjust the size and the location of the inclined line the location of the bottom horizontal edge is also adjusted.

❖   Note that several changes occur as we adjust the size and the location of the inclined line. The location of the bottom horizontal line and the length of the vertical line are adjusted accordingly.

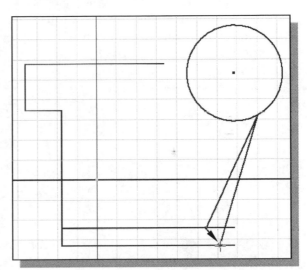

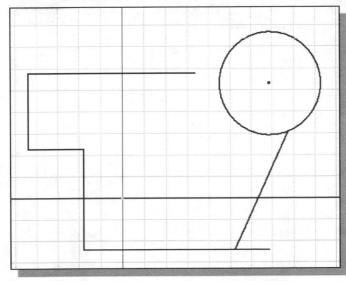

4. On your own, adjust the sketch so that the shape of the sketch appears roughly as shown.

❖   The *Editing by dragging* method is an effective approach that allows designers to explore and experiment with different design concepts.

Add Additional Constraints

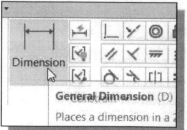

1. Choose **General Dimension** in the *Sketch* panel.

2. Add the horizontal location dimension, from the top left vertical edge to the center of the circle as shown. (Do not be overly concerned with the dimensional value; we are still working on creating a *rough sketch*.)

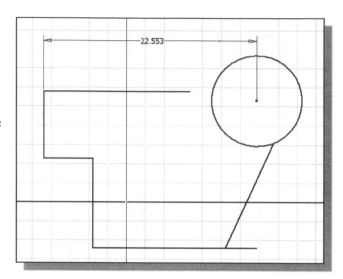

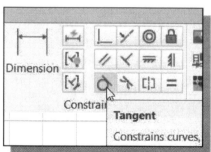

3. Click on the **Tangent** constraint icon in the *Sketch* panel.

4. Pick the inclined line by left-clicking once on the geometry.

5. Pick the circle. The sketched geometry is adjusted as shown below.

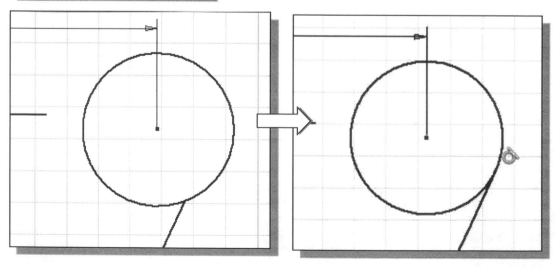

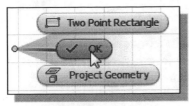

6. Inside the graphics window, right-click to bring up the option menu.

7. In the option menu, select **OK** to end the Tangent command.

8. Use the drag and drop approach, of the circle center, to observe the relations of the sketched geometry.

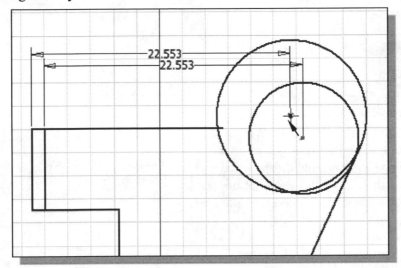

❖ Note that the dimension we added now restricts the horizontal movement of the center of the circle. The tangent relation to the inclined line is maintained.

Use the Trim and Extend Commands

In the following sections, we will illustrate using the Trim and Extend commands to complete the desired 2D profile.

The **Trim** and **Extend** commands can be used to shorten/lengthen an object so that it ends precisely at a boundary. As a general rule, Autodesk Inventor will try to clean up sketches by forming a closed region sketch. Also note that while we are in either **Trim** or **Extend**, we can press the [**Shift**] key to switch to the opposite operation.

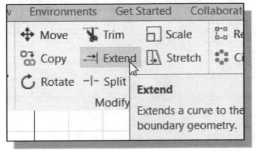

1. Choose **Extend** in the *Modify toolbar* panel. The message "*Extend curves*" is displayed in the prompt area.

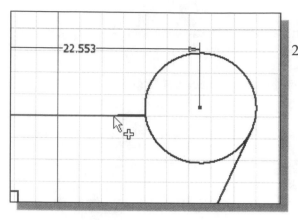

2. We will first extend the top horizontal line to the circle. Move the cursor near the right-hand endpoint of the top horizontal line. Autodesk Inventor will automatically display the possible result of the selection.

3. Next, we will trim the bottom horizontal line to the inclined line. The Trim operation can be activated by selecting the Trim icon in the *Sketch* panel or by pressing down the [**Shift**] key while we are in the Extend command. Move the cursor near the right-hand endpoint of the bottom horizontal line and select the line and press down the [**Shift**] key. Autodesk Inventor will display a dashed line indicating the portion of the line that will be trimmed.

> 3. Press down the [**Shift**] key and move the cursor near the right end of the line.

4. Left-click once on the line to perform the **Trim** operation.

5. On your own, create the vertical location dimension as shown below.

6. Adjust the dimension to **0.0** so that the horizontal line and the center of the circle are aligned horizontally.

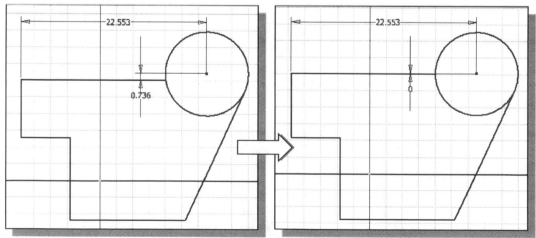

7. On your own, create the **height dimension** to the left and use the **Show Constraints** command to examine the applied constraints. Confirm that a **Perpendicular constraint** is applied to the horizontal and vertical lines as shown.

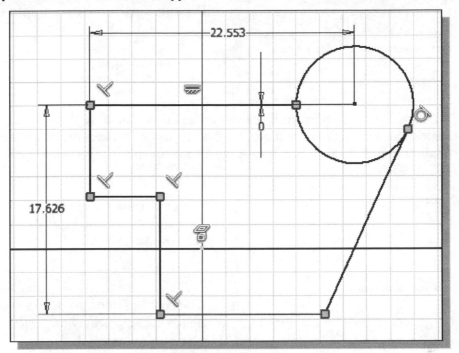

The Auto Dimension Command

In Autodesk Inventor, we can use the **Auto Dimension** command to assist in creating a fully constrained sketch. Fully constrained sketches can be updated more predictably as design changes are implemented. The general procedure for applying dimensions to sketches is to use the **General Dimension** command to add the desired key dimensions, and then use the **Auto Dimension** command as a quick way to calculate all other sketch dimensions necessary. Autodesk Inventor remembers which dimensions were added by the user and which were calculated by the system, so that automatic dimensions do not replace the desired dimensions.

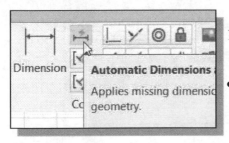

1. Click on the **Auto Dimension** icon in the *Sketch* panel.

- Note that 6 dimensions are needed to fully constrain the sketch, as indicated in the *Status Bar*.

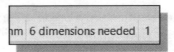

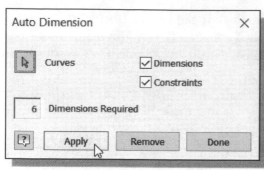

2. The *Auto Dimension* dialog box appears on the screen. Confirm that the **Dimensions** and **Constraints** options are switched *ON* as shown.

3. Click on the **Apply** button to proceed with the Auto Dimension command.

➢ Note that the system automatically calculates additional dimensions for all created geometric entities. Note that at times, the system applied dimensions might not be adequate for the designs. It is best to apply all the key dimensions prior to using the **Auto Dimension** command.

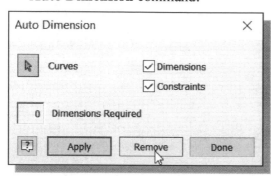

➢ The dimensions created by the **Auto Dimension** command can also be **Removed**.

4. Click on the **Remove** button to undo the dimensions created by the *Auto Dimension* command.

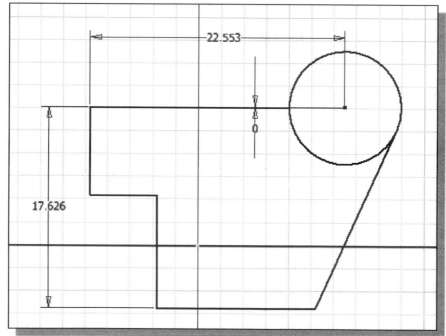

➢ Note the **Remove** option only removes the dimensions created by the **Auto Dimension** command but maintains the dimensions created by the user.

5. On your own, **trim** the circle and add the additional dimensions as shown.

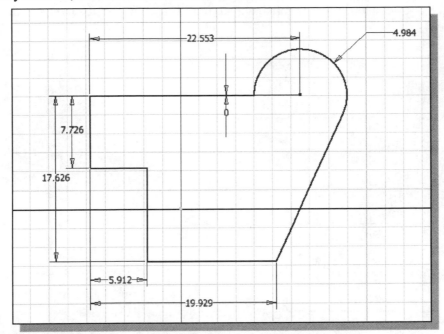

6. Note the display in the *Status Bar* indicates two additional dimensions are still needed to fully constrain the sketch.

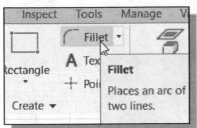

2 dimensions needed   1

Create Fillets and Completing the Sketch

1. Click on the **Fillet** icon in the *Sketch* panel.

2. The *2D Fillet* radius dialog box appears on the screen. Use the **default** radius value and click on the top horizontal line and the arc to create a fillet as shown.

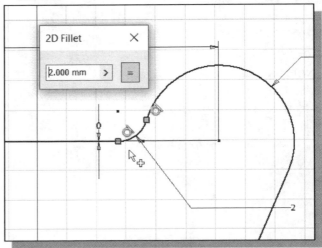

3. On your own, create the three additional fillets as shown in the below figure. Note that the Equal constraint is activated, and all rounds and fillets are created with the constraint.

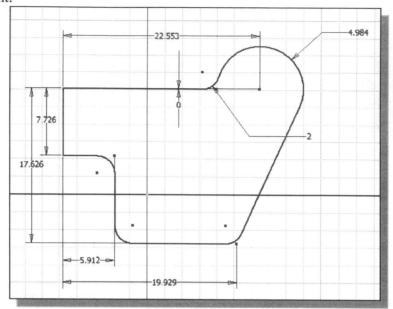

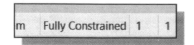

4. Click on the [**X**] icon to close the *2D Fillet* dialog box and end the Fillet command.

Fully Constrained Geometry

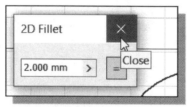

1. Click on the **Fix** constraint icon in the *Sketch Panel*.

2. Apply the **Fix** constraint to the center of the large arc as shown below.

➢ Note that the sketch is now fully constrained.

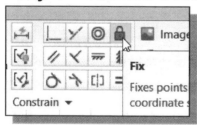

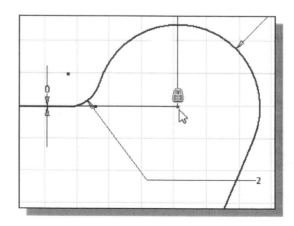

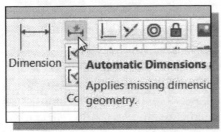

3. Click on the **Auto Dimension** icon in the *Sketch* panel.

➢ Note that the system automatically recalculates if any additional dimension is needed for all created geometric entities. The applied Fix constraint provided the previously missing location dimensions for the created sketches.

4. Click **Done** to exit the Auto Dimension command.

5. On your own, complete the sketch by adjusting the dimensions to the desired values as shown below. (Hint: Change the larger values first.)

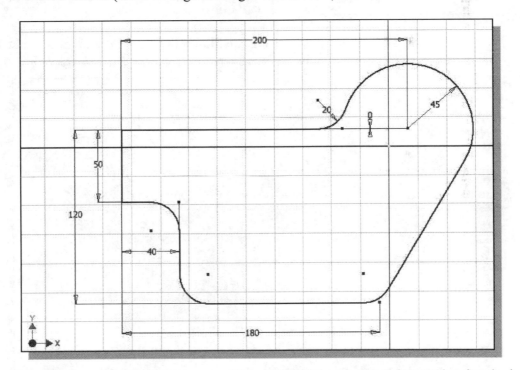

❖ Note that by applying proper geometric and dimensional constraints to the sketched geometry a *constraint network* is created that assures the geometry shape behaves predictably as changes are made.

Profile Sketch

In Autodesk Inventor, ***profiles*** are closed regions that are defined from sketches. Profiles are used as cross sections to create solid features. For example, **Extrude**, **Revolve**, **Sweep**, **Loft**, and **Coil** operations all require the definition of at least a single profile. The sketches used to define a profile can contain additional geometry since the additional geometry entities are consumed when the feature is created. To create a profile we can create single or multiple closed regions, or we can select existing solid edges to form closed regions. A profile cannot contain self-intersecting geometry; regions selected in a single operation form a single profile. As a general rule, we should dimension and constrain profiles to prevent them from unpredictable size and shape changes. Autodesk Inventor does allow us to create under-constrained or non-constrained profiles; the dimensions and/or constraints can be added/edited later.

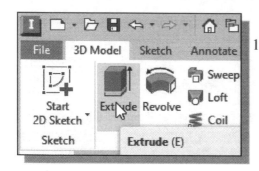

1. In the *3D Model* tab, select the **Extrude** command by left-clicking once on the icon.

❖ Note that Autodesk Inventor automatically highlights the closed region of the sketch. The defining geometry now forms the **profile** required for the *Extrude* operation.

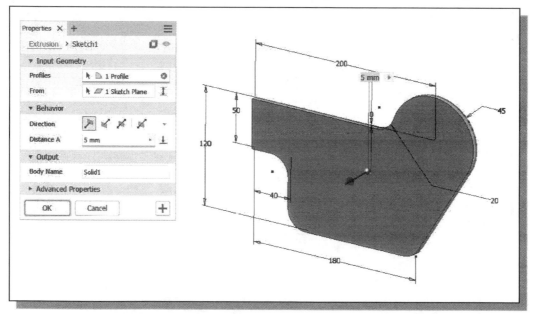

2. In the *Extrude* dialog box, enter **5 mm** as the extrusion distance as shown.

3. Click on the **OK** button to accept the settings and create the solid feature.

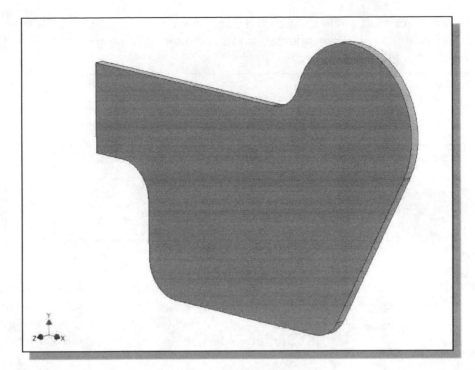

❖ Note that all the sketched geometry entities and dimensions are consumed and have disappeared from the screen when the feature is created.

Redefine the Sketch and Profile

Engineering designs usually go through many revisions and changes. Autodesk Inventor provides an assortment of tools to handle design changes quickly and effectively. We will demonstrate some of the tools available by changing the base feature of the design. The profile used to create the extrusion is selected from the sketched geometry entities. In Autodesk Inventor, any profile can be edited and/or redefined at any time. It is this type of functionality in parametric solid modeling software that provides designers with greater flexibility and the ease to experiment with different design considerations.

1. In the *browser*, **right-click** once on the **Extrusion1** feature to bring up the option menu, and then pick **Edit Sketch** in the pop-up menu.

2. On your own, create a circle and a rectangle of arbitrary sizes, positioned as shown in the figure below. We will intentionally under-constrain the new sketch to illustrate the flexibility of the system.

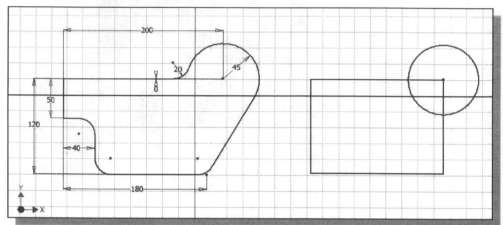

3. Inside the *graphics window*, click once with the right-mouse-button to display the option menu. Select **Finish 2D Sketch** in the pop-up menu to end the Sketch option.

➢ Note that, at this point, the solid model remains the same as before. We will next redefine the profile used for the base feature.

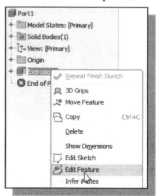

4. In the *browser*, right-click once on the Extrusion1 feature to bring up the option menu, and then pick **Edit Feature** in the pop-up menu.

❖ Autodesk Inventor will now display the 2D sketch of the selected feature in the graphics window. We have literally gone back in time to the point where we defined the extrusion feature.

5. Click on the **Look At** icon in the *Standard* toolbar area.

6. Click on the front face of the gasket design.

• The **Look At** command automatically aligns the *sketch plane* of a selected entity to the screen.

7. Select any line segment of the 2D sketch.

8. Click on the **Zoom All** icon in the *Standard* toolbar.

➢ We have literally gone back to the point where we first defined the 2D profile. The original sketch and the new sketch we just created are wireframe entities recorded by the system as belonging to the same **SKETCH**, but only the **PROFILED** entities are used to create the feature.

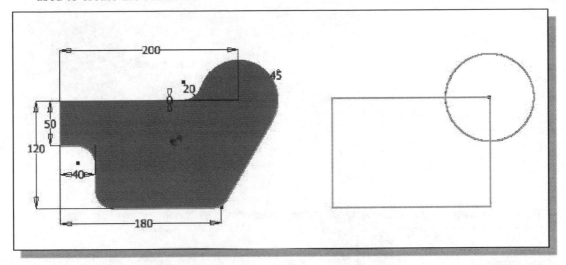

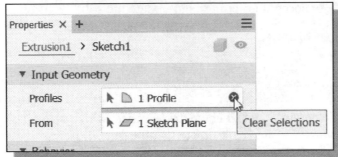

9. In the *Extrude* control panel, click on the **[x]** button to clear the current selected 2D profile.

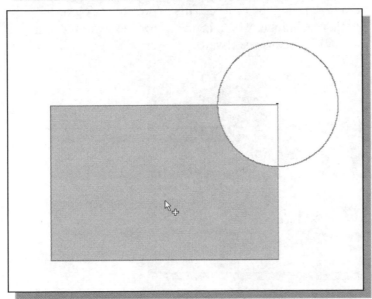

10. Click inside the lower region of the rectangle of the sketch on the right side of the graphics window.

➢ Autodesk Inventor automatically selects the geometry entities that form a closed region to define the profile.

11. Click the inside regions of the circle and complete the profile definition as shown in the figure below.

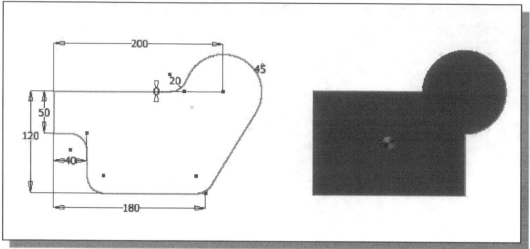

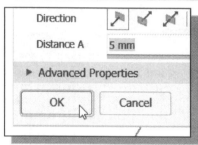

12. In the *Extrude* control panel, click on the **OK** button to accept the settings and update the solid feature.

• The feature is recreated using the newly sketched geometric entities, which are under-constrained and with no dimensions. The profile is created with extra wireframe entities by selecting multiple regions. The extra geometry entities can be used as construction geometry to help in defining the profile. This approach encourages engineering content over drafting technique, which is one of the key features of Autodesk Inventor over other solid modeling software.

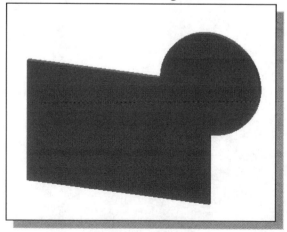

13. On your own, repeat the above steps and reset the profile back to the original gasket sketch.

Create an Offset Cut Feature

To complete the design, we will create a cutout feature by using the Offset command. First, we will set up the sketching plane to align with the front face of the 3D model.

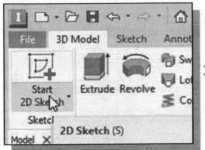

1. In the *3D Model* tab select the **Start 2D Sketch** command by left-clicking once on the icon.

2. In the *Status Bar* area, the message "*Select plane to create sketch or an existing sketch to edit.*" is displayed. Select the **front face** of the 3D model in the graphics window.

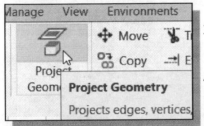

3. Click on the **Project Geometry** icon in the *2D Create* panel.

4. Select the front face of the gasket model, and notice the outline of the design has been projected onto the sketching plane.

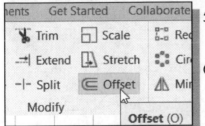

5. Click on the **Offset** icon in the *Sketch* tab on the Ribbon.

6. *Right-click* once to bring up the option menu and activate "**Loop select**" if necessary, as shown.

7. Select any edge of the **front face** of the 3D model. Autodesk Inventor will automatically select all of the connecting geometry to form a closed region.

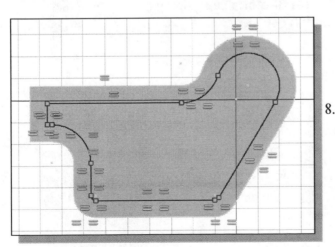

8. Move the cursor toward the center of the selected region and notice an offset copy of the outline is displayed.

9. Left-click once to create the offset profile as shown.

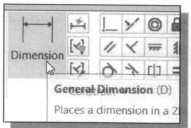

10. On your own, use the **General Dimension** command to create the offset dimension as shown in the figure below.

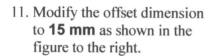

11. Modify the offset dimension to **15 mm** as shown in the figure to the right.

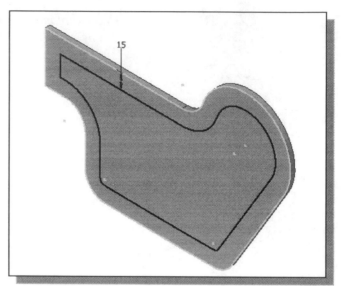

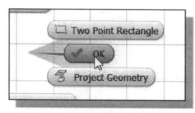

12. Inside the graphics window, click once with the right-mouse-button to display the option menu. Select **OK** in the pop-up menu to end the General Dimension command.

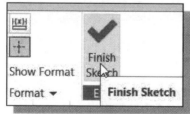

13. Inside the graphics window, click once with the right-mouse-button to display the option menu. Select **Finish Sketch** in the pop-up menu to end the Sketch option.

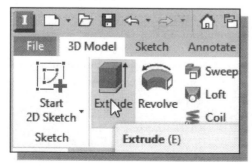

14. In the *3D Model* toolbar (the toolbar that is located to the left side of the graphics window), select the **Extrude** command by left-clicking on the icon.

15. Select the **inside region** of the offset geometry as the profile for the extrusion.

16. Inside the *Extrude* dialog box, select the **Cut** operation, set the *Extents* to **All** and set the direction of the cutout as shown in the figure below.

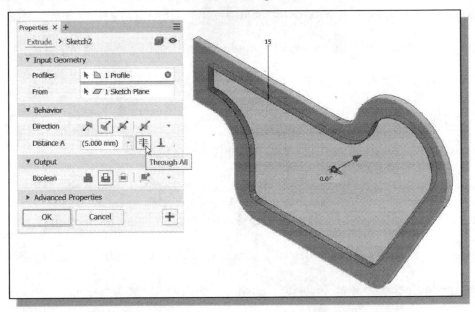

17. In the *Extrude* dialog box, click on the **OK** button to accept the settings and create the solid feature.

➢ The offset geometry is associated with the original geometry. On your own, adjust the overall height of the design to **150 millimeters** and confirm that the offset geometry is adjusted accordingly.

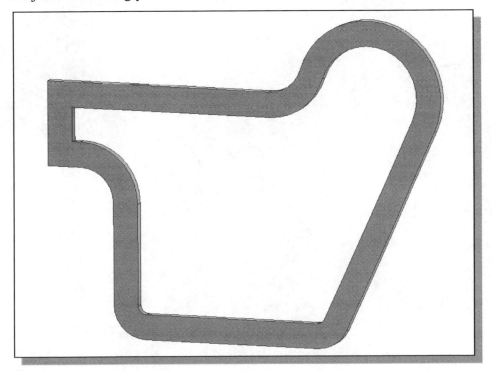

Review Questions: (Time: 25 minutes)

1. What are the two types of wireframe geometry available in Autodesk Inventor?

2. Can we create a profile with extra 2D geometry entities?

3. How do we access the Autodesk Inventor's **Edit Sketch** option?

4. How do we create a *profile* in Autodesk Inventor?

5. Can we build a profile that consists of self-intersecting curves?

6. Describe the procedure to create a copy of a sketched 2D wireframe geometry.

7. Can we create additional entities in a 2D sketch, without using them at all?

8. How do we align the *sketch plane* of a selected entity to the screen?

9. Describe the steps we used to switch existing profiles in the tutorial.

10. Describe the advantages of using the Offset command vs. creating a separate sketch.

Exercises: Create and save the exercises in the Chapter6 folder.
(Time: 180 minutes.)

1. **V-slide Plate** (Dimensions are in inches. Plate Thickness: **0.25**)

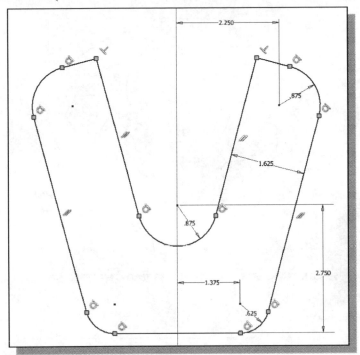

2. **Shaft Support** (Dimensions are in millimeters. Note the two R40 arcs at the base share the same center.)

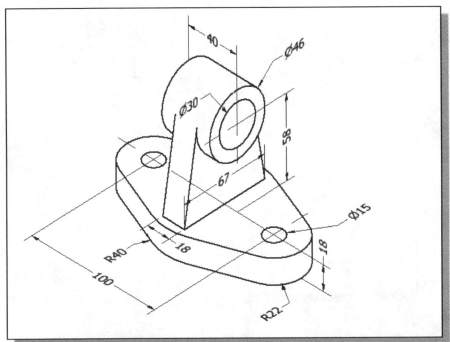

3. **Vent Cover** (Thickness: **0.125** inches. Hint: Use the Ellipse command.)

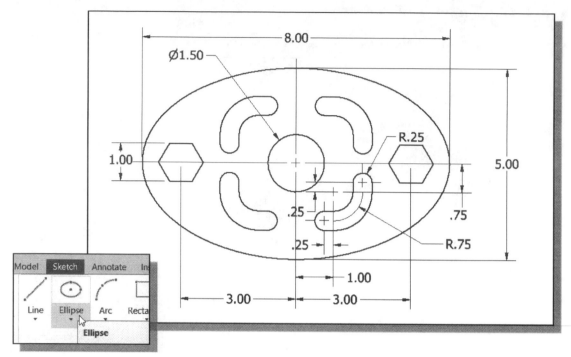

4. **Anchor Base** (Dimensions are in inches.)

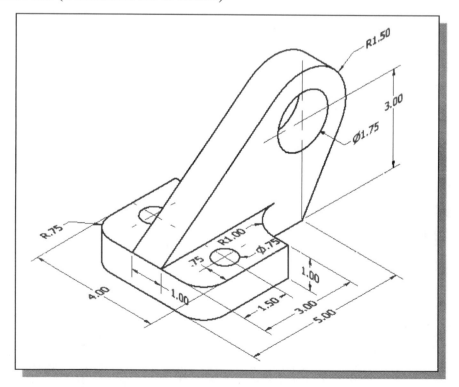

5. **Tube Spacer** (Dimensions are in inches.)

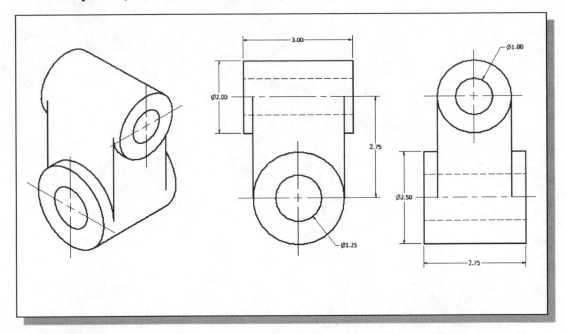

6. **Pivot Lock** (Dimensions are in inches. The circular features in the design are all aligned to the two centers at the base.)

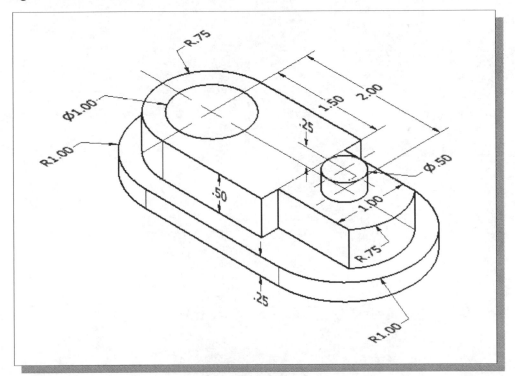

Notes:

Chapter 7
Orthographic Projection and Multiview Constructions

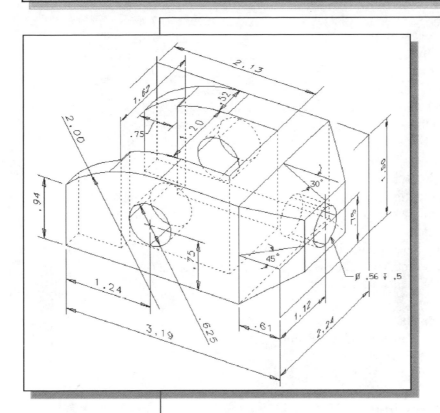

Learning Objectives

- ♦ **Understand the Basic Orthographic Projection Principles**
- ♦ **Be able to Perform 1st and 3rd Angle Projections**
- ♦ **Understand the Concept and Usage of the BORN Technique**
- ♦ **Understand the Importance of Parent/Child Relations in Features**
- ♦ **Use the Suppress Feature Option**
- ♦ **Resolve Undesired Feature Interactions**
- ♦ **Create Drawing Layouts from Solid Models**

Autodesk Inventor Certified User Exam Objectives Coverage

Introduction

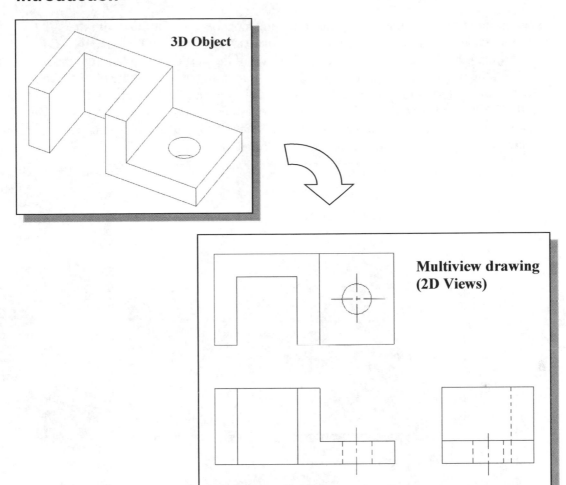

Most drawings produced and used in industry are *multiview drawings*. Multiview drawings are used to provide accurate three-dimensional object information on two-dimensional media, a means of communicating all of the information necessary to transform an idea or concept into reality. The standards and conventions of multiview drawings have been developed over many years, which equip us with a universally understood method of communication.

Multiview drawings usually require several orthographic projections to define the shape of a three-dimensional object. Each orthographic view is a two-dimensional drawing showing only two of the three dimensions of the three-dimensional object. Consequently, no individual view contains sufficient information to completely define the shape of the three-dimensional object. All orthographic views must be looked at together to comprehend the shape of the three-dimensional object. The arrangement and relationship between the views are therefore very important in multiview drawings. Before taking a more in-depth look into the multiview drawings, we will first look at the concepts and principles of projections.

Basic Principles of Projection

To better understand the theory of projection, one must become familiar with the elements that are common to the principles of **projection**. First of all, the **POINT OF SIGHT** (a.k.a. **STATION POINT**) is the position of the observer in relation to the object and the plane of projection. It is from this point that the view of the object is taken. Secondly, the observer views the features of the object through an imaginary PLANE OF PROJECTION (or IMAGE PLANE). Imagine yourself standing in front of a glass window, IMAGE PLANE, looking outward; the image of a house at a distance is sketched onto the glass which is a 2D view of a 3D house.

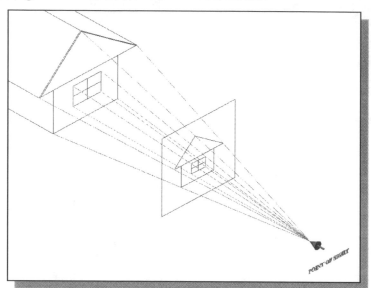

Orthographic Projection

The lines connecting from the *Point of Sight* to the 3D object are called the **Projection Lines** or **Lines of Sight**. Note that in the above figure, the projection lines are connected at the point of sight, and the projected 2D image is smaller than the actual size of the 3D object.

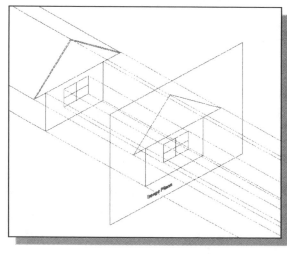

Now, if the *projection lines* are **parallel** to each other and the image plane is also **perpendicular** (*normal*) to the projection lines, the result is what is known as an **orthographic projection**. When the projection lines are parallel to each other, an accurate outline of the visible face of the object is obtained.

The term **orthographic** is derived from the word *orthos* meaning **perpendicular** or **90°**.

In *Engineering Graphics*, the projection of one face of an object usually will not provide an overall description of the object; other planes of projection must be used. To create the necessary 2D views, the *point of sight* is changed to project different views of the same object; hence, each view is from a different point of sight. If the point of sight is moved to the front of the object, this will result in the front view of the object. And then move the point of sight to the top of the object and looking down at the top, and then move to the right side of the object, as the case may be. Each additional view requires a new point of sight.

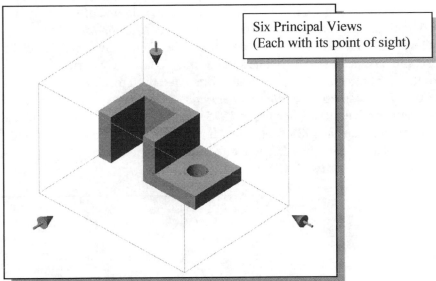

Six Principal Views
(Each with its point of sight)

Multiview Orthographic Projection

In creating multiview orthographic projection, different systems of projection can be used to create the necessary views to fully describe the 3D object. In the figure below, two perpendicular planes are established to form the image planes for a multiview orthographic projection.

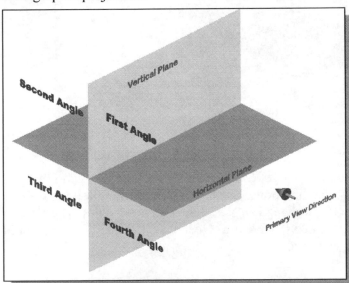

The angles formed between the horizontal and the vertical planes are called the **first**, **second**, **third** and **fourth angles**, as indicated in the figure. For engineering drawings, both **first angle projection** and **third angle projection** are commonly used.

First-Angle Projection

First-angle orthographic projection is commonly used in countries that use the metric (International System of Units: SI) system as the primary units of measurement. The metric system was first developed in 17th century France; most European countries adopted the metric system by the end of the 18th century. Today most countries in the world have adopted and are using the metric SI system as the primary units of measurement. In 1875, the United States solidified its commitment to the development of the internationally recognized metric system by becoming one of the original seventeen signatory nations to the *Metre Convention* or the *Treaty of the Metre*. Over the last 50 years, there has been a quiet ongoing debate in Washington over metric conversion. Many people have said that the United States must adopt the Metric System or face irreparable economic harm, and several efforts have been made to start conversion. Back in 1975 Congress passed the *Metric Conversion Act*. The act mandated a ten-year voluntary conversion period starting in 1975. The full conversion to the metric system was expected to occur in 1985. Yet, today, in spite of years of arguments and Federal efforts, very little has happened. The United States is one of a handful of countries that still use the *Imperial Units* of Measurement.

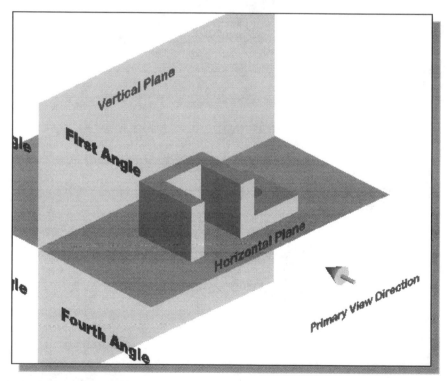

First-angle orthographic projection is a type of *Orthographic Projection*, which is a way of drawing a 3D object from different directions. Usually a front, side and top view are drawn so that a person looking at the drawing can see all the important sides. The first-angle orthographic projection is assumed to have the object placed in the first quadrant of the 3D space as shown in the above figure.

In first-angle projection, the object is placed in **front** of the image planes. And the views are formed by projecting to the image plane located at the back.

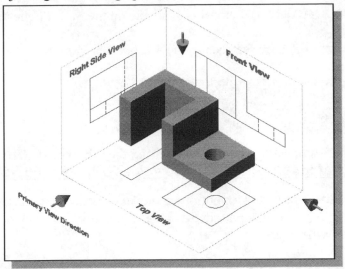

Rotation of the Horizontal and Profile Planes

In order to draw all three views of the object on the same plane, the horizontal (*Top View*) and profile (*Right Side view*) are rotated into the same plane as the primary image plane (*Front View*). Dashed lines are used to indicate edges of the design that are not visible.

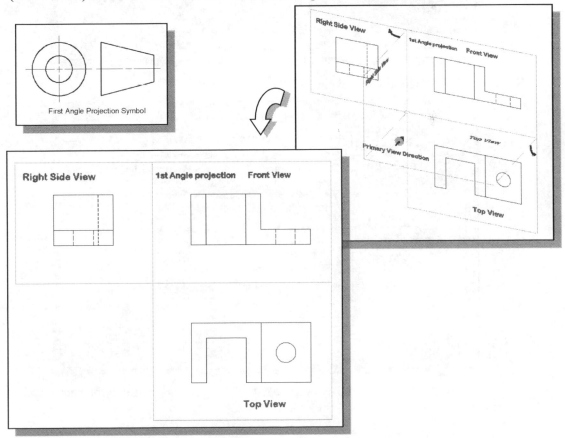

The 3D Adjuster Model and 1st Angle Projection

The **Adjuster1stAngle.ipt** model is available on the SDC Publications website.

1. Launch your internet browser, such as the *MS Internet Explorer* or *Mozilla Firefox* web browsers.

2. Download the Autodesk Inventor part file using the following *URL address*:
   http://www.sdcpublications.com/downloads/978-1-63057-583-0

3. Select the **Autodesk Inventor** option on the *Start* menu, or select the **Autodesk Inventor** icon on the desktop to start Autodesk Inventor. The Autodesk Inventor main window will appear.

4. Select the **Open** icon with a single click of the left-mouse-button on the *Menu Bar* toolbar.

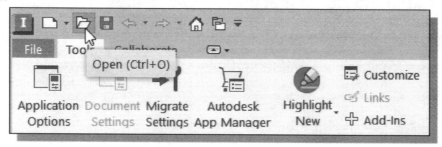

5. Select the downloaded file and click **Open** as shown.

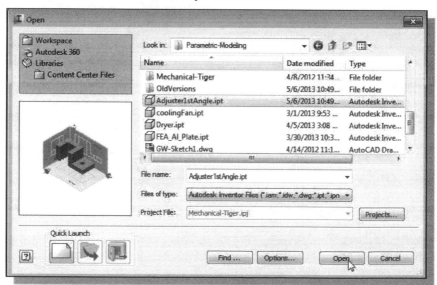

- On your own, use the real-time dynamic rotation feature and examine the relations of the 2D views, projection planes and the 3D object.

General Procedure: 1St Angle Orthographic Projection

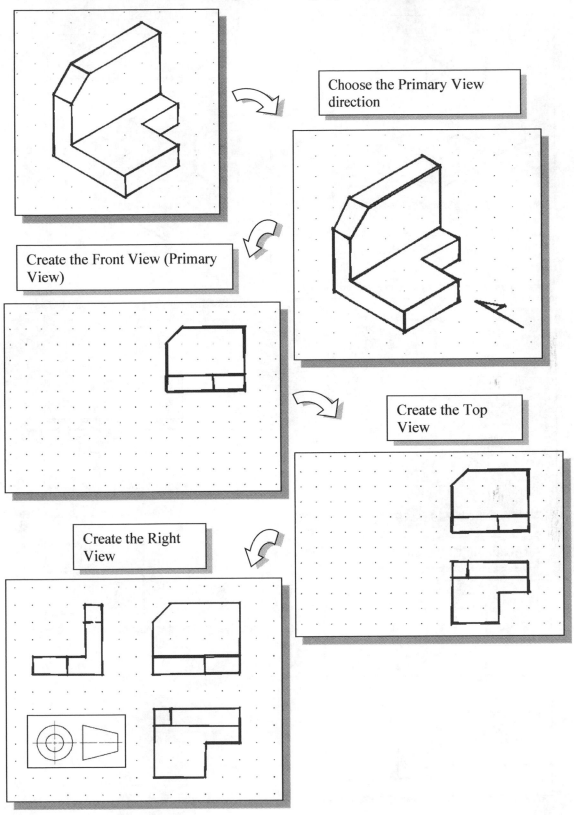

Choose the Primary View direction

Create the Front View (Primary View)

Create the Top View

Create the Right View

Example 1: 1St Angle Orthographic Projection

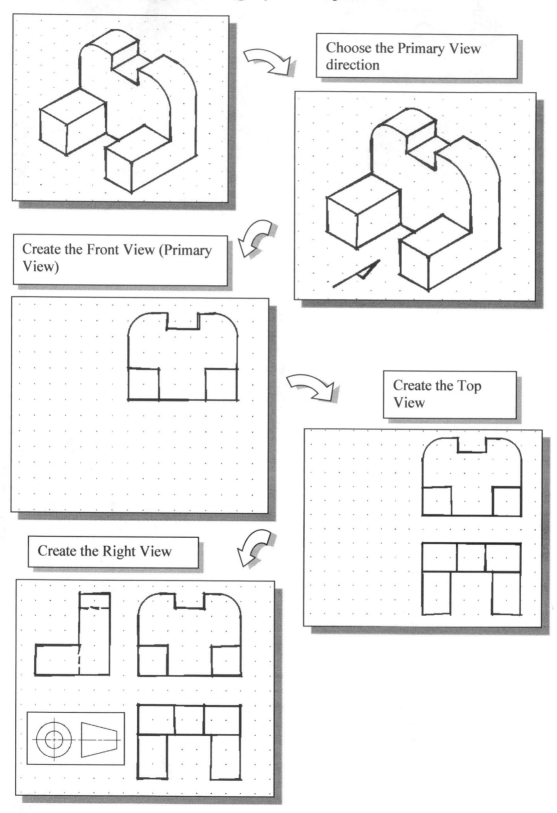

Choose the Primary View direction

Create the Front View (Primary View)

Create the Top View

Create the Right View

Chapter 7 - 1St Angle Orthographic Sketching Exercise 1:

Using the grid to estimate the size of the parts, create the standard three views by freehand rough sketching.
(Note that 3D Models of chapter examples and exercises are available at:
http://www.sdcpublications.com/downloads/978-1-63057-583-0)

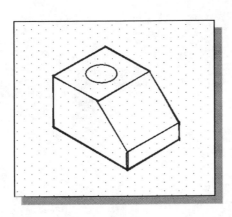

Name: _____   Date: _____

Chapter 7 - 1St Angle Orthographic Sketching Exercise 2:

Using the grid to estimate the size of the parts, create the standard three views by freehand rough sketching.
(Note that 3D Models of chapter examples and exercises are available at:
http://www.sdcpublications.com/downloads/978-1-63057-583-0)

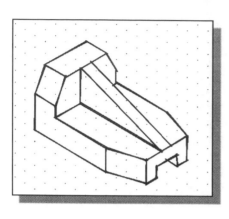

Name: _____   Date: _____

Chapter 7 - 1St Angle Orthographic Sketching Exercise 3:

Using the grid to estimate the size of the parts, create the standard three views by freehand rough sketching.
(Note that 3D Models of chapter examples and exercises are available at: http://www.sdcpublications.com/downloads/978-1-63057-583-0)

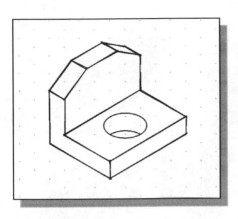

Name: _____   Date: _____

Chapter 7 - 1St Angle Orthographic Sketching Exercise 4:

Using the grid to estimate the size of the parts, create the standard three views by freehand rough sketching.
(Note that 3D Models of chapter examples and exercises are available at: http://www.sdcpublications.com/downloads/978-1-63057-583-0)

Name: _____   Date: _____

Chapter 7 - 1St Angle Orthographic Sketching Exercise 5:

Using the grid to estimate the size of the parts, create the standard three views by freehand rough sketching.
(Note that 3D Models of chapter examples and exercises are available at: http://www.sdcpublications.com/downloads/978-1-63057-583-0)

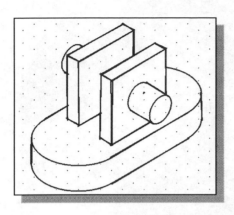

Name: _____   Date: _____

Chapter 7 - 1St Angle Orthographic Sketching Exercise 6:

Using the grid to estimate the size of the parts, create the standard three views by freehand rough sketching.

(Note that 3D Models of chapter examples and exercises are available at: http://www.sdcpublications.com/downloads/978-1-63057-583-0)

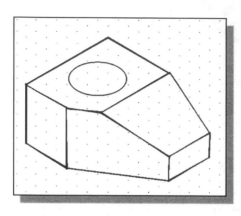

Name: _____   Date: _____

Third-Angle Projection

In the United States, *third-angle projection* is commonly used in Engineering Drawings.

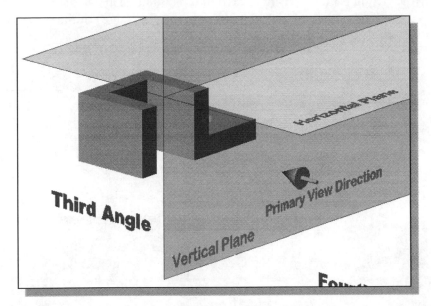

In *third-angle projection*, the image planes are placed in between the object and the observer, and the views are formed by projecting to the image plane located in front of the object.

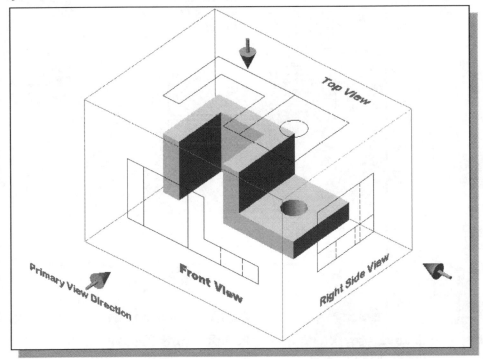

Rotation of the Horizontal and Profile Planes

In order to draw all three views of the object on the same plane, the horizontal (Top View) and profile (Right Side view) are rotated into the same plane as the primary image plane (Front View). Notice the use of dashed lines to indicate edges of the design that are not visible in the specific view direction.

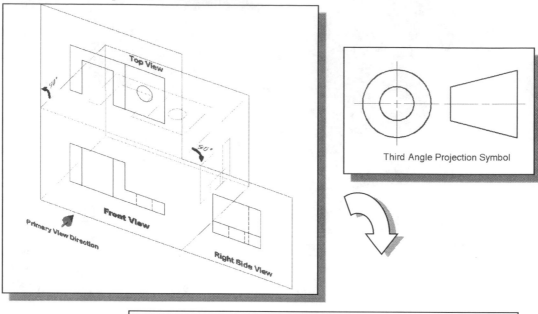

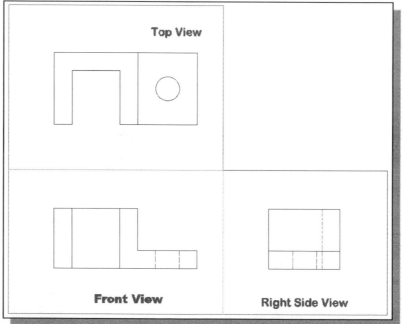

> Using your internet browser, download the **AdjustorRTOP.avi** file to view the rotation of the projection planes:
> ***http://www.sdcpublications.com/downloads/978-1-63057-583-0***

The 3D Adjuster Model and 3rd Angle Projection

The **Adjuster3rdAngle.ipt** model file is available on the SDC Publications website.

1. Launch your internet browser, such as the *MS Internet Explorer* or *Mozilla Firefox* web browsers.

2. Download the Autodesk Inventor part file using the following *URL address*:
   http://www.sdcpublications.com/downloads/978-1-63057-583-0

3. Select the **Autodesk Inventor** option on the *Start* menu or select the **Autodesk Inventor** icon on the desktop to start Autodesk Inventor. The Autodesk Inventor main window will appear.

4. Select the **Open** icon with a single click of the left-mouse-button on the *Menu Bar* toolbar.

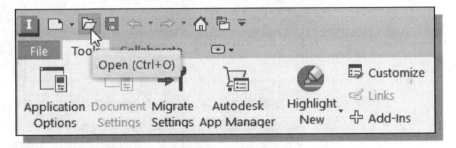

5. Select the downloaded file and click **Open** as shown.

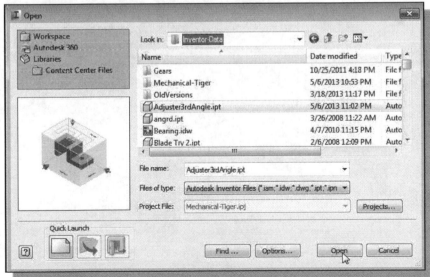

- On your own, use the real-time dynamic rotation feature and examine the relations of the 2D views, projection planes and the 3D object.

The Glass Box and the Six Principal Views

Considering the third angle projection described in the previous section further, we find that the object can be entirely surrounded by a set of six planes, a Glass box. On these planes, views can be obtained of the object as it is seen from the top, front, right side, left side, bottom, and rear.

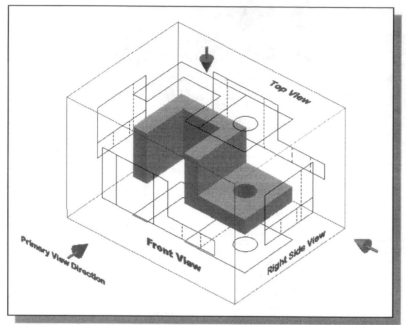

❖ Consider how the six sides of the glass box are being opened up into one plane. The front is the primary plane, and the other sides are hinged and rotated into position.

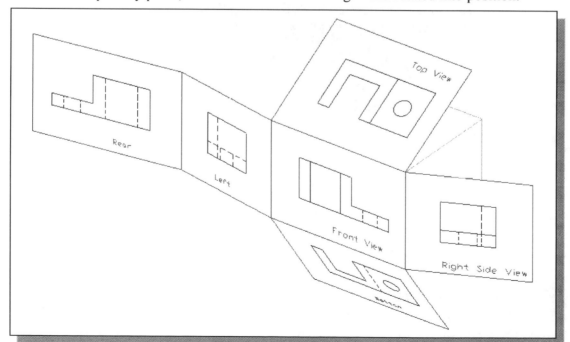

❖ In actual work, there is rarely an occasion when all six principal views are needed on one drawing, but no matter how many are required, their relative positions need to be maintained. These six views are known as the **six principal views**. In performing orthographic projection, each 2D view shows only two of the three dimensions (**height**, **width**, and **depth**) of the 3D object.

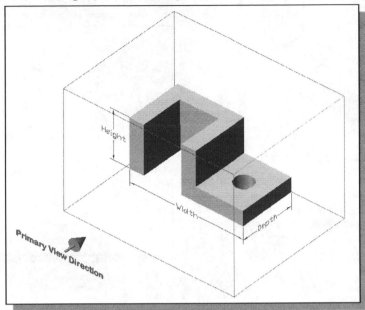

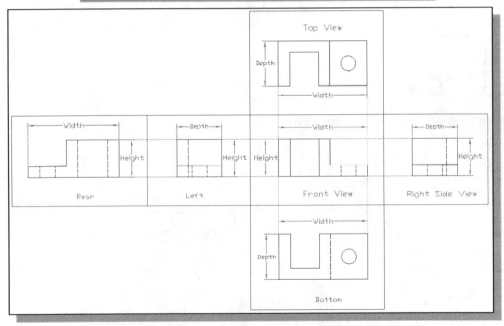

❖ The most usual combination selected from the six possible views consists of the *top*, *front* and *right-side* views. The selection of the *Primary View* is the first step in creating a *multi-view drawing*. The Primary View should typically be (1) the view which describes the main features of the design and (2) considerations of the layout of the other 2D views to properly describe the design.

General Procedure: 3rd Angle Orthographic Projection

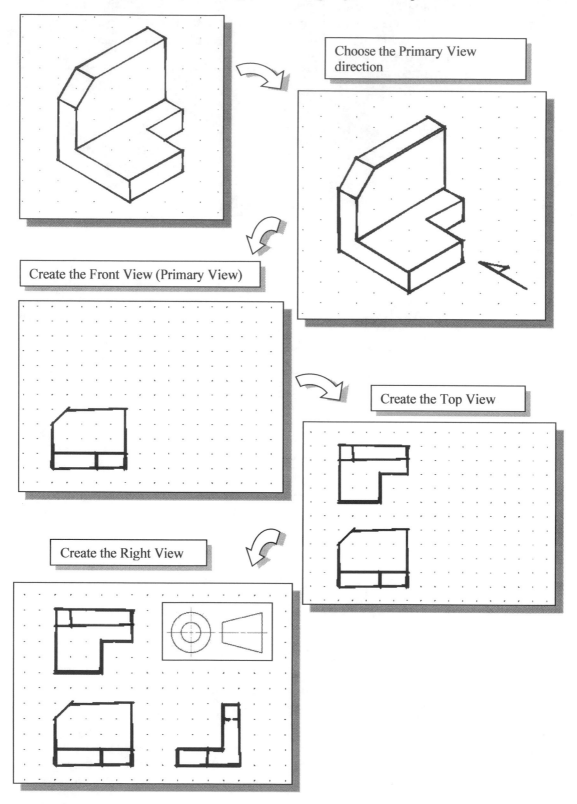

Choose the Primary View direction

Create the Front View (Primary View)

Create the Top View

Create the Right View

Example 2: 3rd Angle Orthographic Projection

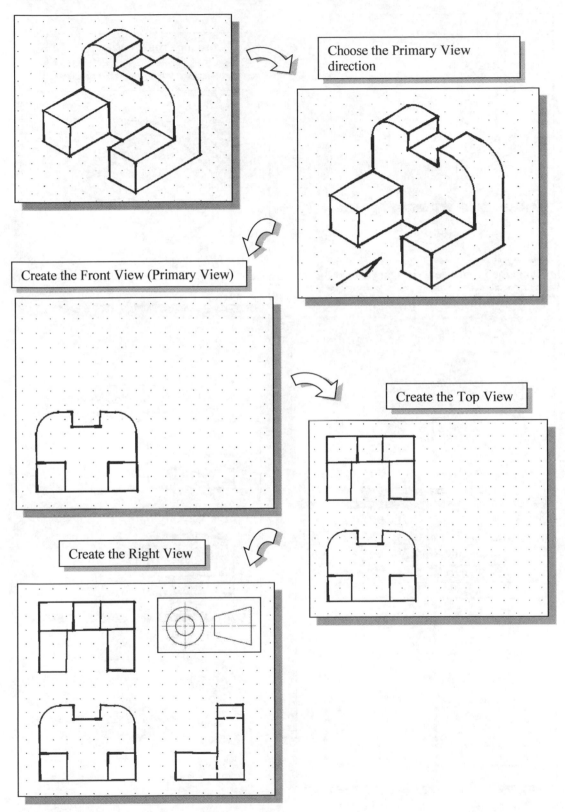

Choose the Primary View direction

Create the Front View (Primary View)

Create the Top View

Create the Right View

Example 3: 3rd Angle Orthographic Projection

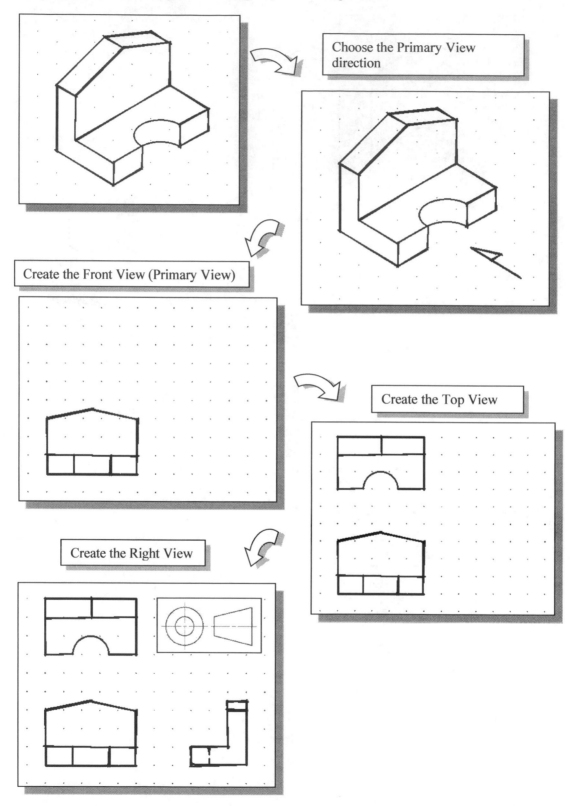

Choose the Primary View direction

Create the Front View (Primary View)

Create the Top View

Create the Right View

Chapter 7 - 3rd Angle Orthographic Sketching Exercise 1:

Using the grid to estimate the size of the parts, create the standard three views by freehand rough sketching.
(Note that 3D Models of chapter examples and exercises are available at:
http://www.sdcpublications.com/downloads/978-1-63057-583-0)

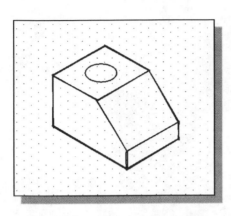

Name: _____   Date: _____

Chapter 7 - 3rd Angle Orthographic Sketching Exercise 2:

Using the grid to estimate the size of the parts, create the standard three views by freehand rough sketching.
(Note that 3D Models of chapter examples and exercises are available at:
http://www.sdcpublications.com/downloads/978-1-63057-583-0)

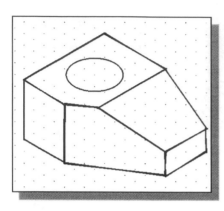

Name: _____   Date: _____

Chapter 7 - 3rd Angle Orthographic Sketching Exercise 3:

Using the grid to estimate the size of the parts, create the standard three views by freehand rough sketching.
(Note that 3D Models of chapter examples and exercises are available at: http://www.sdcpublications.com/downloads/978-1-63057-583-0)

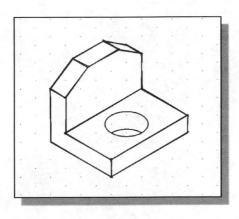

Name: _____   Date: _____

Chapter 7 - 3rd Angle Orthographic Sketching Exercise 4:

Using the grid to estimate the size of the parts, create the standard three views by freehand rough sketching.
(Note that 3D Models of chapter examples and exercises are available at:
http://www.sdcpublications.com/downloads/978-1-63057-583-0)

Name: _____   Date: _____

Chapter 7 - 3rd Angle Orthographic Sketching Exercise 5:

Using the grid to estimate the size of the parts, create the standard three views by freehand rough sketching.
(Note that 3D Models of chapter examples and exercises are available at:
http://www.sdcpublications.com/downloads/978-1-63057-583-0)

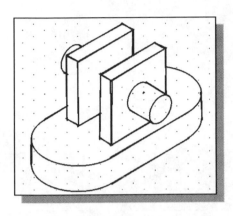

Name: _____   Date: _____

Chapter 7 - 3rd Angle Orthographic Sketching Exercise 6:

Using the grid to estimate the size of the parts, create the standard three views by freehand rough sketching.
(Note that 3D Models of chapter examples and exercises are available at:
http://www.sdcpublications.com/downloads/978-1-63057-583-0)

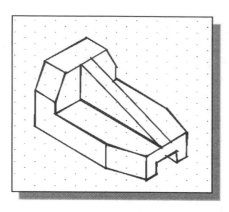

Name: _____   Date: _____

Alphabet of Lines

In *Technical Engineering Drawings*, each line has a definite meaning and is drawn in accordance to the line conventions as illustrated in the figure below. Two widths of lines are typically used on drawings; thick line width should be 0.6 mm and the thin line width should be 0.3 mm.

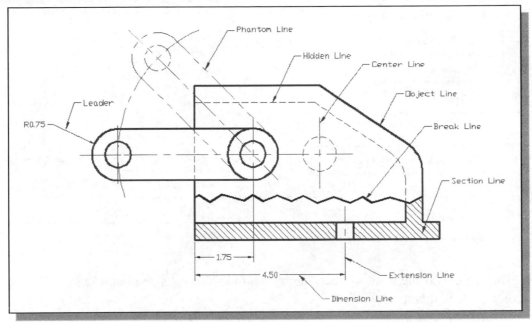

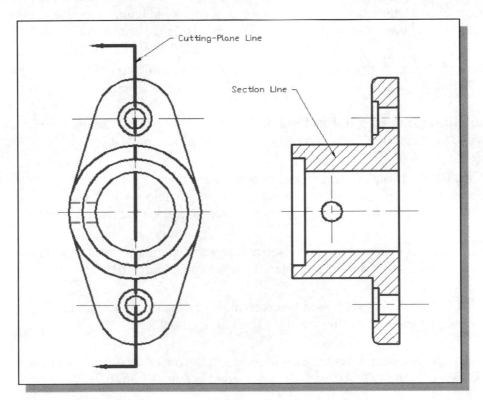

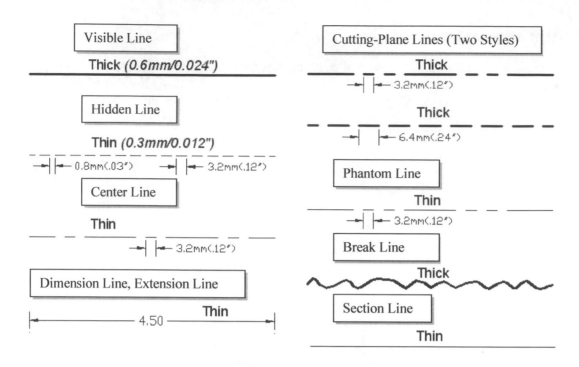

Visible Line Visible lines are used to represent visible edges and boundaries. The line weight is thick, 0.6mm/0.024".

Hidden Line Hidden lines are used to represent edges and boundaries that are not visible from the viewing direction. The line weight is thin, 0.3mm/0.012".

Center Line Center lines are used to represent axes of symmetry. The line weight is thin, 0.3mm/0.012".

Dimension Line, Extension Line and Leader Dimension lines are used to show the sizes and locations of objects. The line weight is thin, 0.3mm/0.012".

Cutting Plane Lines Cutting Plane lines are used to represent where the location of an imaginary cut has been made, so that the interior of the object can be viewed. The line weight is thick, 0.6mm/0.024". (Note that two forms of line type can be used.)

Phantom Line Phantom lines are used to represent imaginary features or objects, such as a rotated position of a part. The line weight is thin, 0.3mm/0.012".

Break Line Break lines are used to represent an imaginary cut, so that the interior of the object can be viewed. The line weight is thick, 0.6mm/0.024".

Section Line Section lines are used to represent the regions that have been cut with the *break lines* or *cutting plane lines*. The line weight is thin, 0.3mm/0.012".

Precedence of Lines

In multiview drawings, coincidence lines may exist within the same view. For example, hidden features may project lines to coincide with the visible object lines. And center lines may occur where there is a visible or hidden outline.

In creating a multiview drawing, the features of the design are to be represented, therefore object and hidden lines take precedence over all other lines. And since the visible outline is more important than hidden features, the visible object lines take precedence over hidden lines, as shown in the below figure.

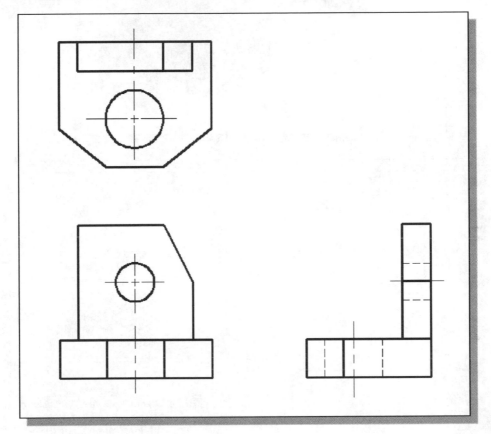

The following list gives the order of precedence of lines:
1. **Visible object lines**
2. **Hidden lines**
3. **Center line or cutting-plane line**
4. **Break lines**
5. **Dimension and extension lines**
6. **Crosshatch/section lines**

In the following sections, the general procedure of creating a 3rd angle three-view orthographic projection from a solid model using Autodesk Inventor is presented.

The U-Bracket Design

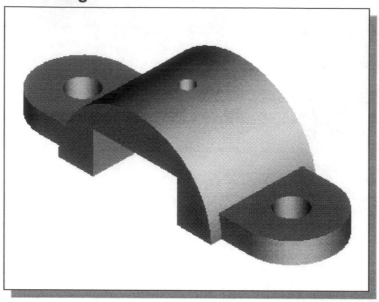

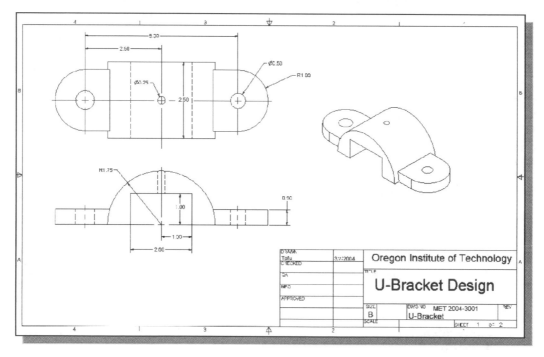

> ➢ Based on your knowledge of Autodesk Inventor so far, how many features would you use to create the model? What is your choice for arranging the order of the features? Would you organize the features differently if the rectangular cut at the center is changed to a circular shape?

The BORN Technique

In *parametric modeling*, the **base feature** (the *first solid feature* of the solid model) is the center of all features and is considered the key feature of the design. All subsequent features are built by referencing the first feature. A great deal of emphasis is placed on the selection of the *base feature*.

A more advanced technique of creating solid models is what is known as the "**Base Orphan Reference Node**" (**BORN**) technique. The basic concept of the BORN technique is to use a *Cartesian coordinate system* as the first feature prior to creating any solid features. With the *Cartesian coordinate system* established, we then have three mutually perpendicular datum planes (namely the *XY*, *YZ*, and *ZX planes*), three datum axes and a datum point available to use as sketching planes. The three datum planes can also be used as references for dimensions and geometric constructions. Using this technique, the first node in the history tree is called an "orphan," meaning that it has no history to be replayed. The technique of using the reference geometry in this "base node" is therefore called the "Base Orphan Reference Node" (BORN) technique.

Autodesk Inventor automatically establishes a set of reference geometry when we start a new part, namely a *Cartesian coordinate system* with three work planes, three work axes, and a work point. All subsequent solid features can then use the coordinate system and/or reference geometry as sketching planes. The *base feature* is still important, but the *base feature* is no longer the ONLY choice for creating subsequent solid features. This approach provides us with more options while we are creating parametric solid models. More importantly, this approach provides greater flexibility for part modifications and design changes. This approach is also very useful in creating assembly models, which will be illustrated in the later chapters of this text.

Starting Autodesk Inventor

1. Select the **Autodesk Inventor** option on the *Start* menu to start Autodesk Inventor. The Autodesk Inventor main window will appear on the screen.

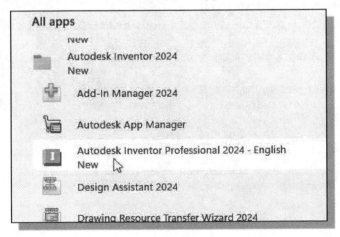

Sketch Plane Settings

1. Select the Autodesk Inventor option on the *Start* menu or select the Autodesk Inventor icon on the desktop to start Autodesk Inventor. The Autodesk Inventor main window will appear on the screen.

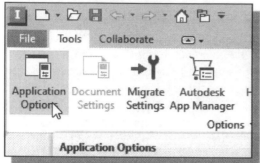

2. Select **Application Options** in the **Tools** pull-down menu.

- The **Application Options** menu allows us to set behavioral options, such as *color*, *file locations*, etc.

3. Click on the **Part** tab to display and/or modify the default sketch plane settings.

- Note that a sketch plane can be aligned to one of the work planes during new part creation. Confirm the **No new sketch** option is set as shown.

4. Click on the **Sketch** tab to examine/modify the default sketching settings.

5. Turn *OFF* the *Look at sketch plane on sketch creation - In Part environment* option as shown.

6. Click on the **OK** button to accept the setting.

- Note that the new settings will take effect when a new part file is created.

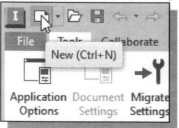

7. Click on the **New** icon in the *Standard* toolbar.

8. On your own, open a new *English* units **standard (in) part** file.

Apply the BORN Technique

1. In the *Part Browser* window, click on the [**+**] symbol in front of the ***Origin*** feature to display more information on the feature.

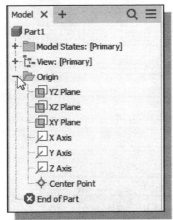

❖ In the *Part Browser* window, notice a new part name appeared with seven work features established. The seven work features include three *work planes*, three *work axes*, and a *work point*. By default, the three work planes and work axes are aligned to the **world coordinate system** and the work point is aligned to the *origin* of the **world coordinate system**.

2. Inside the *browser* window, move the cursor on top of the third work plane, the **XY Plane**. Notice a rectangle, representing the work plane, appears in the graphics window.

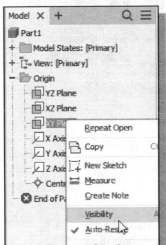

3. Inside the *browser* window, click once with the right-mouse-button on XY Plane to display the option menu. Click on **Visibility** to toggle on the display of the plane.

4. On your own, repeat the above steps and toggle *ON* the display of all of the *work planes* and the *center point* on the screen.

5. On your own, use the *Dynamic Viewing* options (ViewCube, 3D Orbit, Zoom and Pan) to view the default work features.

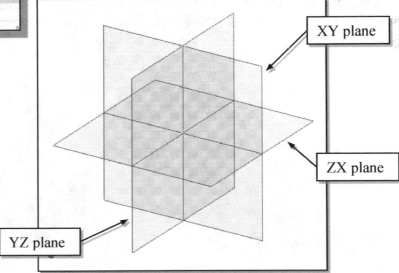

❖ By default, the basic set of work planes is aligned to the world coordinate system; the work planes are the first features of the part. We can now proceed to create solid features referencing the three mutually perpendicular datum planes.

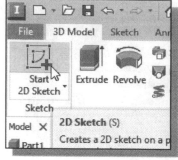

6. In the *Sketch* toolbar select the **Start 2D Sketch** command by left-clicking once on the icon.

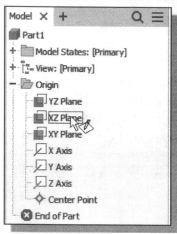

7. In the *Status Bar* area, the message "*Select plane to create sketch or an existing sketch to edit.*" is displayed. Autodesk Inventor expects us to identify a planar surface where the 2D sketch of the next feature is to be created. Move the graphics cursor on top of the **XZ Plane**, inside the *browser* window as shown, and notice that Autodesk Inventor will automatically highlight the corresponding plane in the graphics window. Left-click once to select the XZ Plane as the sketching plane.

❖ Autodesk Inventor allows us to identify and select features in the graphics window as well as in the *browser* window.

● Note that since both of the ***Look at sketch plane on sketch creation*** options are turned ***OFF***, the view will be adjusted back to the default ***isometric view***.

➢ Note the alignment of the sketch plane is set to the XZ plane as shown.

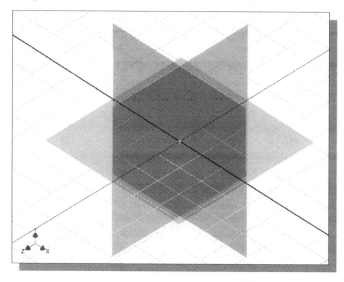

Create the 2D Sketch for the Base Feature

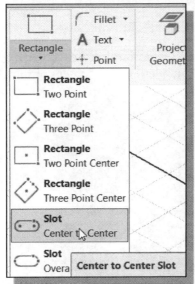

1. Select the **Center to Center Slot** command by clicking once with the **left-mouse-button** on the icon in the *Draw* toolbar.

2. Create a horizontal line, representing center to center distance of the slot, in front of the work planes as shown below. (Note the **Horizontal** symbol indicates the alignment of the sketched line.)

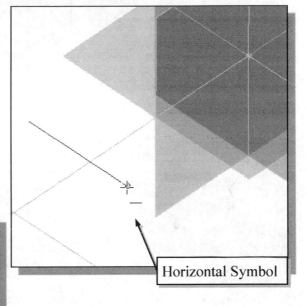

Horizontal Symbol

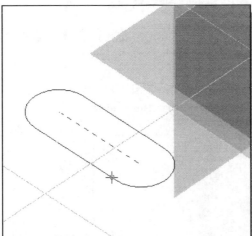

3. Move the cursor outward to define the size of the center to center slot; click once with the left-mouse-button to create the shape.

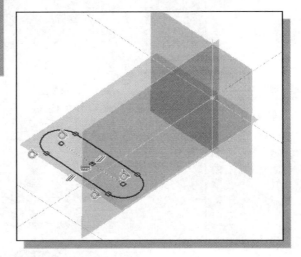

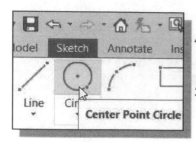

4. Choose **Center Point Circle** in the *Draw* toolbar.

5. On your own, create the two inner circles; the center points of the circles are coincident to the centers of the overall slot.

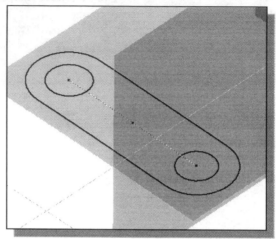

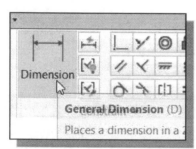

6. Left-click once on the **General Dimension** icon to activate the General Dimension command.

7. On your own, set the display orientation to the Top view position as shown in the below figure.

8. On your own, create the dimensions, referencing the origin point, to fully constrain the sketch as shown. (Do not be overly concerned with the actual numbers displayed; the dimensions will be adjusted in the next steps.)

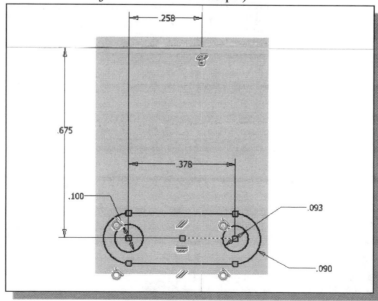

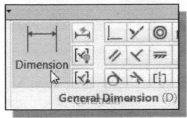

9. Choose **General Dimension** in the *Constrain* panel.

10. On your own, adjust the dimensions as shown in the figure.

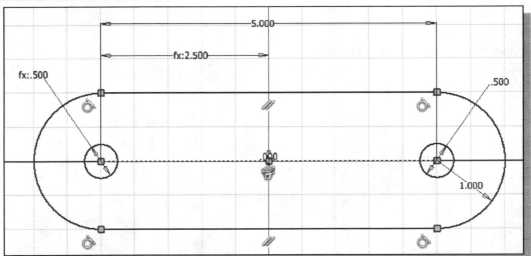

11. Inside the graphics window, click once with the right-mouse-button to display the option menu. Select **OK** in the pop-up menu to end the General Dimension command.

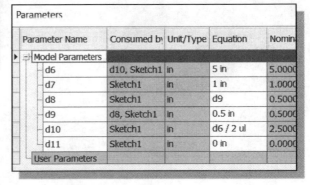

12. On your own, use the **Parameters** command to examine the two parametric equations used.

Parameter Name	Consumed by	Unit/Type	Equation	Nomin
Model Parameters				
d6	d10, Sketch1	in	5 in	5.0000
d7	Sketch1	in	1 in	1.0000
d8	Sketch1	in	d9	0.5000
d9	d8, Sketch1	in	0.5 in	0.5000
d10	Sketch1	in	d6 / 2 ul	2.5000
d11	Sketch1	in	0 in	0.0000
User Parameters				

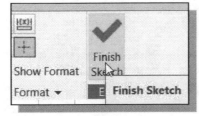

13. In the *Ribbon toolbar* area, select **Finish Sketch** in the pop-up menu to end the Sketch option.

Create the First Extrude Feature

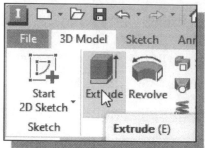

1. In the *Create* toolbar, select the **Extrude** command by clicking once with the left-mouse-button on the icon.

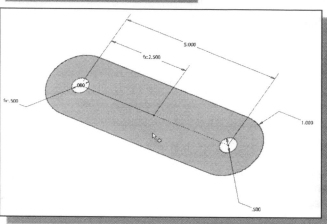

2. Select the inside regions of the sketch to define the profile of the extrusion as shown.

3. In the *Distance* option box, enter **0.5** as the extrusion distance.

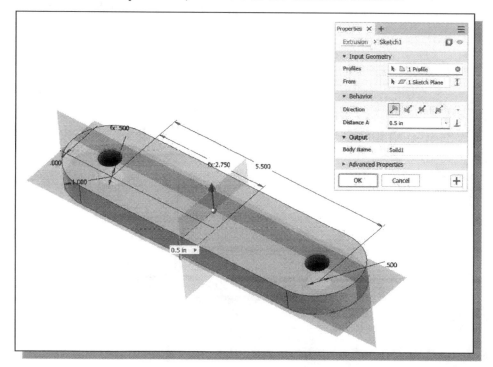

4. In the *Extrude* pop-up control, click on the **OK** button to create the base feature.

The Implied Parent/Child Relationships

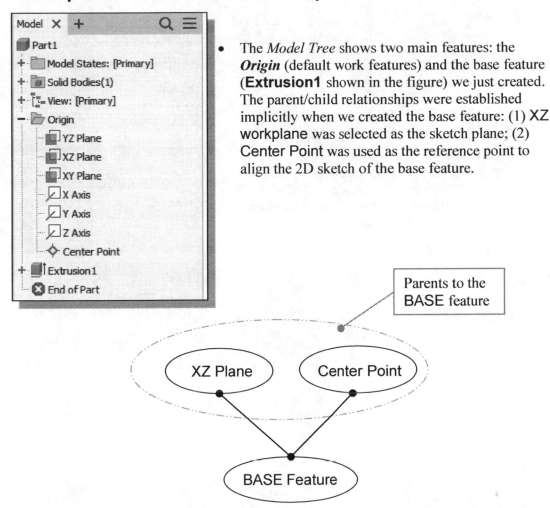

- The *Model Tree* shows two main features: the **Origin** (default work features) and the base feature (**Extrusion1** shown in the figure) we just created. The parent/child relationships were established implicitly when we created the base feature: (1) XZ workplane was selected as the sketch plane; (2) Center Point was used as the reference point to align the 2D sketch of the base feature.

Create the Second Solid Feature

For the next solid feature, we will create the top section of the design. Note that the center of the base feature is aligned to the Center Point of the default work features. This was done intentionally so that additional solid features can be created referencing the default work features. For the second solid feature, the XY workplane will be used as the sketch plane.

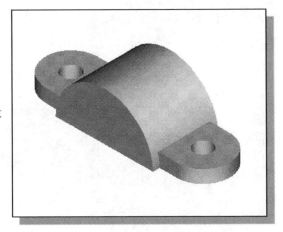

1. In the *Sketch* toolbar select the **Start 2D Sketch** command by left-clicking once on the icon.

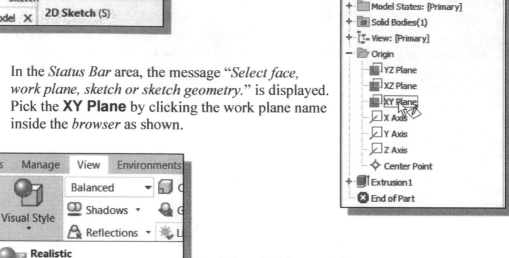

2. In the *Status Bar* area, the message "*Select face, work plane, sketch or sketch geometry.*" is displayed. Pick the **XY Plane** by clicking the work plane name inside the *browser* as shown.

3. Select the **View** tab in the *Ribbon* panel as shown.

4. Select **Wireframe** display mode under the *Visual Style* icon as shown.

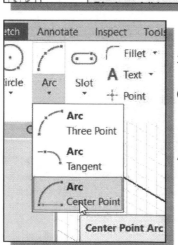

5. Select the **Sketch** tab in the *Ribbon*.

6. Select the **Center point arc** command by clicking once with the left-mouse-button on the icon in the *icon stack* as shown.

7. Pick the **center point** and watch for the Green dot for alignment as the center location of the new arc.

8. On your own, create a semi-circle of arbitrary size, with the center point aligned to the origin and both endpoints aligned to the X-axis, as shown below.

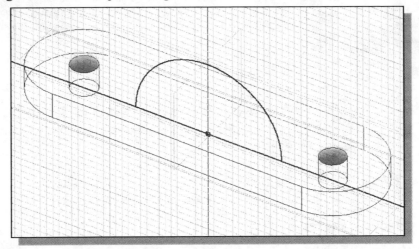

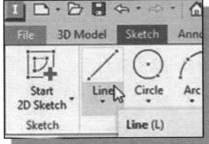

9. Move the cursor on top of the **General Dimension** icon. Left-click once on the icon to activate the General Dimension command.

10. On your own, create and adjust the radius of the arc to **1.75**.

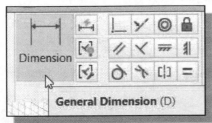

11. Select the **Line** command in the *Draw* toolbar.

12. Create a line connecting the two endpoints of the arc as shown in the figure below. (Hint: Use the **Coincident** constraint to help align the line to the origin if necessary.)

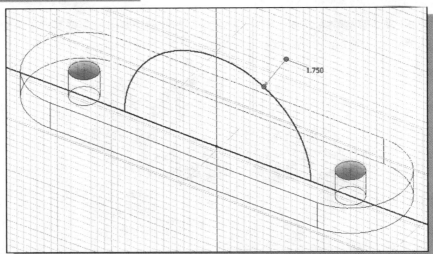

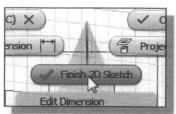

13. Inside the graphics window, click once with the right-mouse-button to display the option menu. Select **Finish 2D Sketch** in the pop-up menu to end the Sketch option.

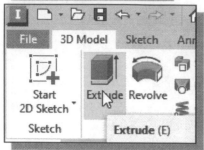

14. In the *Create* toolbar, select the **Extrude** command by clicking once with the left-mouse-button on the icon.

15. Select the inside region of the sketched arc-line curves as the profile to be extruded.

16. In the *Extrude* dialog box, set to the **Symmetric** option.

17. In the *Distance* value box, set the extrusion distance to **2.5**.

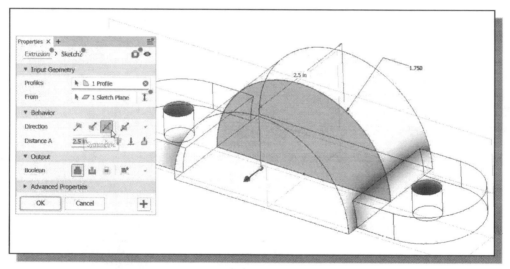

18. Click on the **OK** button to proceed with the **Extrude** operation.

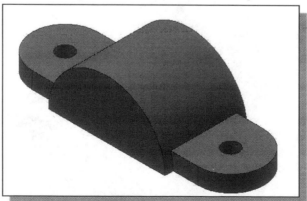

Create a Cut Feature

A rectangular cut will be created as the next solid feature.

1. In the *Sketch* toolbar select the **Start 2D Sketch** command by left-clicking once on the icon.

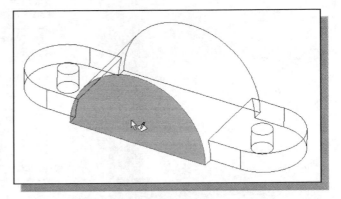

2. In the *Status Bar* area, the message "*Select plane to create sketch or an existing sketch to edit*" is displayed. Pick the front vertical face of the solid model shown.

3. On your own, create a rectangle and apply the dimensions as shown below. (Hint: Use the **YZ Plane** to assure the proper alignment of the 2D sketch.)

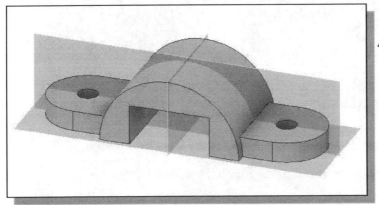

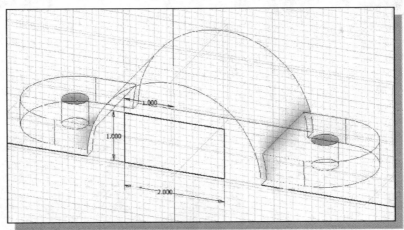

4. On your own, use the **Extrude** command and create a cutout that cuts through the entire 3D solid model as shown.

The Second Cut Feature

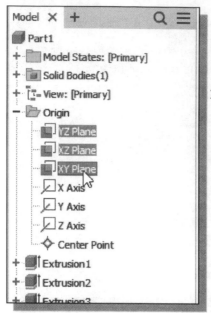

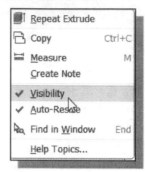

1. In the *Model Tree area*, select all of the **Datum Planes** while holding down the [**SHIFT**] key.

2. *Right-click* on any of the selected items to display the option menu and turn **OFF** the visibility of the selected items as shown.

3. In the *Sketch* toolbar select the **Start 2D Sketch** command by left-clicking once on the icon.

4. In the *Status Bar* area, the message "*Select face, work plane, sketch or sketch geometry*" is displayed. Select the horizontal face of the last cut feature as the *sketching plane*.

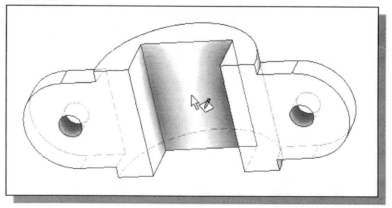

5. In the *Standard* toolbar area, click on the **Look At** button.

6. Select one of the edges of the highlighted sketching plane.

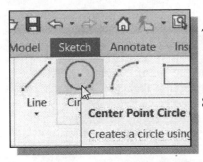

7. Select the **Center Point Circle** command by clicking once with the left-mouse-button on the icon in the *Draw* panel.

8. Pick the **center point**, located at the center of the part, to align the center of the new circle.

9. On your own, create a circle of arbitrary size.

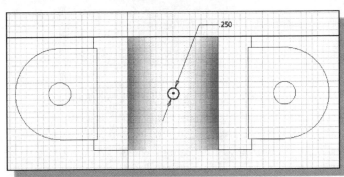

10. On your own, add the size dimension of the circle and set the dimension to **0.25**.

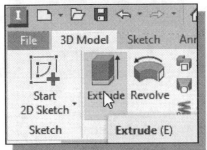

11. Inside the graphics window, click once with the right-mouse-button to display the option menu. Select **Finish 2D Sketch** in the pop-up menu to end the Sketch option.

12. In the *Create toolbar*, select the **Extrude** command by clicking once with the left-mouse-button on the icon.

13. Select the inside region of the sketched circle as the profile to be extruded.

14. In the *Extrude* control panel, set to the **Cut – Through All** option.

15. Click on the **OK** button to proceed with creating the cut feature.

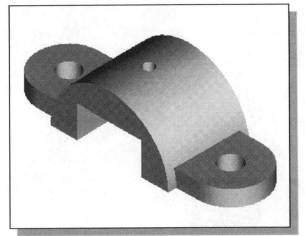

Examine the Parent/Child Relationships

1. On your own, rename the feature names to **Base**, **MainBody**, **Rect_Cut** and **Center_Drill** as shown in the figure.

The *Model Tree* window now contains seven items: the ***Origin*** (default work features) and four solid features. All of the parent/child relationships were established implicitly as we created the solid features. As more features are created, it becomes much more difficult to make a sketch showing all the parent/child relationships involved in the model. On the other hand, it is not really necessary to have a detailed picture showing all the relationships among the features. In using a feature-based modeler, the main emphasis is to consider the interactions that exist between the **immediate features**. Treat each feature as a unit by itself and be clear on the parent/child relationships for each feature. Thinking in terms of *features* is what distinguishes *feature-based modeling* and the previous generation solid modeling techniques. Let us take a look at the last feature we created, the **Center_Drill** feature. What are the parent/child relationships associated with this feature? (1) Since this is the last feature we created, it is not a parent feature to any other features. (2) Since we used one of the surfaces of the rectangular cutout as the sketching plane, the **Rect_Cut** feature is a parent feature to the **Center_Drill** feature. (3) We also used the Origin as a reference point to align the center; therefore, the ***Origin*** is also a parent to the **Center_Drill** feature.

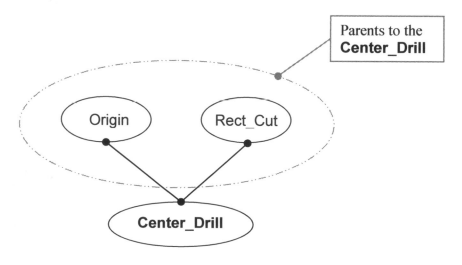

Modify a Parent Dimension

Any changes to the parent features will affect the child feature. For example, if we modify the height of the **Rect_Cut** feature from 1.0 to 0.75, the depth of the child feature (**Center_Drill** feature) will be affected.

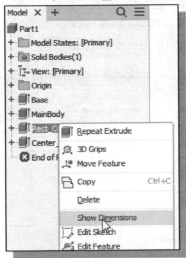

1. In the *Model Tree* window, right-click on **Rect_Cut** to bring up the option menu.

2. In the option menu, select **Show Dimensions**.

3. Select the height dimension (1.0) by double-clicking on the dimension.

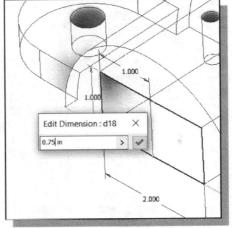

4. Enter **0.75** as the new height dimension as shown.

5. Click on the **Update** button in the *Quick Access Toolbar* area to proceed with updating the solid model.

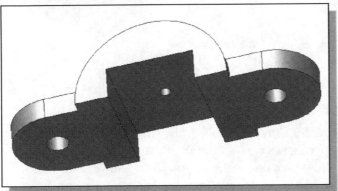

➤ Note that the position of the **Center_Drill** feature is also adjusted as the placement plane is lowered. The drill-hole still goes through the main body of the *U-Bracket* design. The parent/child relationship assures the intent of the design is maintained.

6. On your own, adjust the height of the **Rect_Cut** feature back to **1.0** inch before proceeding to the next section.

A Design Change

Engineering designs usually go through many revisions and changes. For example, a design change may call for a circular cutout instead of the current rectangular cutout feature in our model. Autodesk Inventor provides an assortment of tools to handle design changes quickly and effectively. In the following sections, we will demonstrate some of the more advanced tools available in Autodesk Inventor, which allow us to perform the modification of changing the rectangular cutout (2.0 × 1.0 inch) to a circular cutout (radius: 1.25 inch).

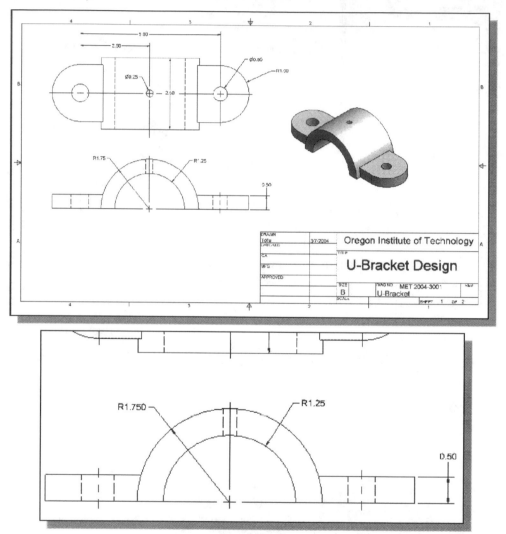

❖ Based on your knowledge of Autodesk Inventor so far, how would you accomplish this modification? What other approaches can you think of that are also feasible? Of the approaches you came up with, which one is the easiest to do and which is the most flexible? If this design change was anticipated right at the beginning of the design process, what would be your choice in arranging the order of the features? You are encouraged to perform the modifications prior to following through the rest of the tutorial.

Feature Suppression

With Autodesk Inventor, we can take several different approaches to accomplish this modification. We could (1) create a new model, or (2) change the shape of the existing cut feature using the **Redefine** command, or (3) perform **feature suppression** on the rectangular cut feature and add a circular cut feature. The third approach offers the most flexibility and requires the least amount of editing to the existing geometry. **Feature suppression** is a method that enables us to disable a feature while retaining the complete feature information; the feature can be reactivated at any time. Prior to adding the new cut feature, we will first suppress the rectangular cut feature.

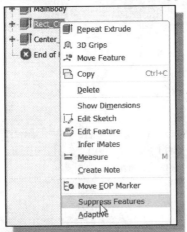

1. Move the cursor inside the *Model Tree* window. Click once with the right-mouse-button on **Rect_Cut** to bring up the option menu.

2. Pick **Suppress** in the pop-up menu.

❖ With the *Suppress* command, the Rect_Cut and Center_Drill features have disappeared in the display area. The child feature cannot exist without its parent(s), and any modification to the parent (Rect_Cut) influences the child (Center_Drill).

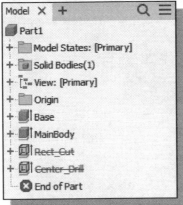

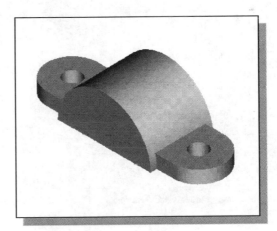

3. Move the cursor inside the *Model Tree* window. Click once with the right-mouse-button on top of **Center_Drill** to bring up the option menu.

4. Pick **Unsuppress Features** in the pop-up menu.

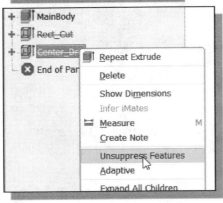

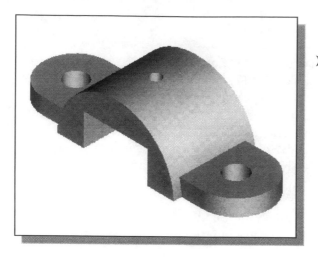

> ➤ In the display area and the *Model Tree* window, both the **Rect_Cut** feature and the **Center_Drill** feature are re-activated. The child feature cannot exist without its parent(s); the parent (Rect_Cut) must be activated to enable the child (Center_Drill).

A Different Approach to the Center_Drill Feature

The main advantage of using the BORN technique is to provide greater flexibility for part modifications and design changes. In this case, the Center_Drill feature can be placed on the XZ workplane and therefore not be linked to the Rect_Cut feature.

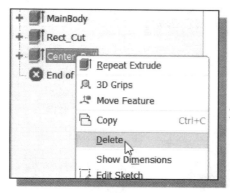

1. Move the cursor inside the *Model Tree* window. Click once with the **right-mouse-button** on top of **Center_Drill** to bring up the option menu.

2. Pick **Delete** in the pop-up menu.

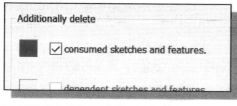

3. In the *Delete Features* window, confirm the **consumed sketches and features** option is switched *ON*.

4. Click **OK** to proceed with the Delete command.

5. In the *Sketch* toolbar select the **Start 2D Sketch** command by left-clicking once on the icon.

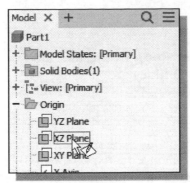

6. In the *Status Bar* area, the message "*Select plane to create sketch or an existing sketch to edit*" is displayed. Pick the **XZ Plane** in the *Model Tree* window as shown.

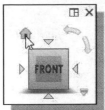

7. Inside the graphics area, single left-click to activate the **Home View** option as shown. The view will be adjusted back to the default *isometric view*.

➢ Note the alignment of the sketch plane is set to the XZ plane as shown.

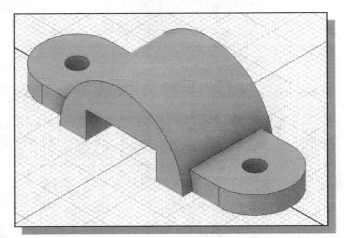

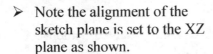

8. Select the **Center Point Circle** command by clicking once with the left-mouse-button on the icon in the *Draw* panel.

9. Pick the center point to align the center of the new circle. Select another location to create a circle of arbitrary size.

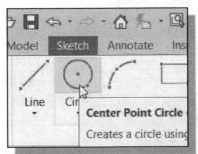

10. On your own, add the size dimension of the circle and set the dimension to **0.25**.

11. On your own, complete the extrude cut feature, cutting upward through the main body of the design.

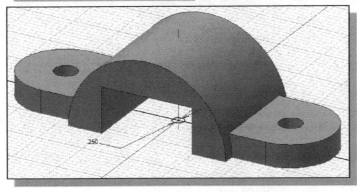

Suppress the Rect_Cut Feature

Now the new **Center_Drill** feature is no longer a child of the **Rect_Cut** feature, any changes to the **Rect_Cut** feature do not affect the **Center_Drill** feature.

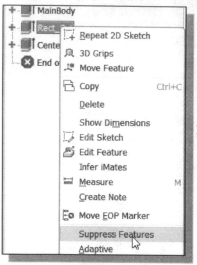

1. Move the cursor inside the *Model Tree* window. Click once with the right-mouse-button on top of **Rect_Cut** to bring up the option menu.

2. Pick **Suppress Features** in the pop-up menu.

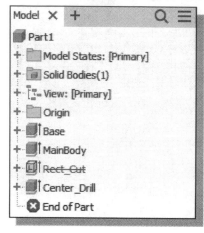

❖ The **Rect_Cut** feature is now disabled without affecting the *new* **Center_Drill** feature.

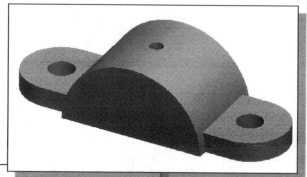

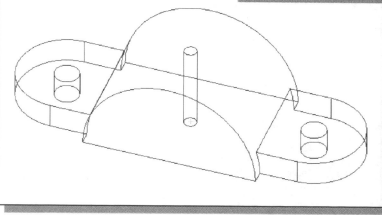

Create a Circular Cut Feature

1. In the *Sketch* toolbar select the **Start 2D Sketch** command by left-clicking once on the icon.

2. In the *Status Bar* area, the message "*Select face, work plane, sketch or sketch geometry*" is displayed. Pick the **XY Plane** in the *Model Tree* window as shown.

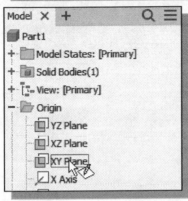

3. Activate the **View** tab in the *Ribbon* panel as shown.

4. Select **Wireframe** display mode under the *Visual Style* icon as shown.

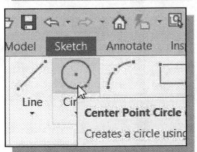

5. Select the **Center Point Circle** command by clicking once with the left-mouse-button on the icon in the *Draw* panel.

6. Pick the **center point** at the origin to align the center of the new circle.

7. On your own, create a circle of arbitrary size.

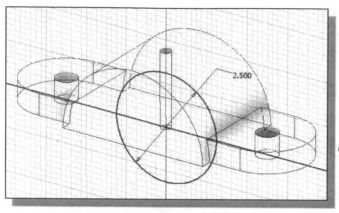

8. On your own, create the size dimension of the circle and set the dimension to **2.5** as shown in the figure.

9. On your own, complete the **cut** feature using the **Symmetric - All** command.

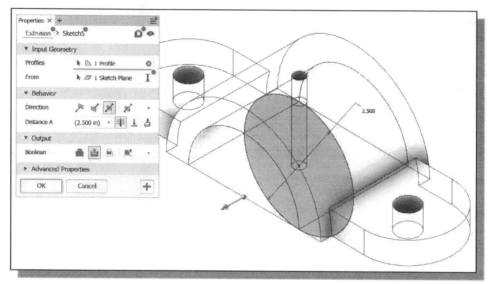

❖   Note that the parents of the Circular_Cut feature are the XY Plane and the Center Point.

➢   On your own, save the model as **U-Bracket**; this model will be used again in the next chapter.

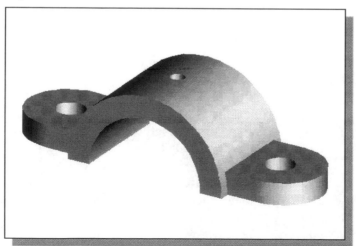

A Flexible Design Approach

In a typical design process, the initial design will undergo many analyses, tests, reviews and revisions. Autodesk Inventor allows the users to quickly make changes and explore different options of the initial design throughout the design process.

The model we constructed in this chapter contains two distinct design options. The *feature-based parametric modeling* approach enables us to quickly explore design alternatives and we can include different design ideas into the same model. With parametric modeling, designers can concentrate on improving the design, and the design process becomes quicker and requires less effort. The key to successfully using parametric modeling as a design tool lies in understanding and properly controlling the interactions of features, especially the parent/child relations.

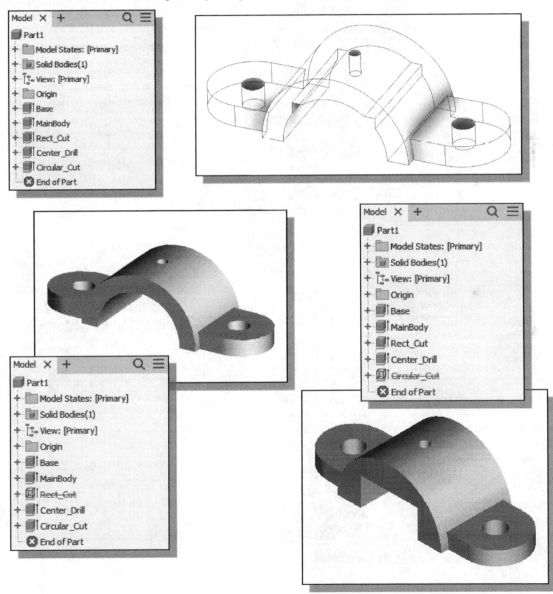

View and Edit Material Properties

The *Inventor Material Library* provides many commonly used materials. The material properties listed in the *Material Library* can be edited and new materials can be added.

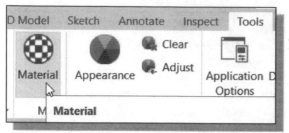

1. In the *Tools* tab, click **Material** as shown.

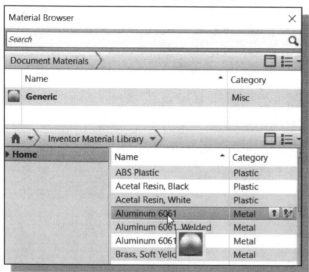

2. In the *Inventor Material* library group, select **Aluminum-6061**.

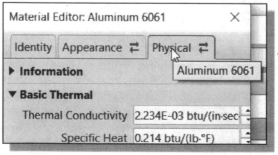

3. Click the **Add Material to document and display in editor** icon as shown.

4. Click on the **Physical Aspect** tab to view the associated material properties.

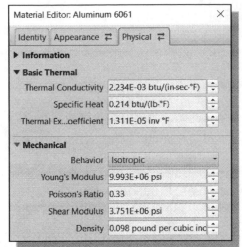

5. Expand the **Mechanical** properties list as shown.

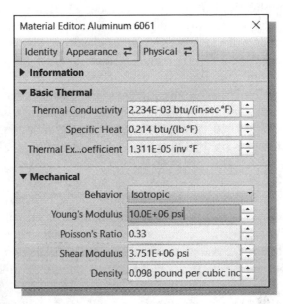

6. Enter **10.0E6 psi** as the new value for *Young's Modulus*.

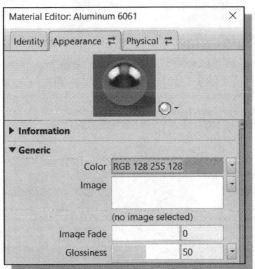

7. On your own, edit the material **Appearance - Glossiness** to **50** and color to **light green** as shown in the figure.

8. Click **OK** to accept the changes and close the editor.

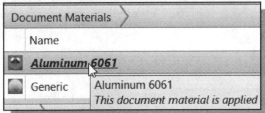

❖ Note the **Aluminum-6061** material in the document material list is highlighted showing the applied material properties.

9. Click **Close** to exit the **Material Browser**.

• Note the new material setting is assigned to the model which can be shown under any of the shaded display modes.

Drawings from Parts and Associative Functionality

With the software/hardware improvements in solid modeling, the importance of two-dimensional drawings is decreasing. Drafting is considered one of the downstream applications of using solid models. In many production facilities, solid models are used to generate machine tool paths for *computer numerical control* (CNC) machines. Solid models are also used in *rapid prototyping* to create 3D physical models out of plastic resins, powdered metal, etc. Ideally, the solid model database should be used directly to generate the final product. However, the majority of applications in most production facilities still require the use of two-dimensional drawings. Using the solid model as the starting point for a design, solid modeling tools can easily create all the necessary two-dimensional views. In this sense, solid modeling tools are making the process of creating two-dimensional drawings more efficient and effective.

Autodesk Inventor provides associative functionality in the different Autodesk Inventor modes. This functionality allows us to change the design at any level, and the system reflects it at all levels automatically. For example, a solid model can be modified in the *Part Modeling* mode and the system automatically reflects that change in the *Drawing* mode. We can also modify a feature dimension in the *Drawing* mode, and the system automatically updates the solid model in all modes.

In this lesson, the general procedure of creating multi-view drawings is illustrated. The *U_Bracket* design from the last chapter is used to demonstrate the associative functionality between the model and drawing views.

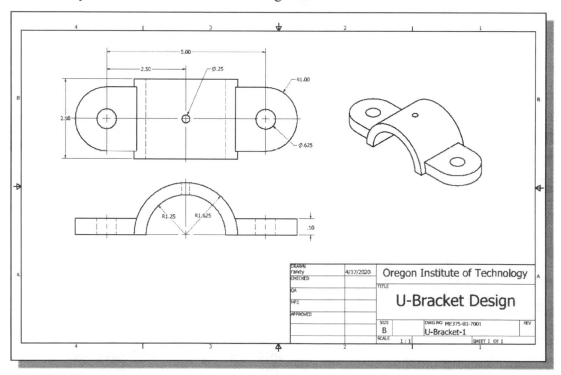

Drawing Mode

Autodesk Inventor allows us to generate 2D engineering drawings from solid models so that we can plot the drawings to any exact scale on paper. An engineering drawing is a tool that can be used to communicate engineering ideas/designs to manufacturing, purchasing, service, and other departments. Until now we have been working in *Model* mode to create our design in *__full size__*. We can arrange our design on a two-dimensional sheet of paper so that the plotted hardcopy is exactly what we want. This two-dimensional sheet of paper is saved in a *drawing* file in Autodesk Inventor. We can place borders and title blocks, objects that are less critical to our design, in the *drawing*. In general, each company uses a set of standards for drawing content based on the type of product and also on established internal processes. The appearance of an engineering drawing varies depending on when, where, and for what purpose it is produced. However, the general procedure for creating an engineering drawing from a solid model is fairly well defined. In Autodesk Inventor, creation of 2D engineering drawings from solid models consists of four basic steps: drawing sheet formatting, creating/positioning views, annotations, and printing/plotting.

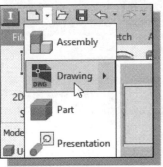

1. Click on the **drop-down arrow** next to the **New File** icon in the *Quick Access* toolbar area to display the available New File options.

2. Select **Drawing** from the option list.

• Note that a new graphics window appears on the screen. We can switch between the solid model and the drawing by clicking the corresponding tabs or graphics windows.

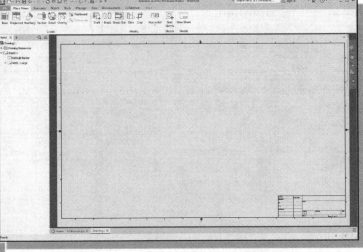

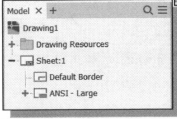

• In the *browser* area, the Drawing1 icon is displayed at the top, which indicates we are in the ***Drawing Mode***. **Sheet1** is the current drawing-sheet that is displayed in the graphics window.

Drawing Sheet Format

1. Choose the **Manage** tab in the *Ribbon* toolbar.

2. Click **Styles Editor** in the *Styles and Standards* toolbar as shown.

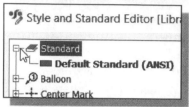

3. Click on the [+] sign in front of **Standard** to display the current active standard. Note that there can only be **one active standard** for each drawing.

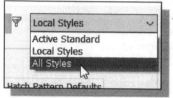

4. Set the *Filter Styles Setting* to display **All Styles**. Note that besides the default ANSI drafting standard, other standards, such as ISO, GB, BSI, DIN, and JIS, are also available.

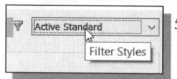

5. Set the *Filter Styles Setting* to display **Active Standard** and note only the active standard is displayed.

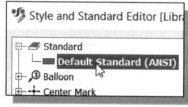

6. Click **Default Standard (ANSI)** to toggle the display of detailed settings for the current standard.

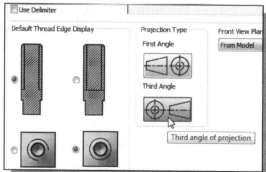

7. In the *View Preferences* page, confirm that the *Projection Type* is set to **Third Angle of projection**.

❖ Notice the different settings available in the *General* option window, such as the *Units* setting and the *Line Weight* setting.

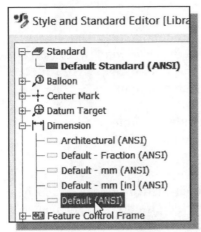

8. Choose **Default (ANSI)** in the **Dimension** list as shown.

➢ Note that the default *Dimension Style* in Inventor is based on the ANSI Y14.5-1994 standard.

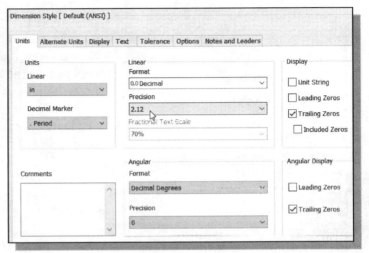

9. The **Units** tab contains the settings for linear/angular units. Note that the *Linear* units are set to Decimal and the *Precision* for linear dimensions is set to two digits after the decimal point.

10. Click on the **Text** tab to display and examine the settings for dimension text. Note that the default *Dimension Style*, DEFAULT-ANSI, cannot be modified. However, new *Dimension Styles* can be created and modified.

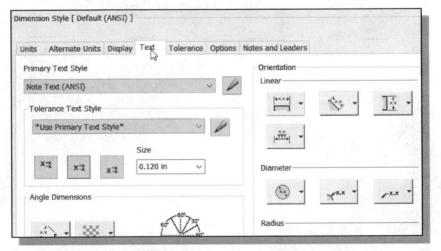

➢ On your own, click on the other tabs and examine the other *Settings* available.

11. Click on the **Cancel** button to exit the *Style and Standard Editor* dialog box.

Using the Pre-defined Drawing Sheet Formats

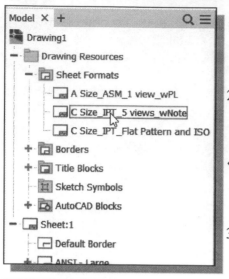

1. Inside the *Drawing Browser* window, click on the [**+**] symbol in front of **Drawing Resources** to display the available options.

2. Click on the [**+**] symbol in front of **Sheet Formats** to display the available pre-defined sheet formats.

❖ Notice several pre-defined *sheet formats*, each with a different view configuration, are available in the *browser* window.

3. Inside the *browser* window, **double-click** on the **C size IPT 5 view** sheet format.

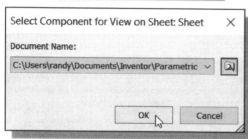

4. Click on the **OK** button to accept the default part file and generate the 2D views.

➢ The ***U-Bracket*** model is the only model opened. By default, all of the 2D drawings will be generated from this model file.

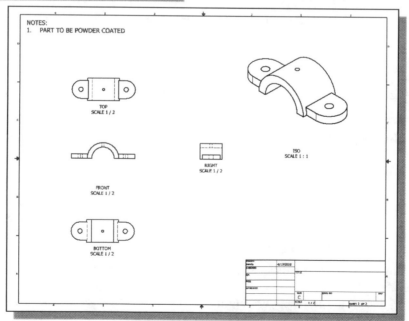

❖ We have created a C-size drawing of the *U-Bracket* model. Autodesk Inventor automatically generates and positions five of the pre-defined views of the model inside the title block.

Activate, Delete and Edit Drawing Sheets

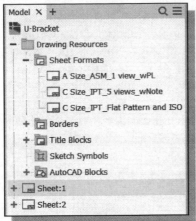

❖ Note that we have created two drawing sheets, displayed in the *Drawing Browser* window as Sheet1, and Sheet2. Autodesk Inventor allows us to create multiple 2D drawings from the same model file, which can be used for different purposes.

➢ In most cases, the pre-defined *sheet formats* can be used to quickly set up the views needed. However, it is also important to understand the concepts and principles involved in setting up the views. In the next sections, the procedures to set up drawing sheets and different types of views are illustrated.

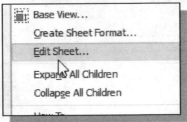

1. Inside the *Drawing Browser* window, double-click on **Sheet1** to activate this drawing sheet.

   • Alternatively, use the right-mouse-click and select **Activate** in the option menu.

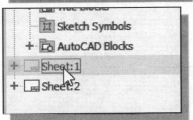

2. Inside the *Drawing Browser* window, right-click on **Sheet2** to display the option menu.

3. Select **Delete Sheet** in the option menu to remove the Sheet2 drawing.

4. In the *warning window*, click on the **OK** button to proceed with deleting the drawing.

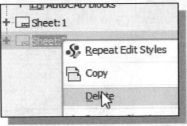

5. Inside the *Drawing Browser* window, **right-click** on **Sheet1** to display the *option menu*.

6. Select **Edit Sheet** in the option menu to display the settings for the *Sheet1* drawing.

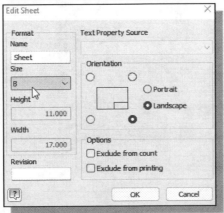

7. Set the sheet size to **B-size** and click on the **OK** button to exit the *Edit Sheet* dialog box.

Add a Base View

In *Autodesk Inventor Drawing Mode*, the first drawing view we create is called a **base view**. A *base view* is the primary view in the drawing; other views can be derived from this view. When creating a *base view*, Autodesk Inventor allows us to specify the view to be shown. By default, Autodesk Inventor will treat the *world XY plane* as the front view of the solid model. Note that there can be more than one *base view* in a drawing.

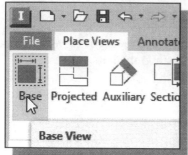

1. Click on the **Base View** in the *Place Views* panel to create a base view.

2. In the *Drawing View* dialog box, set the **Scale** to **1 : 1** and Style to **Hidden Line** as shown in the figure below.

3. In the *graphics area*, click on the arrows to switch to different views; set it to the **Front View** as shown in the below figure.

4. Inside the *graphics window,* drag and place the *base* view near the left center of the graphics window as shown below. Click **OK** to place the *base view* and close the *Drawing View* dialog box.

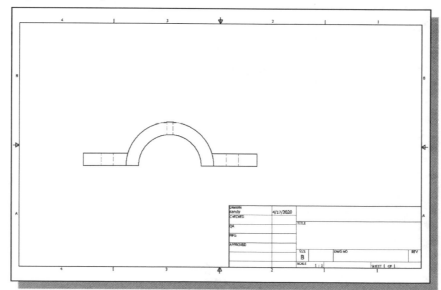

Create Projected Views

In *Autodesk Inventor Drawing Mode*, **projected views** can be created with a first-angle or third-angle projection, depending on the drafting standard used for the drawing. We must have a base view before a projected view can be created. Projected views can be orthographic projections or isometric projections. Orthographic projections are aligned to the base view and inherit the base view's scale and display settings. Isometric projections are not aligned to the base view.

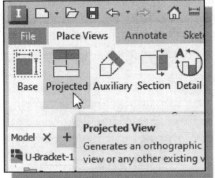

1. Click the **Projected View** button in the *Place Views* panel; this command allows us to create projected views.

2. Select the **base view** as the main view for the projected views.

3. Move the cursor **above** the *base view* and select a location to position the projected side view of the model.

4. Move the cursor toward the upper right corner of the title block and select a location to position the isometric view of the model as shown below.

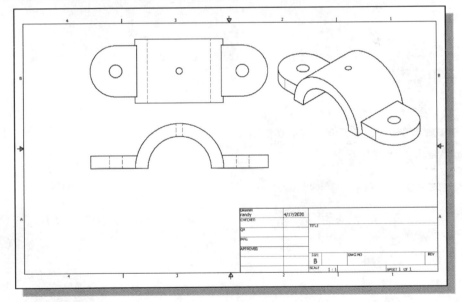

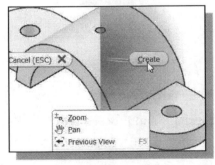

5. Inside the *graphics window*, right-click once to bring up the **option menu**.

6. Select **Create** to proceed with creating the two projected views.

Adjust the View Scale

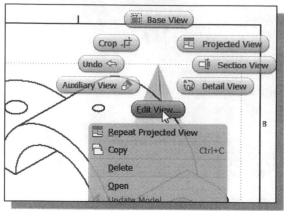

1. Move the cursor on top of the *isometric view* and watch for the box around the entire view indicating the view is selectable as shown in the figure. **Right-click** once to bring up the *option menu*.

2. Select **Edit View** in the option menu.

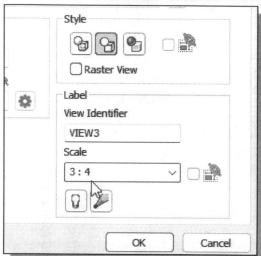

3. Inside the *Drawing View* dialog box set the *Scale* to **3:4** as shown in the figure.

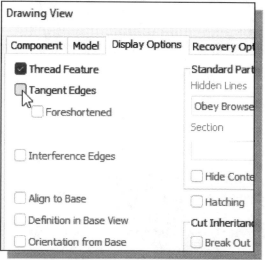

4. Click on the **Display Options** tab and turn *OFF* the **Tangent Edges** option as shown.

5. Click on the **OK** button to accept the settings and proceed with updating the drawing views.

Repositioning Views

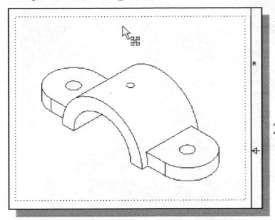

1. Move the cursor on top of the isometric view and watch for the **four-arrow Move symbol** as the cursor is near the border indicating the view can be dragged to a new location as shown in the figure.

2. Press and hold down the left-mouse-button and reposition the view to a new location.

3. On your own, reposition the views we have created so far. Note that the top view can be repositioned only in the vertical direction. The top view remains aligned to the base view, the front view.

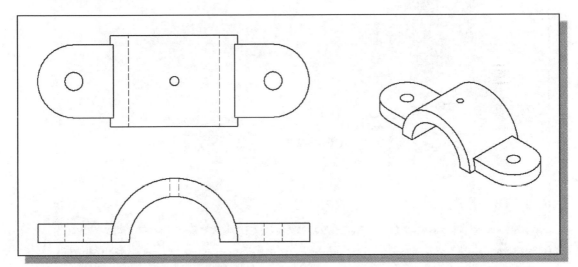

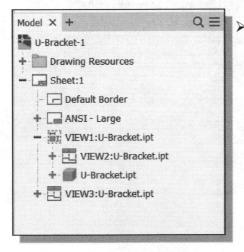

➢ Note that in the *Drawing Browser* area, a hierarchy of the created views is displayed under Sheet1. The base view, View1, is listed as the first view created, with View2 linked to it. The top view, View2, is projected from the base view, View1. The implied parent/child relationship is maintained by the system. Drawing views are associated with the model and the drawing sheets. As we create views from the base view, they are nested beneath the base view in the *browser*.

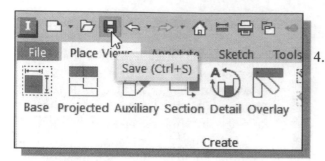

4. Click **Save** to save the drawing file.

5. On your own, save the U-Bracket.dwg file to the appropriate folder.

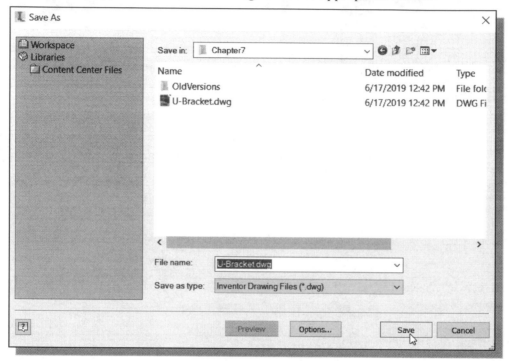

➤ Note that we have completed the creation of the necessary 2D orthographic views of the U-Bracket design for a multiview engineering drawing. In the next chapter, we will look into the general procedure in providing detailed descriptions of the created 2D views to document the design.

Review Questions:

1. Explain what an orthographic view is and why it is important to engineering graphics.

2. What is the purpose of creating 2D orthographic views?

3. What information is missing in the front view? Top view? Right view?

4. Which dimension stays the same in between the Top view and the front view?

5. How do you decide which views are necessary for a multi-view drawing?

6. Explain the creation of 2D views using the 1st Angle Projection method.

7. Explain the creation of 2D views using the 3rd Angle Projection method.

8. How do you modify the sheet size in Autodesk Inventor?

Exercises: (Unless otherwise specified, dimensions are in inches.)

1. **Swivel Yoke** (Material: **Cast Iron**)

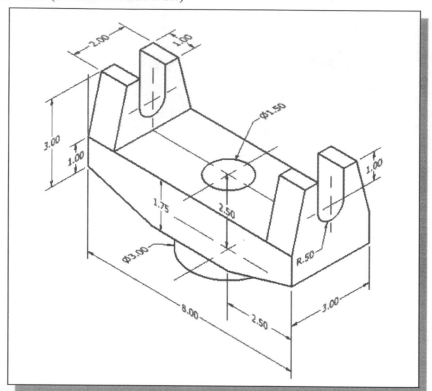

2. **Angle Bracket** (Material: **Carbon Steel**)

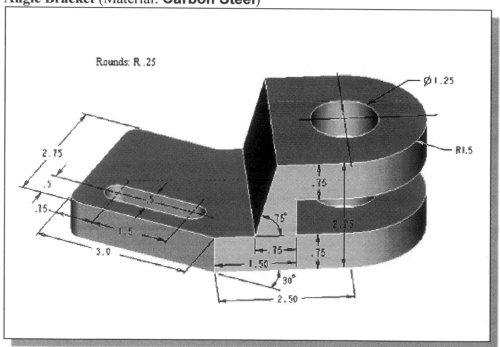

3. **Connecting Rod** (Material: **Carbon Steel**)

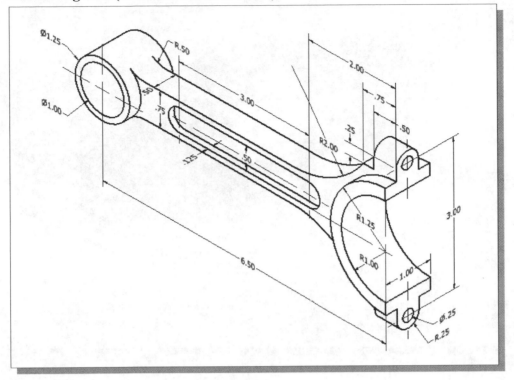

4. **Tube Hanger** (Material: **Aluminum 6061**)

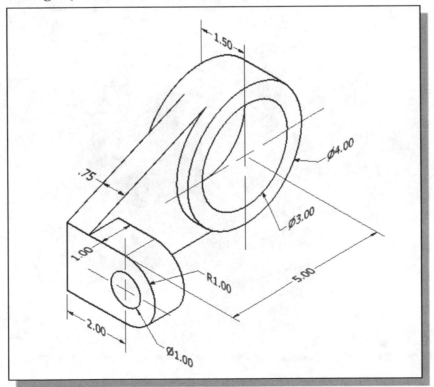

5. **Angle Latch** (Dimensions are in millimeters. Material: **Brass**)

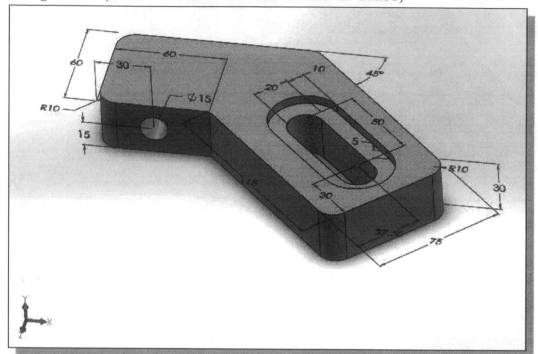

6. **Inclined Lift** (Material: **Mild Steel**)

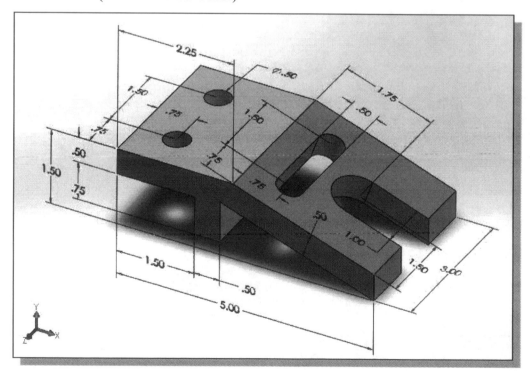

Chapter 8
Dimensioning and Notes

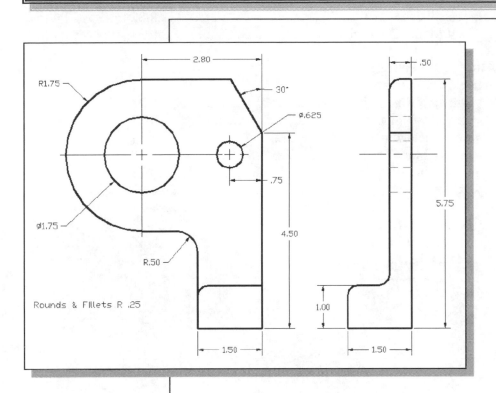

Learning Objectives

- **Understand Dimensioning Nomenclature and Basics**
- **Understand Associative Functionality**
- **Use the Default Sheet Formats**
- **Arrange and Manage 2D Views in Drawing Mode**
- **Display and Hide Feature Dimensions**
- **Create Reference Dimensions**
- **Understand and Create 3D Model-Based Definition on Solid Models**

Certified Autodesk Inventor User Exam Objectives Coverage

Autodesk Inventor Certified User Reference Guide

Introduction

In engineering graphics, a detailed drawing is a drawing consisting of the necessary orthographic projections, complete with all the necessary dimensions, notes, and specifications needed to manufacture the design. The dimensions put on the drawing are not necessarily the same ones used to create the drawing but are those required for the proper functioning of the part, as well as those needed to manufacture the design. It is therefore important, prior to dimensioning the drawing, to study and understand the design's functional requirements; then consider the manufacturing processes that are to be performed by the pattern-maker, die-maker, machinist, etc., in determining the dimensions that best give the information. In engineering graphics, a **detailed drawing** is a drawing consisting of the necessary orthographic projections, complete with all the necessary dimensions, notes, and specifications needed to manufacture the design.

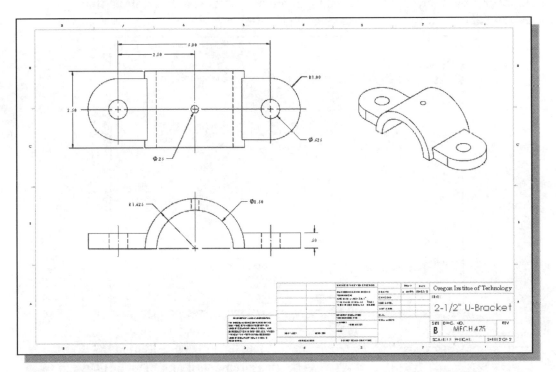

Considerable experience and judgment are required for accurate size and shape description. Dimensioning a design correctly requires conformance to many rules. For example, detail drawings should contain only those dimensions that are necessary to make the design. Dimensions for the same feature of the design should be given only once in the same drawing. Nothing should be left to chance or guesswork on a drawing. Drawings should be dimensioned to avoid any possibility of questions. Dimensions should be carefully positioned, preferably near the profile of the feature being dimensioned. The designer and the CAD operator should also be as familiar as possible with materials, methods of manufacturing and shop processes.

Dimensioning Standards and Basic Terminology

The first step to learning to dimension is a thorough knowledge of the terminology and elements required for dimensions and notes on engineering drawings. Various industrial standards exist for the different industrial sectors and countries; the appropriate standards must be used for the particular type of drawing that is being produced. Some commonly used standards include **ANSI** (American National Standard Institute), **BSI** (British Standards Association), **DIN** (Deutsches Institut für Normüng), **ISO** (International Standardization Organization), and **JIS** (Japanese Industry Standards). In this text, the discussions of dimensioning techniques are based primarily on the standards of the *American National Standard Institute* (ANSI/ASME Y14.5); other commonsense practices are also illustrated.

In engineering graphics, two basic methods are used to give a description of a measurement on a drawing: **Dimensions** and **Notes**.

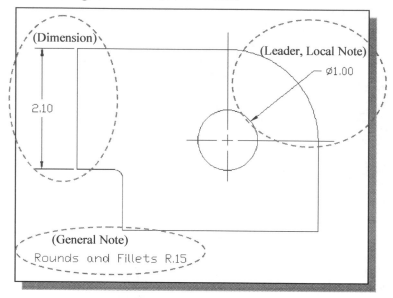

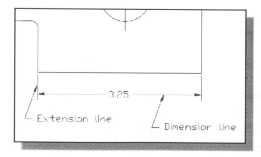

A dimension is used to give the measurement of an object or between two locations on a drawing. The **numerical value** gives the actual measurement, and the **dimension lines** and the **arrowheads** indicate the direction in which the measurement applies. **Extension lines** are used to indicate the locations where the dimension is associated.

Two types of notes are generally used in engineering drawings: **Local Notes** and **General Notes**. A *Local Note* is done with a leader and an arrowhead referring the statement of the note to the proper place on the drawing. A *General Note* refers to descriptions applying to the design as a whole and it is given without a leader.

Selection and Placement of Dimensions and Notes

One of the most important considerations to assure proper manufacturing and functionality of a design is the selection of dimensions to be given in the drawing. The selection should always be based upon the **functionality of the part**, as well as the **production requirements in the shop**. The method of manufacturing can affect the type of detailed dimensions and notes given in the drawing. It would be very beneficial to have a good knowledge of the different shop practices to give concise information on an engineering drawing.

After the dimensions and notes to be given have been selected, the next step is the actual placement of the dimensions and notes on the drawing. In reading a drawing, it is natural to look for the dimensions of a given feature wherever that feature appears most characteristic. One of the views of an object will usually describe the shape of some detailed feature better than with other views. Placing the dimensions in those views will promote clarity and ease of reading. This practice is known as the **Contour Rule**.

The two methods of positioning dimension texts on a dimension line are the **aligned** system and the **unidirectional** system. The *unidirectional* system is more widely used as it is easier to apply, and easier to read, the texts horizontally. The texts are oriented to be read from the **bottom** of the drawing with the *unidirectional* system. The *unidirectional* system originated in the automotive and aircraft industries and is also referred to as the "horizontal system."

For the *aligned* system, the texts are oriented to be aligned to the dimension line. The texts are oriented to be read from the **bottom or right side** of the drawing with the *aligned* system.

General Notes must be oriented horizontally and be read from the bottom of the drawing in both systems.

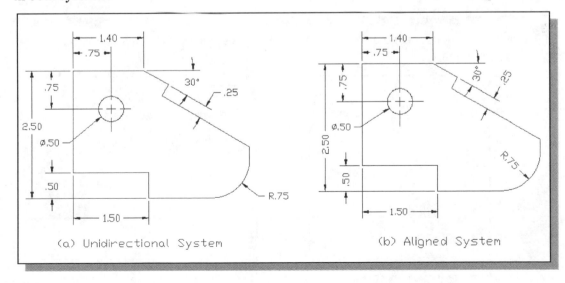

The following rules need to be followed for clearness and legibility when applying dimensions or notes to a design:

1. Always dimension **features**, not individual geometric entities. Consider the dimensions related to the **sizes** and **locations** of each feature.

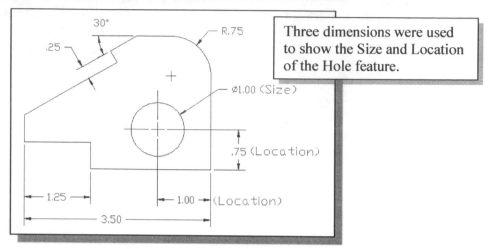

2. Choose the view that best describes the feature to place the User dimensions; this is known as the **Contour Rule**.

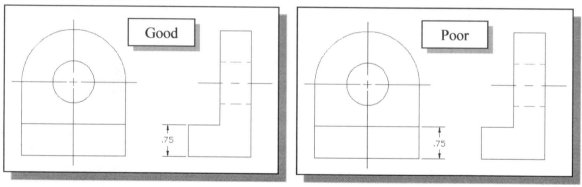

3. Place the dimensions next to the features being described; this is also known as the **Proximity Rule**.

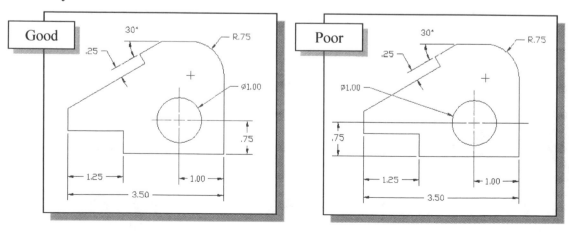

4. Dimensions should be placed outside the view if possible; only place dimensions on the inside if they add clarity, simplicity, and ease of reading.

5. Dimensions common to two views are generally placed in between the views.

6. Center lines may also be used as extension lines; therefore, there should be **no gap** in between an extension line and a center line.

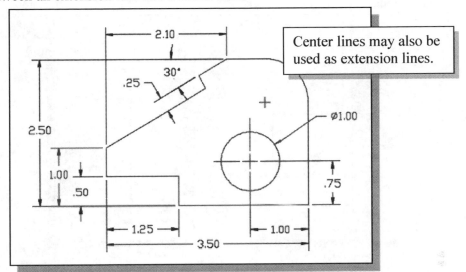

Center lines may also be used as extension lines.

7. Dimensions should be applied to one view only. When dimensions are placed between views, the extension lines should be **drawn from one view**, not from both views.

8. Center lines are also used to indicate the **symmetry** of shapes, and frequently eliminate the need for a positioning dimension. They should extend about 6 mm (0.25 in.) beyond the shape for which they indicate symmetry unless they are carried further to serve as extension lines. Center lines should not be continued between views.

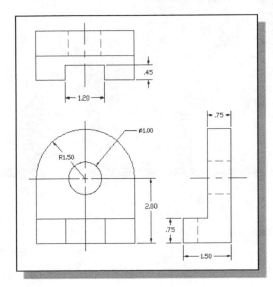

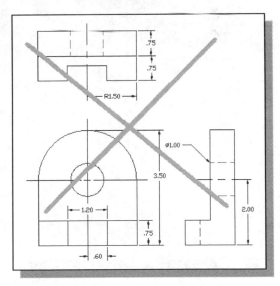

9. Dimensions should be placed only on the view that shows the measurement in its **true length**.

10. Always place a shorter dimension line inside a longer one to avoid crossing dimension lines with the extension lines of other dimensions. Thus an overall dimension, the maximum size of part in a given direction, will be placed outside all other dimensions.

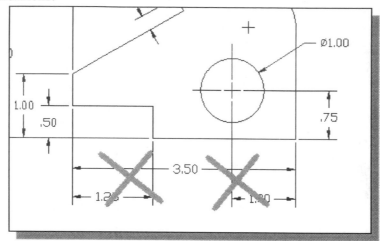

11. As a general rule, local notes (leaders) and general notes are created and placed after the regular dimensions.

12. Dimensions should never be crowded. If the space is small and crowded, an enlarged view or a partial view may be used.

13. Do not dimension to hidden lines; if necessary, create sectional views to convert the hidden features into visible features.

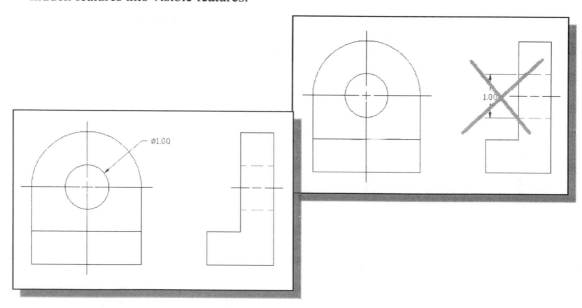

14. The spacing of dimension lines should be uniform throughout the drawing. Dimension lines should be spaced, in general, **10 mm** (0.4 inches) away from the outlines of the view. This applies to a single dimension or to the first dimension of several in a series. The space between subsequent dimension lines should be at least **6 mm** (0.25 inches) and uniformly spaced.

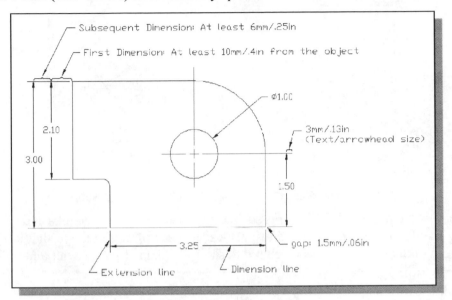

15. The **Overall Width**, **Height**, and **Depth** of the part should generally be shown in a drawing unless curved contours are present.

16. Never allow the crossing of dimension lines.

17. Dimension texts should be placed midway between the arrowheads, except when several parallel dimensions are present, where the texts should be staggered.

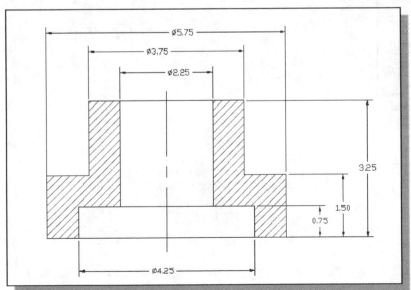

18. A **leader** should always be placed to allow its extension to pass through the **center** of all round features.

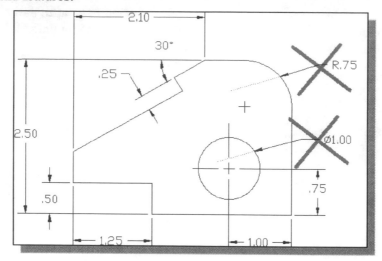

19. **Symbols** are preferred for features such as counter-bores, countersinks and spot faces; they should also be dimensioned using a leader. Each of these features has a special dimensioning symbol that can be used to show (a) Shape (b) Diameter and (c) Depth.

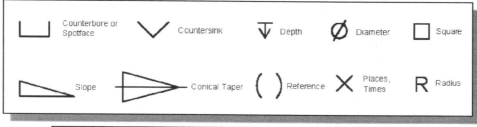

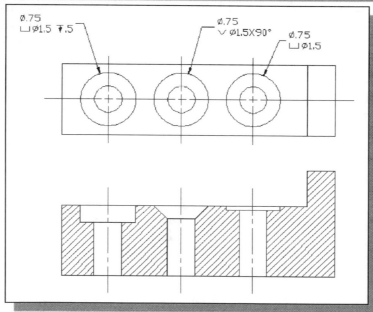

20. For leaders, avoid crossing leaders, avoid long leaders, avoid leaders in a horizontal or vertical direction, and avoid leaders parallel to adjacent dimension lines, extension lines, or section lines.

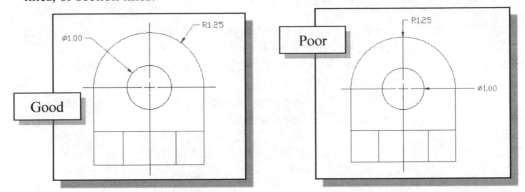

21. When dimensioning a circular feature, full cylinders (holes and bosses) must always be measured by their diameters. Arcs should always be shown by specifying a radius value. If a leader is used, the leader should meet the surface at an angle between 30° to 60° if possible.

22. Dual dimensions may be used, but they must be consistent and clearly noted. The dual-dimensioning can be shown by placing the alternate-units values below or next to the primary-units dimension, as shown in the figure.

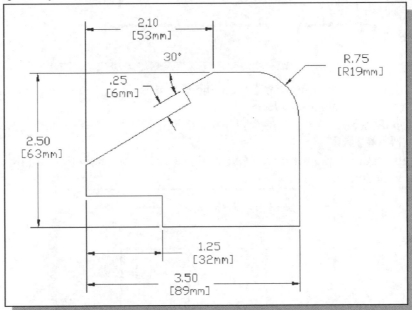

23. Never pass through a dimension text with any kind of line.

Metric Dimensioning versus English Dimensioning

In general, for Mechanical Engineering drawings, **Metric system** are given in **millimeters** and **English** system are given in **inches**.

- Millimeter (mm): common International System of Units (SI) unit of measure, generally **rounded to the nearest whole number**.
- Note on drawings: "UNLESS OTHERWISE SPECIFIED, ALL DIMENSIONS ARE IN MILLIMETERS"
- Omit the decimal point and trailing zeros when dimension is a whole number.
- Show a leading zero for values less than 1 unit.

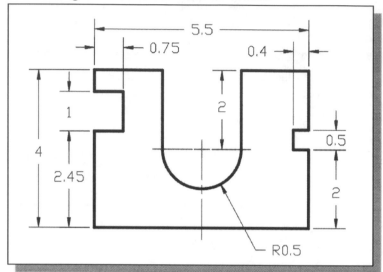

- Inch (IN): commonly used on US engineering drawings, and they are typically shown with **two decimal places**.
- Note on drawings: "UNLESS OTHERWISE SPECIFIED, ALL DIMENSIONS ARE IN INCHES"
- Omit the decimal point for values less than 1 unit, also show the trailing zeros after the decimal point.

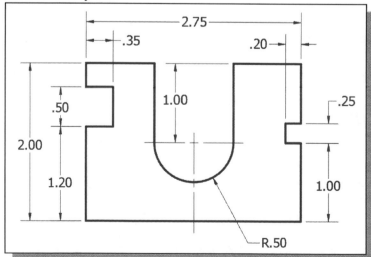

Machined Holes

In machining, a hole is a cylindrical feature that is cut from the work piece by a rotating cutting tool. The diameter of the machined hole will be the same as the cutting tool with matching the geometry. Non-cylindrical features, or pockets, can also be machined, but these features require end milling operations not the typical hole-making operations. While all machined holes have the same basic form, they can still differ in many ways to best suit a given application.

A machined hole can be characterized by several different parameters or features which will determine the hole machining operation and tool that is required.

Diameter - Holes can be machined in a wide variety of diameters, determined by the selected tool. The cutting tools used for hole-making are available in standard sizes that can be as small as 0.0019 inches and as large as 3 inches. Several standards exist including fractional sizes, letter sizes, number sizes, and metric sizes. A custom tool can be created to machine a non-standard diameter, but it is generally more cost effective to use the standard sized tool.

Tolerance - In any machining operation, the precision of a cut can be affected by several factors, including the sharpness of the tool, any vibration of the tool, or the buildup of chips of material. The specified tolerance of a hole will determine the method of hole-making used, as some methods are suited for tight-tolerance holes. Refer to the next section and Chapter 9 for more details on tolerances.

Depth - A machined hole may extend to a point within the work piece, known as a blind hole, or it may extend completely through the work piece, known as a through hole. A blind hole may have a flat bottom, but typically ends in a point due to the pointed end of the tool. When specifying the depth of a hole, one may reference the depth to the point or the depth to the end of the full diameter portion of the hole. The total depth of the hole is limited by the length of the cutting tool. For a through hole, the depth information is typically omitted in a drawing.

Recessed top - A common feature of machined holes is to recess the top of the hole into the work piece. This is typically done to accommodate the head of a fastener and allow it to sit flush with the work piece surface. Two types of recessed holes are typically used: a counterbore, which has a cylindrical recess, and a countersink, which has a cone-shaped recess.

Threads - Threaded holes are machined to accommodate a threaded fastener and are typically specified by their outer diameter and pitch. The pitch is a measure of the spacing between threads and may be expressed in the English standard, as the number of threads per inch (TPI), or in the metric standard, as the distance in between threads. For more details on threads, refer to Chapter 13 of this text.

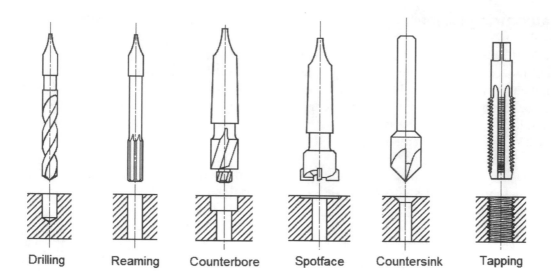

| Drilling | Reaming | Counterbore | Spotface | Countersink | Tapping |

Drilling - A drill bit enters the work piece in the axial direction and cuts a blind hole or a through hole with a diameter equal to that of the tool. A drill bit is a multi-point tool and typically has a pointed end. A twist drill is the most commonly used, but other types of drill bits, such as center drill, spot drill, or tap drill can also be used to start a hole that will be completed by additional operation.

Reaming - A reamer enters the work piece in the axial direction and enlarges an existing hole to the diameter of the tool. A reamer is a multi-point tool that has many flutes, which may be straight or in a helix. Reaming removes a minimal amount of material and is often performed after drilling to obtain both a more accurate diameter and a smoother internal finish.

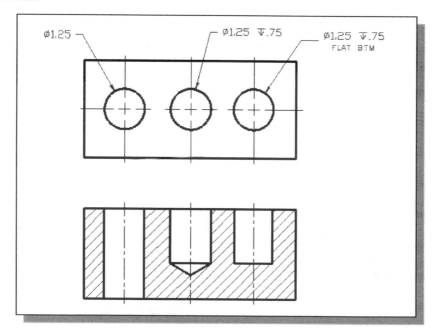

Counterbore - A counterbore tool enters the work piece axially and enlarges the top portion of an existing hole to the diameter of the tool. Counterboring is often performed after drilling to provide space for the head of a fastener, such as a bolt, to sit flush with the work piece surface. The counterboring tool has a pilot on the end to guide it straight into the existing hole. The depth of the counterbored portion is typically identified and dimensioned as a second operation on a drawing.

Spotface - A spotfaced hole is similar to a counterbored hole, but a spotfaced hole has a relatively shallow counterbored depth. The depth of the counterbored portion is left to the judgment of the machinist and thus, not dimensioned in the drawing. The smoothing is necessary in cast or forged surfaces to provide better contact between assembled pieces. For drawing purposes, the depth of the counterbored portion for a spotfaced hole is typically 1/16" or 2 mm.

Countersink - A countersink tool enters the work piece axially and enlarges the top portion of an existing hole to a cone-shaped opening. Countersinking is often performed after drilling to provide space for the head of a fastener, such as a screw, to sit flush with thc work piece surface. Common included angles for a countersink include 60, 82, 90, 100, 118, and 120 degrees.

Tapping - A tap enters the work piece in the axial direction and cuts internal threads inside an existing hole. The existing hole is typically drilled with a drill size that will accommodate the desired tap. The tap is selected based on the major diameter and pitch of the threaded hole. Threads may be cut to a specified depth inside the hole (bottom tap) or the complete depth of a through hole (through tap).

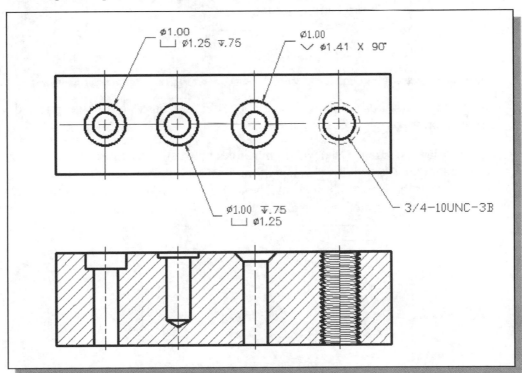

Baseline and Chain Dimensioning

Frequently the need arises to dimension a feature which involves a series of dimensions. Two methods should be considered for this type of situation: **Baseline dimensioning** and **Chain dimensioning**. The baseline dimensioning method creates dimensions by measuring from a common baseline. The chain dimensioning method creates a series or a chain of dimensions that are placed one after another. It is important to always consider the functionality and the manufacturing process of the specific features to be dimensioned. Note that both methods can be used to different features within the same drawing as shown in the below figure.

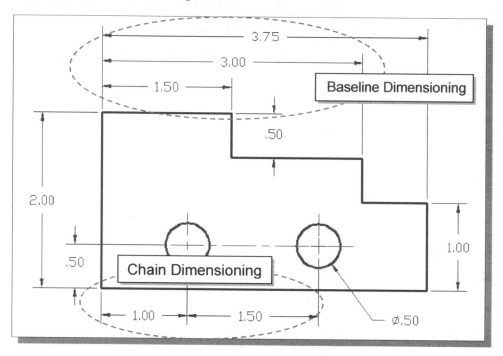

- **Baseline dimensioning:** used when the **location** of features must be controlled from a common reference point or plane.

- **Chain dimensioning:** Used when **tolerances** between adjacent features is more important than the overall tolerance of the feature. For example, dimensioning mating parts, hole patterns, slots, etc. Note that more in-depth discussions on tolerances are presented in the next section and also in the next chapter.

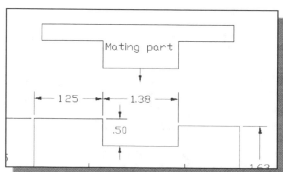

Dimensioning and Tolerance Accumulation

Drawings with dimensions and notes often serve as construction documents to ensure the proper functioning of a design. When dimensioning a drawing, it is therefore critical to consider the *precision* required for the part. *Precision* is the degree of accuracy required during manufacturing. Different machining methods will produce different degrees of precisions. However, one should realize it is nearly impossible to produce any dimension to an absolute, accurate measurement; some variation must be allowed in manufacturing. **Tolerance** is the allowable variation for any given size and provides a practical means to achieve the precision necessary in a design. With **tolerancing**, each dimension is allowed to vary within a specified zone. The general machining tolerances of some of the more commonly used manufacturing processes are listed in the figure below.

International Tolerance Grade (IT)							
4	5	6	7	8	9	10	11
Lapping or Honing							
	Cylindrical Grinding						
	Surface Grinding						
	Diamond Turning or Boring						
	Broaching						
	Powder Metal-sizes						
		Reaming					
			Turning				
			Powder Metal-sintered				
			Boring				
						Milling, Drilling	
						Planing & Shaping	
						Punching	
							Die Casting

As it was discussed in the previous sections, the selections and placements of dimensions and notes can have a drastic effect on the manufacturing of the designs. Furthermore, a poorly dimensioned drawing can create problems that may invalidate the design intent and/or become very costly to the company. One of the undesirable results in poorly dimensioned drawings is **Tolerance Accumulation**. *Tolerance Accumulation* is the effect of cumulative tolerances caused by the relationship of tolerances associated with relating dimensions. As an example, let's examine the *tolerance accumulation* that can occur with two mating parts, as shown in the below figure, using a standard tolerance of ± 0.01 for all dimensions.

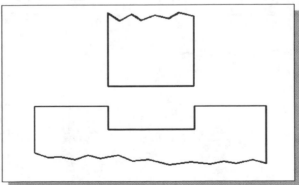

(1) Tolerance Accumulation - Baseline Dimensioning

Consider the use of the *Baseline dimensioning* approach, as described in the previous sections, to dimension the parts. At first glance, it may appear that the tolerance of the overall size is maintained and thus this approach is a better approach in dimensioning the parts!?

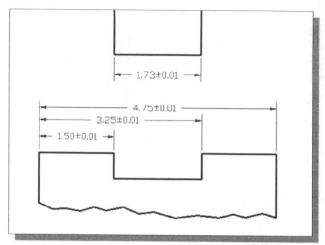

Now let's examine the size of the center slot, which is intended to fit the other part:

Upper limit of the slot size: 3.26-1.49= 1.77
Lower limit of the slot size: 3.24-1.51= 1.73

The tolerance of the center slot has a different range of tolerance than the standard ± 0.01. With this result, the two parts will not fit as well as the tolerance has been loosened.

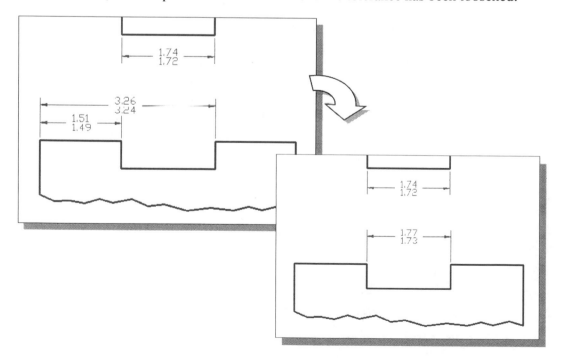

(2) Tolerance Accumulation - Chain Dimensioning

Now consider the use of the *chain dimensioning* approach, as described in the previous sections, to dimension the parts.

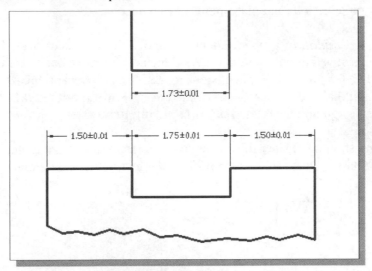

The *tolerance accumulation* occurred since the overall size is determined with the three dimensions that are chained together. Considering the upper and lower limits of the overall size of the bottom part:

Upper limit of the overall size: 1.51+1.76+1.51= 4.78
Lower limit of the overall size: 1.49+1.74+1.49= 4.72

With the chaining of the dimensions, the tolerance of the overall size has been changed, perhaps unintentionally, to a wider range.

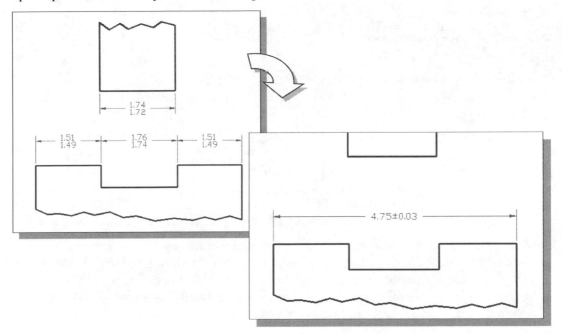

(3) Avoid Tolerance Accumulation – Dimensioning Features

The effect of the *tolerance accumulation* is usually undesirable. We could be using fairly precise and expensive manufacturing processes but resulting in lower quality products simply due to the poor selections and placements of dimensions.

To avoid *tolerance accumulation,* proper dimensioning of a drawing is critical. The first rule of dimensioning states that one should always dimension **features**, not an individual geometric entity. Always consider the dimensions related to the **sizes** and **locations** of each feature. One should also consider the associated **design intent** of the **feature**, the **functionality** of the design and the related **manufacturing processes**.

For our example, one should consider the center slot as a feature, where the other part will be fit into, and therefore the **size** and **location** of the feature should be placed as shown in the figure below.

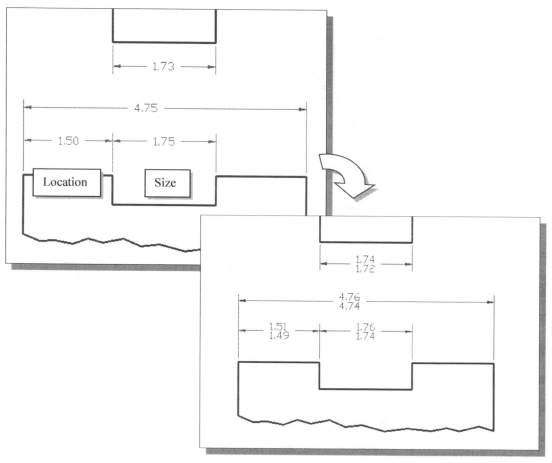

> More detailed discussions on tolerances are covered in the next chapter; however, the effects of *tolerance accumulation* should be avoided even during the initial selection and placement of dimensions. Note that there are also other options, generally covered under the topics of **Geometric Dimensioning and Tolerancing (GD&T)**, to assure the accuracy of designs through productions.

Dimensioning Tools in Autodesk Inventor

Thus far, the use of **Autodesk Inventor** to define the *shape* of designs has been illustrated. For the rest of the chapter, the general procedures to convey the *size* definitions of designs using **Autodesk Inventor** are discussed. The *tools of size description* are known as *dimensions* and *notes*.

Considerable experience and judgment are required for accurate size description. As it was outlined in the previous sections, detail drawings should contain only those dimensions that are necessary to make the design. Dimensions for the same feature of the design should be given only once in the same drawing. Nothing should be left to chance or guesswork on a drawing. Drawings should be dimensioned to avoid any possibility of questions. Dimensions should be carefully positioned, preferably near the profile of the feature being dimensioned. The designer and the CAD operator should be as familiar as possible with materials, methods of manufacturing, and shop processes.

Traditionally, detailing a drawing is the biggest bottleneck of the design process; and when doing board drafting, dimensioning is one of the most time consuming and tedious tasks. Today, most CAD systems provide what is known as an **auto-dimensioning feature**, where the CAD system automatically creates the extension lines, dimensional lines, arrowheads, and dimension values. Most CAD systems also provide an **associative dimensioning feature** so that the system automatically updates the dimensions when the drawing is modified.

The U-Bracket Design

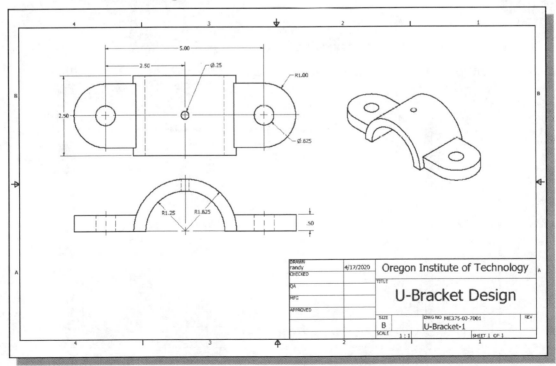

Starting Autodesk Inventor

1. Select the **Autodesk Inventor** option on the *Start* menu or select the **Autodesk Inventor** icon on the desktop to start Autodesk Inventor. The Autodesk Inventor main window will appear on the screen.

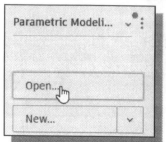

2. Select the **Open** icon on the *Menu Bar* with a single click of the left-mouse-button.

3. In the *Open* window select the **U-Bracket.dwg** file. Use the *browser* to locate the file if it is not displayed in the *File name* list box.

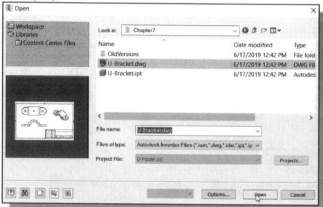

4. Click on the **Open** button in the *Open* dialog box to accept the selected settings.

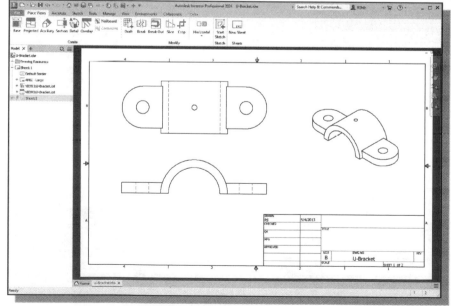

Display Feature Dimensions

By default, feature dimensions are not displayed in 2D views in Autodesk Inventor. We can change the default settings while creating the views or switch on the display of the parametric dimensions using the option menu.

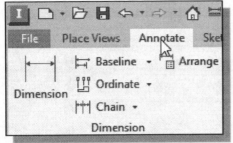

1. Select **Annotate** by left-clicking once in the *Ribbon* toolbar system.

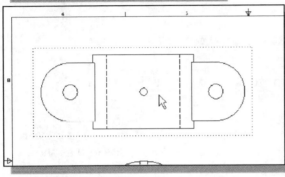

2. Move the cursor on top of the *top* view of the model and watch for the box around the entire view indicating the view is selectable as shown in the figure.

3. Inside the graphics window, **right-click** once on the top view to bring up the option menu.

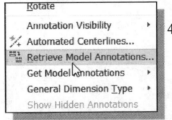

4. Select **Retrieve Model Annotations** to display the parametric dimensions used to create the model. Note the command can also be accessed through the *Ribbon* toolbar.

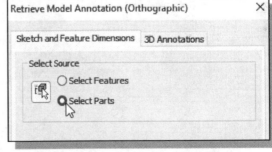

5. In the *Sketch and Features Dimensions* tab, set the *Select Source* option to **Select Parts** as shown.

6. Move the *Retrieve Model annotations* dialog box by pressing the title of the box and dragging with the left-mouse button to the right side of the view.

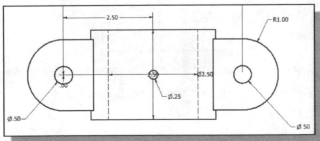

➤ Note that many of the dimensions used to create the part are now displayed in the selected view.

➢ The system now expects us to select the dimensions to retrieve.

7. On your own, select the dimensions to retrieve by left-clicking once on the dimensions as shown. (Note that only selected dimensions will be retrieved.)

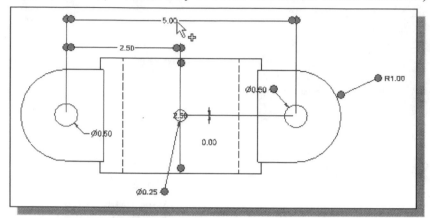

 8. Click on the **Apply** button to proceed with retrieving the selected dimensions.

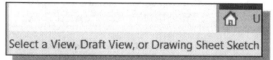 9. In the *message* area, the Inventor system expects us to **Select a View** as shown.

10. Select the *front* view.

11. On your own, retrieve the three dimensions as shown.

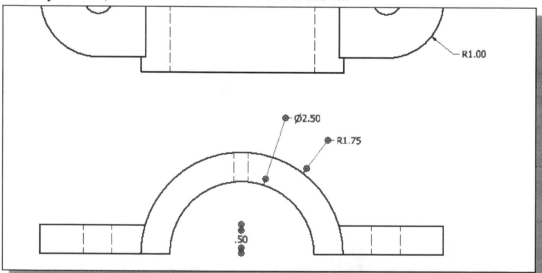

12. Click on the **OK** button to end the Retrieve Model Annotations command.

Repositioning and Hiding Feature Dimensions

1. Move the cursor on top of the width dimension text **5.00**, and watch for when the dimension text becomes highlighted with the four-arrow symbol indicating the dimension is selectable.

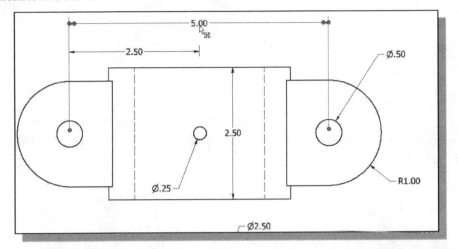

2. Reposition the dimension by using the left-mouse-button and drag the dimension text to a new location.

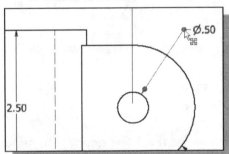

3. Move the cursor on top of the diameter dimension **0.5** and drag the grip point green dot associated with the dimension to reposition the dimension. Note that we can also drag on the dimension text, which only repositions the text.

4. On your own, reposition the dimensions displayed in the *top* view as shown in the figure below.

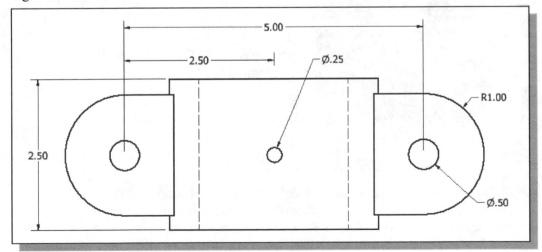

5. Move the cursor on top of the radius dimension **R 1.75** and notice two green grip points appear. The grip points can be used to reposition the dimension.

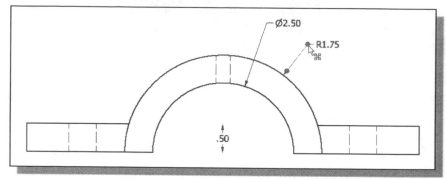

6. Use the left-mouse-button and drag the grip point near the center of the view and notice the arrowhead is automatically adjusted to the inside, as shown in the figures below.

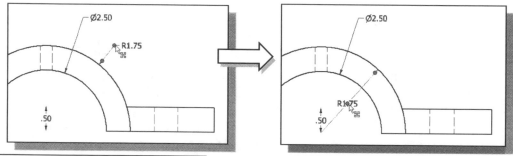

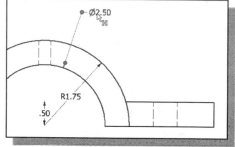

7. Move the cursor on top of the diameter dimension **2.50** and left-mouse-click once to select the dimension.

8. Right-click once on the dimension text to bring up the **option menu**.

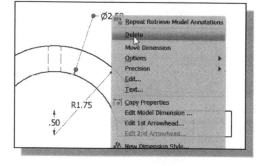

9. Select **Delete** to remove the dimension from the display.

➢ Note that the feature dimension is deleted from the display, but the removed dimension still remains in the database. In other words, the feature dimension is turned off or is hidden. Any feature dimensions can be removed from the display just as they can be displayed.

Add Additional Dimensions – Reference Dimensions

Besides displaying the **feature dimensions**, dimensions used to create the features, we can also add additional **reference dimensions** in the drawing. *Feature dimensions* are used to control the geometry, whereas *reference dimensions* are controlled by the existing geometry. In the drawing layout, therefore, we can ***add* or *delete*** *reference dimensions*, but we can only *hide* the *feature dimensions*. One should try to use as many *feature dimensions* as possible and add *reference dimensions* only if necessary. It is also more effective to use *feature dimensions* in the drawing layout since they are created when the model was built. Note that additional *Drawing Mode* entities, such as lines and arcs, can be added to drawing views. Before *Drawing Mode* entities can be used in a reference dimension, they must be associated to a *drawing view*.

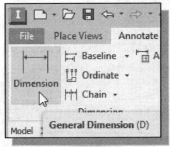

1. Click on the **General Dimension** button.

➤ Note the **General Dimension** command is similar to the **Smart Dimensioning** command in the *3D Modeling Mode*.

2. In the prompt area, the message "*Select first object:*" is displayed. Select the smaller arc of the front view.

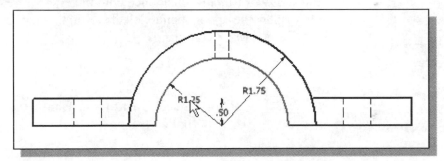

3. Place the dimension text on the inside of the arc as shown in the above figure.

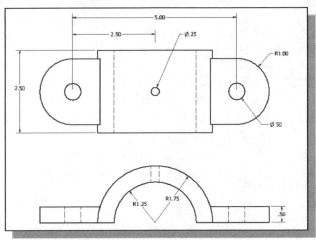

4. On your own, position the necessary dimensions for the design as shown in the figure.

➤ Note the extension lines can also be repositioned by dragging the associated grip points.

Add Center Marks and Center Lines

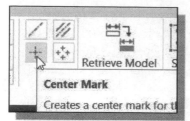

1. Click on the **Center Mark** button in the *Drawing Annotation* window.

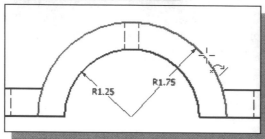

2. Click on the larger arc in the front view to add the center mark.

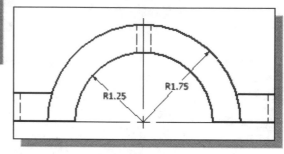

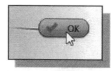

3. Inside the *graphics window*, click once with the **right-mouse-button** to display the option menu. Select **OK** in the pop-up menu to end the Center Mark command.

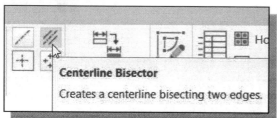

4. Select **Centerline Bisector** from the option list.

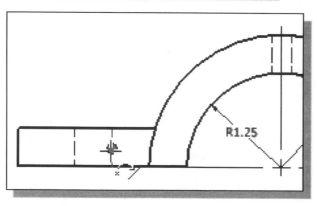

5. Click on the two hidden edges of one of the *drill* features of the front view as shown in the figure.

6. On your own, repeat the above step and create another centerline on the right side of the front view as shown.

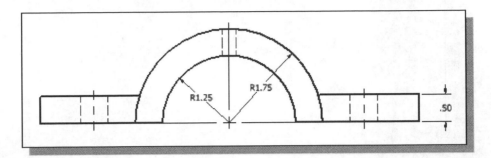

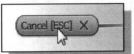

7. Inside the graphics window, click once with the right-mouse-button to display the option menu. Select **Cancel [ESC]** in the pop-up menu to end the Centerline Bisector command.

8. On your own, repeat the above steps and create additional centerlines as shown in the figure below.

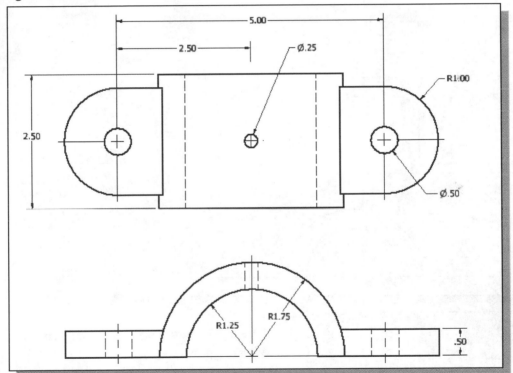

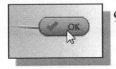

9. Inside the graphics window, click once with the right-mouse-button to display the option menu. Select **OK** in the pop-up menu to end the Centerline command.

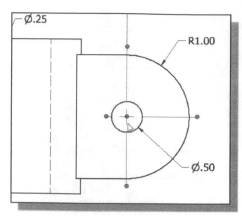

10. Click on the right centerlines in the *top* view as shown.

11. Adjust the length of the horizontal centerline by dragging on one of the grip points as shown.

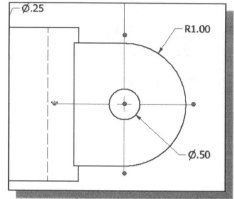

12. On your own, repeat the above steps and adjust the dimensions/centerlines as shown below.

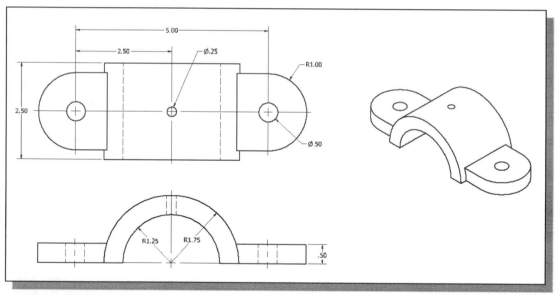

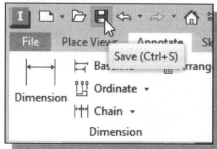

13. Click on the **Save File** icon in the *Standard* toolbar as shown.

14. Click the **Save** button to use the default drawing name.

Complete the Drawing Sheet

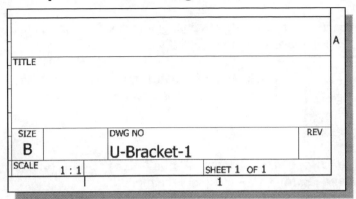

1. On your own, use the **Zoom** and the **Pan** commands to adjust the display as shown; this is so that we can complete the title block.

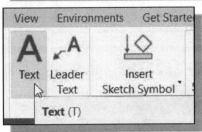

2. In the *Drawing Annotation* window, click on the **Text** button.

3. Pick a location that is inside the top block area as the location for the new text to be entered.

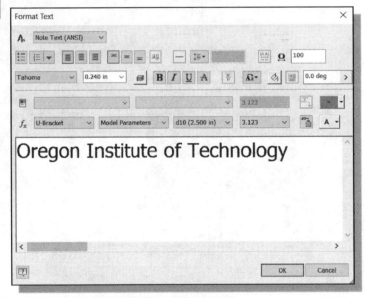

4. In the *Format Text* dialog box, enter the name of your organization. Also note the different settings available.

5. Click **OK** to proceed.

6. On your own, repeat the above steps and complete the title block.

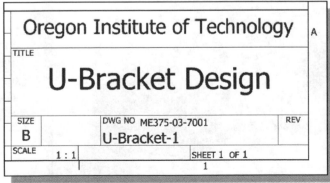

Associative Functionality – Modifying Feature Dimensions

Autodesk Inventor's *associative functionality* allows us to change the design at any level, and the system reflects the changes at all levels automatically.

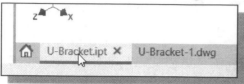

1. Click on the **U-Bracket** part window or tab to switch to the *Solid Model*.

2. In the *browser* window, click once on the [+] sign in front of the **Base** (Extrusion1) to expand the menu list.

3. Select **Sketch1** to display the sketch and dimensions in the graphics area.

4. Choose **Make Sketch visible** as shown.

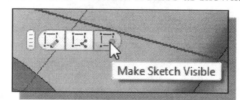

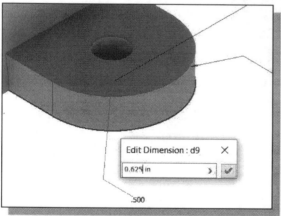

5. Double-click on the diameter dimensions (**0.50**) of the drill feature on the base feature as shown in the figure.

6. In the *Edit Dimension* dialog box, enter **0.625** as the new diameter dimension.

7. Click on the **check mark** button to accept the new setting.

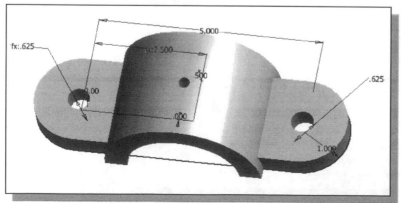

8. On your own, confirm the diameter of the other drill feature is also set to **0.625** as shown.

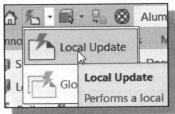

9. Click on the **Update** button in the *Standard* toolbar area to proceed with updating the solid model.

10. Click on the **U-Bracket-1** drawing graphics window or tab to switch to the *Multi-View Drawing*.

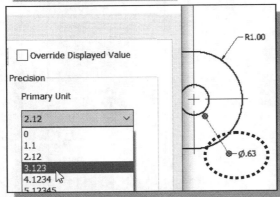

11. Inside the *graphics window*, double-click on the **0.63** dimension in the *top* view to bring up the *Precision and Tolerance* dialog box.

12. Set the *Precision* option to **3 digits after the decimal point** as shown.

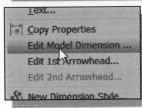

13. Inside the graphics window, right-click once on the **R 1.75** dimension in the *front* view to bring up the option menu.

14. Select **Edit Model Dimension** in the pop-up menu.

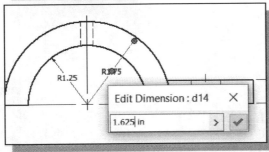

15. Change the dimension to **1.625**.

16. Click on the **check mark** button to accept the setting.

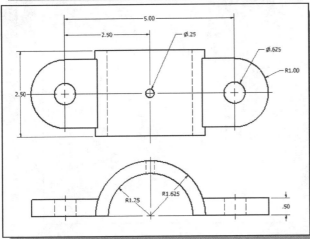

➤ Note the geometry of the cut feature is updated in all views automatically.

❖ On your own, switch to the *Part Modeling Mode* and confirm the design is updated as well.

➤ The completed multi-view drawing should appear as shown on the next page.

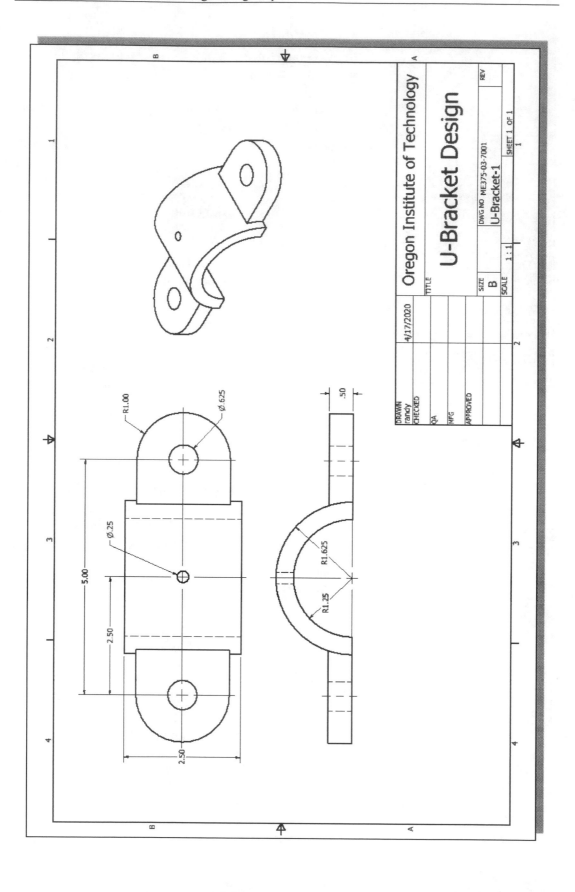

3D Model-Based Definition

The 3D Model-Based Definition (MBD) functionality is also available in Autodesk Inventor 2024. Modern 3D CAD applications allow for the insertion of engineering information such as dimensions, GD&T, notes and other product details directly within the 3D digital data set for components and assemblies. MBD uses such capabilities to establish the 3D digital data set as the source of these specifications and design authority for the product. The 3D digital data set may contain enough information to manufacture and inspect a product without the need for 2D engineering drawings.

Autodesk has partnered with *Sigmetrix/Advanced Dimensional Management*, leading providers of tolerance analysis and GD&T software and training, to help users apply semantic GD&T properly per the rules of the ASME Y14.5 and applicable ISO GPS standards. Autodesk Inventor 2024 supports both ASME and ISO standards. Inventor 2024 provides API access to all the MBD data, which means that downstream software can directly access the Product Manufacturing Information from the model.

The new Annotate ribbon provides a suite of tools for annotating models and exporting data sets to 3D PDF and other 3D formats. The 3D Annotation tools are divided into three categories: Geometric Annotation, General Annotation, and Notes.

Geometric Annotation tools are related to GD&T and GPS. The **Tolerance Feature** tool is used to apply geometric tolerances to features. The **DRF** tool is used to define Datum Reference Frames and datum systems. The **Tolerance Advisor** tool provides feedback on completeness of the GD&T scheme, error messages and warnings as the GD&T is applied.

General Annotation includes the **Dimension** tool, the **Hole/Thread Note** tool, and the **Surface Texture** tool. The *Dimension* tool is used to apply dimensions and directly-toleranced dimensions. The *Hole/Thread Note* is used to annotate complex holes and threaded holes. The *Surface Texture* tool is used to define surface texture requirements.

Notes are used to apply notes and general profile tolerances. Local notes may be defined using the *Leader Text* tool. General notes may be defined by the *General Note* tool. Default profile tolerances may be defined by the *General Profile Note* tool.

With this set of MBD tools, the mental translation of 3D models to two-dimensional (2D) drawings for communicating information and then back to 3D models for manufacturing is replaced by a simplified approach. The 3D annotation approach means faster and more precise product communication, which can improve the product development process.

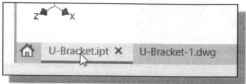

1. Click on the **U-Bracket** part window or tab to switch to the *Solid Model*.

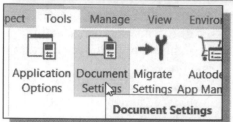

2. Click on the **Tools** tab and pick the *Document Settings* as shown.

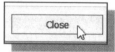

3. Confirm the *Annotations Standard* is set to **ASME** as the *annotation* standard as shown.

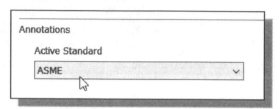

4. Click **Close** to accept the settings.

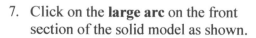

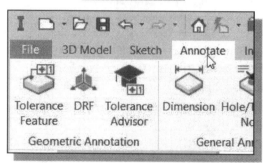

5. Click on the **Annotate** tab to switch to the *MBD toolbar set* as shown.

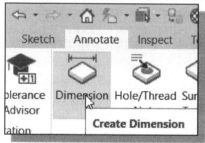

6. Activate the **Create Dimension** command in the *General Annotation* toolbar as shown.

7. Click on the **large arc** on the front section of the solid model as shown.

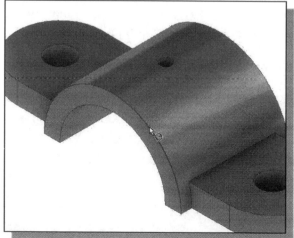

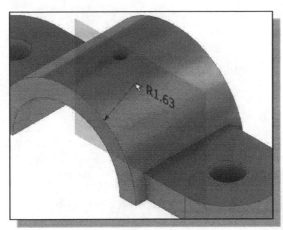

8. On your own, place the dimension above the arc as shown in the figure.

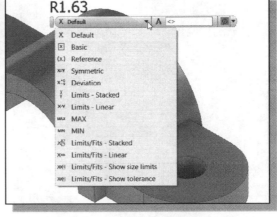

9. Display the first option list and note the different GD&T options available for the associated dimension.

10. On your own, use the **Edit Dimension** icon and change the number of digits displayed.

11. Click **OK** to accept the setting and create the dimension as shown.

12. On your own, repeat the above steps and add the other arc dimension as shown.

➢ Note that the radius and diameter dimensions are placed on the same plane of the selected arc and circle. The 3D annotation command will automatically select the placement plane if the selected geometry lies on a specific plane.

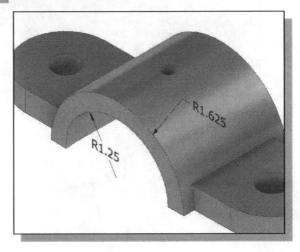

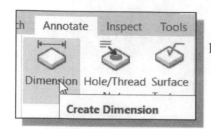

13. Activate the **Create Dimension** command in the *General Annotation* toolbar as shown.

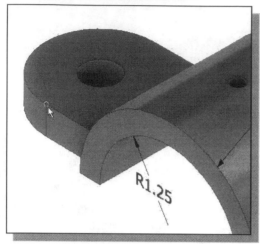

14. Click on the arc endpoint on the left section of the solid model as shown.

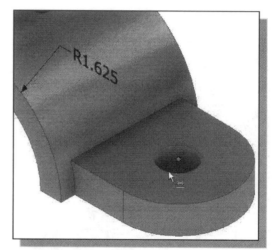

15. Click on the top circle on the right section of the solid model as shown.

➢ Note that by selecting a circle, we also set the placement plane for the dimension.

16. On your own, place the dimension in front of the solid model as shown.

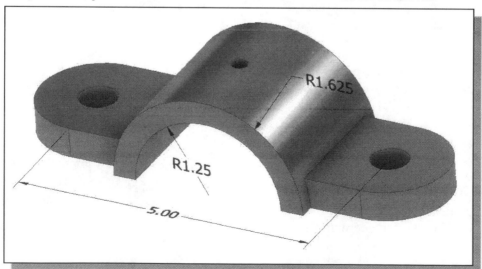

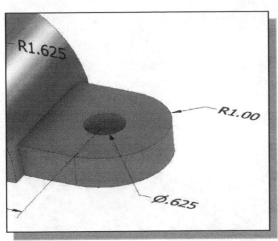

17. On your own, repeat the above steps and create the arc and hole dimensions as shown.

18. On your own, use the dynamic rotation option to view the bottom of the design as shown.

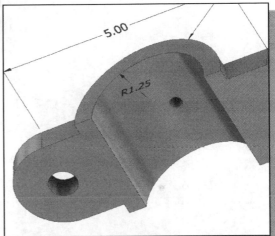

19. Activate the **Create Dimension** command in the *General Annotation* toolbar as shown.

20. On your own, select the straight edge and create the dimension as shown.

➤ Note the **General Dimension** command in *3D Annotation* behaves similarly to the **Smart Dimensioning** command in the *3D Modeling Mode*. *Inventor* will automatically create the proper dimensions based on the selected geometry.

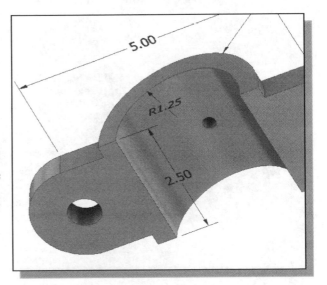

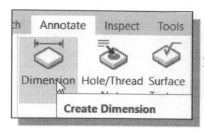

21. Activate the **Create Dimension** command in the General Annotation toolbar as shown.

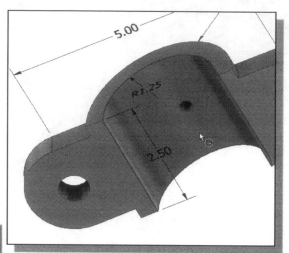

22. Select the inside **cylindrical surface** as shown.

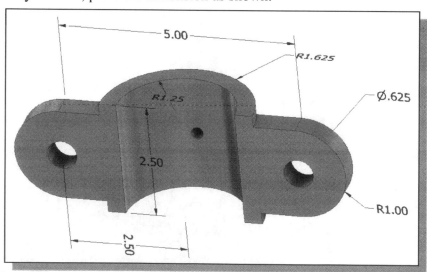

23. Click on the **bottom circle** on the left section of the solid model as shown.

24. On your own, place the dimension as shown.

25. Press the function key **F6** once or select **Home View** in the **ViewCube** to change the display back to the default isometric view.

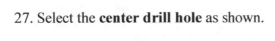

26. Activate the **Hole/Thread Note** command in the *General Annotation* toolbar as shown.

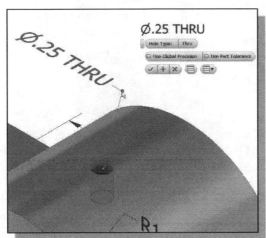

27. Select the **center drill hole** as shown.

28. Place the dimension toward the left side as shown. Note the different options available of the selected hole feature.

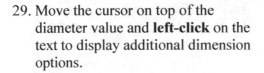

29. Move the cursor on top of the diameter value and **left-click** on the text to display additional dimension options.

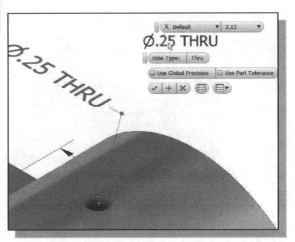

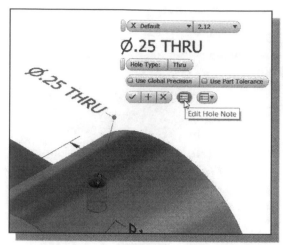

30. Click on the **Edit Hole Note** icon as shown. We will use this option to identify the related features for the small drill hole at the center.

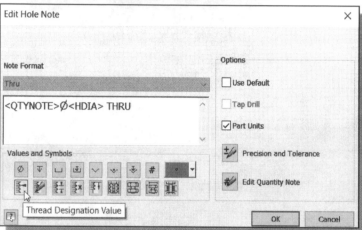

31. Note that we can add additional threads information to the *Hole Note*.

32. Click **OK** to close the *Editor*.

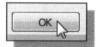

33. Click **OK** to accept the settings and create the dimension and complete the addition of the 3D model-based definition.

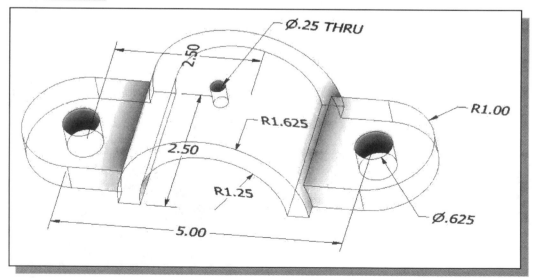

Review Questions:

1. What does Autodesk Inventor's *associative functionality* allow us to do?

2. How do we move a view on the *Drawing Sheet*?

3. How do we display feature/model dimensions in the drawing mode?

4. What is the difference between a *feature dimension* and a *reference dimension*?

5. How do we reposition dimensions?

6. What is a *base view*?

7. Can we delete a drawing view? How?

8. Can we adjust the length of centerlines in the drafting mode of *Inventor*? How?

9. Describe the purpose and usage of the *Leader Text* command.

10. Describe the advantages of using the *3D annotations in isometric drawing views* approach as a documentation tool that is available in Autodesk Inventor.

Exercises:

Create the Solid models and the associated 2D drawings.

1. **Shaft Guide** (Design is symmetrical and dimensions are in inches.)

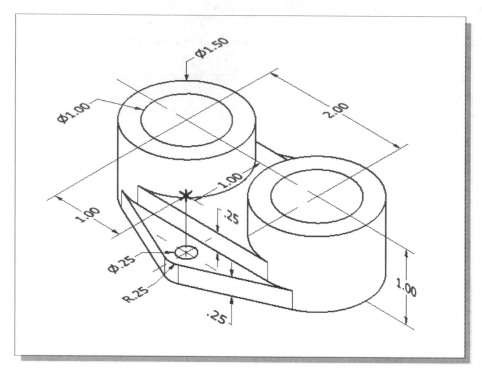

2. **Shaft Guide** (Dimensions are in inches.).

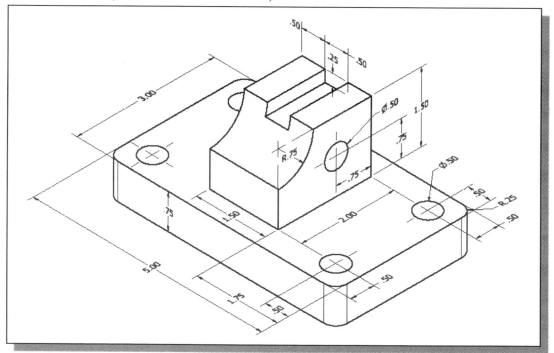

3. **Cylinder Support** (Dimensions are in inches.)

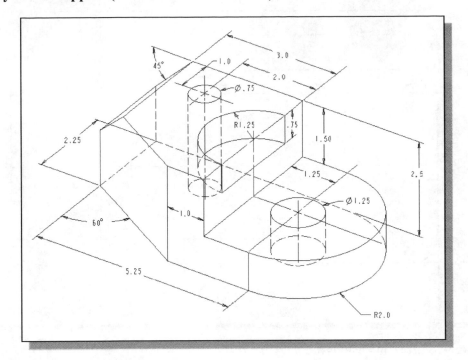

4. **Swivel Base** (Dimensions are in inches.)

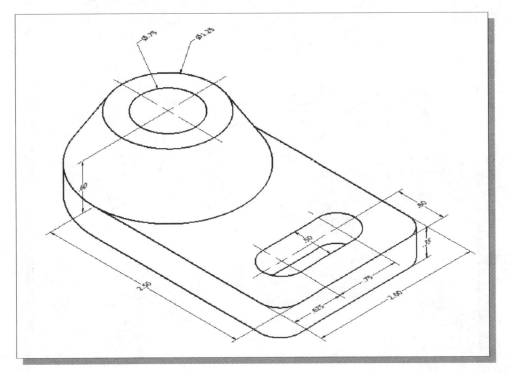

5. **Block Base** (Dimensions are in inches. Plate Thickness: 0.25)

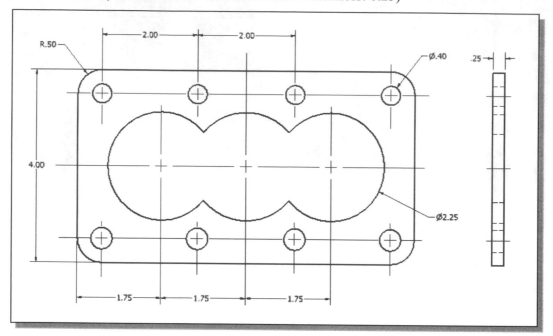

6. **Slide Mount** (Dimensions are in inches.)

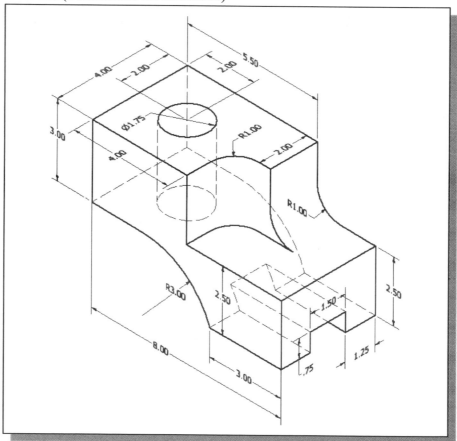

Chapter 9
Tolerancing and Fits

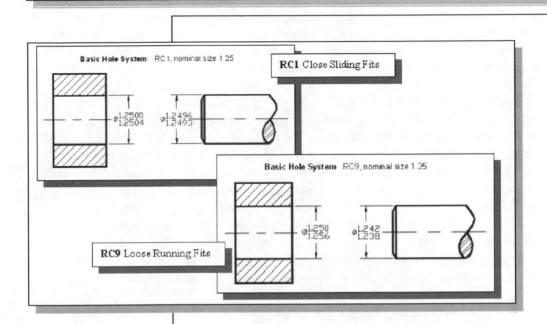

Basic Hole System: RC1, nominal size 1.25

RC1 Close Sliding Fits

ø1.2500
1.2504

ø1.2496
1.2493

Basic Hole System: RC9, nominal size 1.25

RC9 Loose Running Fits

ø1.250
1.256

ø1.242
1.238

Learning Objectives

- ♦ **Understand Tolerancing Nomenclature**
- ♦ **The Tolerancing Designation Method**
- ♦ **Understand the Basics of ANSI Standard Fits**
- ♦ **Use of the ISO Metric Standard Fits**
- ♦ **Use the Open Files Autodesk Inventor Option**
- ♦ **Set up the Tolerancing Option in Autodesk Inventor**

Precision and Tolerance

In the manufacturing of any product, quality and cost are always the two primary considerations. Drawings with dimensions and notes often serve as construction documents and legal contracts to ensure the proper functioning of a design. When dimensioning a drawing, it is therefore critical to consider the *precision* required for the part. *Precision* is the degree of accuracy required during manufacturing. However, it is impossible to produce any dimension to an absolute, accurate measurement; some variation must be allowed in manufacturing. Specifying higher precision on a drawing may ensure better quality of a product but doing so may also raise the costs of the product. And requiring unnecessary high precision of a design, resulting in high production costs, may cause the inability of the design to compete with similar products in the market. As an example, consider a design that contains cast parts. A cast part usually has two types of surfaces: mating surfaces and non-mating surfaces. The mating surfaces, as they will interact with other parts, are typically machined to a proper smoothness and require higher precision on all corresponding dimensions. The non-mating surfaces are usually left in the original rough-cast form as they have no important relationship with the other parts. The dimensions on a drawing must clearly indicate which surfaces are to be finished and provide the degree of precision needed for the finishing.

The method of specifying the degree of precision is called **Tolerancing**. **Tolerance** is the allowable variation for any given size and provides a practical means to achieve the precision necessary in a design. Tolerancing also ensures interchangeability in manufacturing; parts can be made by different companies in different locations and still maintain the proper functioning of the design. With tolerancing, each dimension is allowed to vary within a specified zone. By assigning as large a tolerance as possible, without interfering with the functionality of a design, the production costs can be reduced. The smaller the tolerance zone specified, the more expensive it is to manufacture.

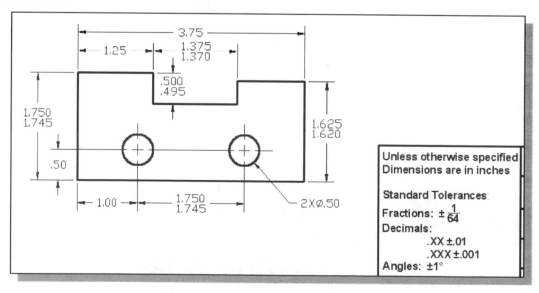

Methods of Specifying Tolerances – English System

Three methods are commonly used in specifying tolerances: **Limits**, **Unilateral tolerances** and **Bilateral tolerances**. Note that the unilateral and bilateral methods employ the use of a **Base Dimension** in specifying the variation range.

- **Limits** – Specifies the range of size that a part may be.
  This is the preferred method as approved by ANSI/ASME Y14.5M; the maximum and minimum limits of size and locations are specified. The maximum value is placed above the minimum value. In single-line form, a dash is used to separate the two values.

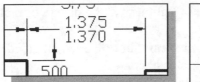 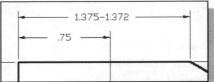

- **Bilateral** – Variation of size is in both directions.
  This method uses a basic size and is followed by a plus-and-minus expression of tolerance. The bilateral tolerances method allows variations in both directions from the basic size. If an equal variation in both directions is desired, the combined plus-and minus symbol is used with a single value.

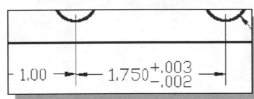 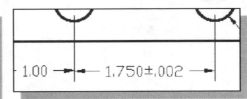

- **Unilateral** – Variation of size is in only one direction.
  This method uses a basic size and is followed by a plus-or-minus expression of tolerance. The unilateral tolerances method allows variations only in one direction from the basic size.

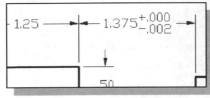 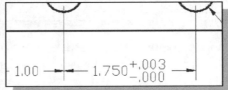

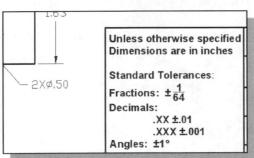

- General tolerances – General tolerances are typically covered in the title block; the general tolerances are applied to the dimensions in which tolerances are not given.

Nomenclature

The terms used in tolerancing and dimensioning should be clearly understood before more detailed drawings are studied. The following are definitions as defined in the ANSI/ASME Y 14.5M standard.

- **Nominal size** is the designation used for the purpose of general identification.
- **Basic size, or basic dimension,** is the theoretical size from which limits of size are derived. It is the size from which limits are determined for the size or location of a feature in a design.
- **Actual size** is the measured size of the manufactured part.
- **Fit** is the general term used to signify the range of tightness in the design of mating parts.
- **Allowance** is the minimum clearance space or maximum interference intended between two mating parts under the maximum material condition.
- **Tolerance** is the total permissible variation of a size. The tolerance is the difference between the limits of size.
- **Maximum Material Condition (MMC)** is the size of the part when it consists of the most material.
- **Least Material Condition (LMC)** is the size of the part when it consists of the least material.

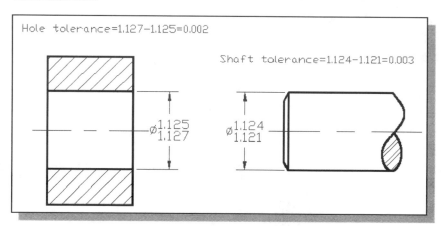

In the figure above, the tightest fit between the two parts will be when the largest shaft is fit inside the smallest hole. The allowance between the two parts can be calculated as:

Allowance = (MMC Hole)-(MMC Shaft)=1.125-1.124 = 0.001

The loosest fit between the above two parts will be when the smallest shaft is fit inside the largest hole. The maximum clearance between the two parts can be calculated as:

Max. Clearance = (LMC Hole)-(LMC Shaft)=1.127-1.121 = 0.006

- ➢ Note that the above dimensions will assure there is always a space between the two mating parts.

Example 9.1

The size limits of two mating parts are as shown in the below figure. Determine the following items, as described on the previous page: nominal size, tolerance of the shaft, tolerance of the hole, allowance and maximum clearance between the parts.

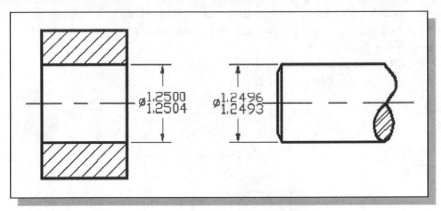

Solution:

      Nominal size = 1.25

      Tolerance of the shaft = Max. Shaft - Min. Shaft
                      = 1.2496-1.2493 = 0.0003

      Tolerance of the hole = Max. Hole - Min. Hole
                      = 1.2504-1.2500 = 0.0004

      Allowance = Min. Hole (MMC Hole) - Max. Shaft (MMC Shaft)
                      = 1.2500-1.2496 = 0.0004

      Max. Clearance = Max. Hole (LMC Hole) - Min. Shaft (LMC Shaft)
                    =1.2504-1.2493 = 0.0011

Exercise: Given the size limits of two mating parts, determine the following items: nominal size, tolerance of the shaft, tolerance of the hole, allowance and maximum clearance between the parts.

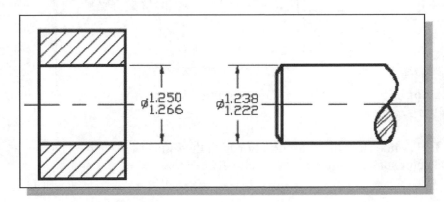

Fits between Mating Parts

Fit is the general term used to signify the range of tightness in the design of mating parts. In ANSI/ASME Y 14.5M, four general types of fits are designated for mating parts.

- **Clearance Fit:** A clearance fit is the condition in which the internal part is smaller than the external part and always leaves a space or clearance between the parts.

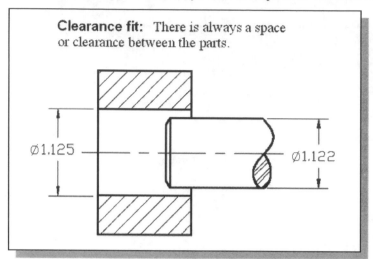

- **Interference Fit:** An interference fit is the condition in which the internal part is larger than the external part and there is always an interference between the parts.

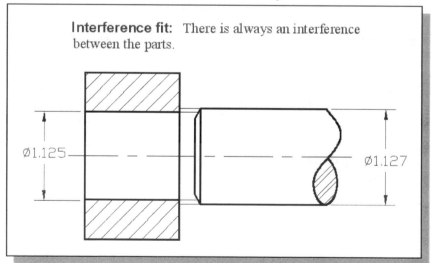

- **Transition Fit:** A transition fit is the condition in which there is either a *clearance* or *interference* between the parts.

- **Line Fit:** A line fit is the condition in which the limits of size are such that a clearance or surface contact may result between the mating parts.

Selective Assembly

By specifying the proper allowances and tolerances, mating parts can be completely interchangeable. But sometimes the fit desired may require very small allowances and tolerances, and the production cost may become very high. In such cases, either manual or computer-controlled selective assembly is often used. The manufactured parts are then graded as small, medium and large based on the actual sizes. In this way, very satisfactory fits may be achieved at much lower cost than to manufacture all parts to very accurate dimensions. Interference and transition fits often require the use of selective assembly to get the desirable interference or clearance.

Basic Hole and Basic Shaft Systems

In manufacturing, standard tools (such as reamers and drills) are often used to produce holes. In the **basic hole system**, the **minimum size of the hole** is taken as a base, and an allowance is assigned to derive the necessary size limits for both parts. The hole can often be made with a standard size tool and the shaft can easily be machined to any size desired.

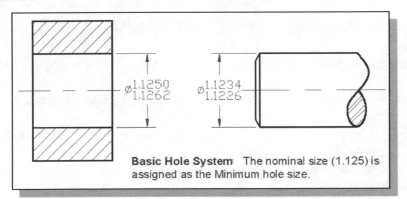

Basic Hole System   The nominal size (1.125) is assigned as the Minimum hole size.

On the other hand, in some branches of manufacturing industry, where it is necessary to manufacture the shafts using standard sizes, the **basic shaft system** is often used. This system should only be used when there is a special need for it; for example, if a shaft is to be assembled with several other parts using different fits. In the **basic shaft system**, the **maximum shaft** is taken as the basic size, and all size limits are derived from this basic size.

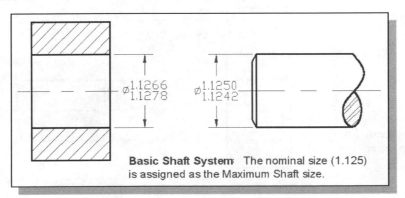

Basic Shaft System   The nominal size (1.125) is assigned as the Maximum Shaft size.

American National Standard Limits and Fits – Inches

The standard fits designated by the American National Standards Institute are as follows:

- **RC** Running or Sliding Clearance Fits (See appendix A for the complete table.)
- **LC** Locational Clearance Fits
- **LT** Transition Clearance or Interference Fits
- **LN** Locational Interference Fits
- **FN** Force or Shrink Fits

These letter symbols are used in conjunction with numbers for the class of fit. For example, **RC5** represents a class 5 *Running or Sliding Clearance Fits*. The limits of size for both of the mating parts are given in the standard.

1. **Running or Sliding Clearance Fits:** RC1 through RC9

   These fits provide limits of size for mating parts that require sliding and running performance, with suitable lubrication allowance.

   RC1, **close sliding fits**, are for mating parts to be assembled without much play in between.

   RC2, **sliding fits**, are for mating parts that need to move and turn easily but not run freely.

   RC3, **precision running fits**, are for precision work running at low speed and light journal pressure. This class provides the closest running fits.

   RC4, **close running fits**, are for moderate running speed and journal pressure in accurate machinery.

   RC5, RC6 and RC7, **medium running fits**, are for running at higher speed and/or higher journal pressure.

   RC8 and RC9, **loose running fits**, are for general purpose running condition where wide commercial tolerances are sufficient; typically used for standard stock parts.

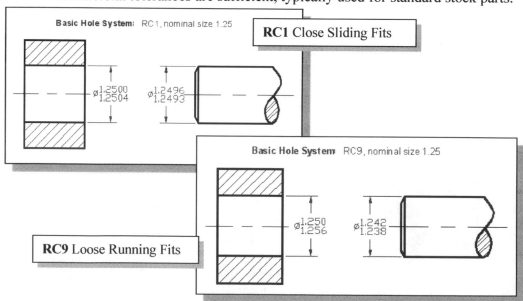

2. Locational Clearance Fits: LC1 through LC11

These fits provide limits of size for mating parts that are normally stationary and can be freely assembled and disassembled. They run from snug fits for parts requiring accuracy location, through the medium clearance fits for parts (such as spigots) to the looser fastener fits where freedom of assembly is of prime importance.

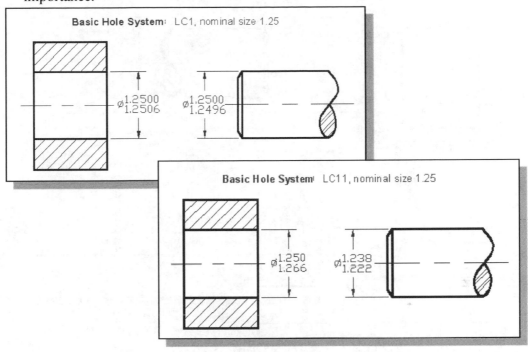

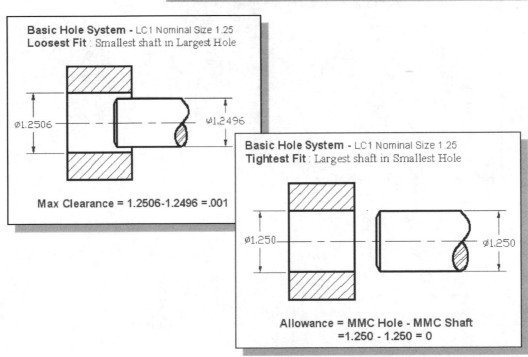

3. Transition Clearance or Interference Fits: LT1 through LT6

These fits provide limits of size for mating parts requiring accurate location but either a small amount of clearance or interference is permissible.

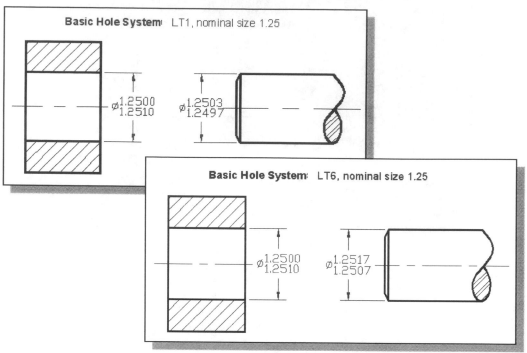

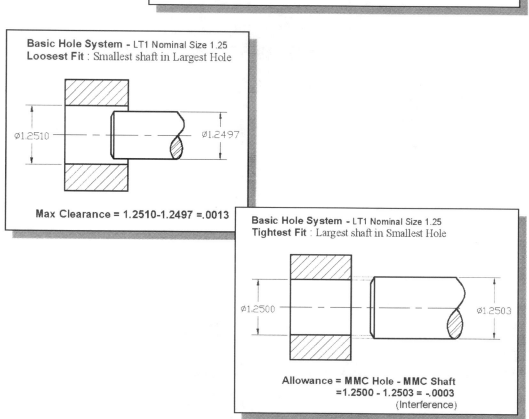

4. Locational Interference Fits: LN1 through LN3

These fits provide limits of size for mating parts requiring accuracy location and for parts needing rigidity and alignment. Such fits are not for parts that transmit frictional loads from one part to another by the tightness of the fit.

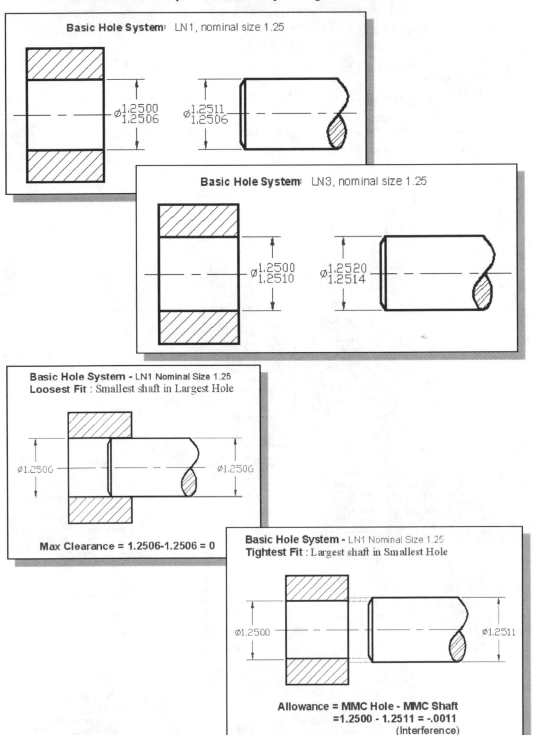

5. Force or Shrink Fits: FN1 through FN5

These fits provide limits of size for mating parts requiring constant bore pressure in between the parts. The interference amount varies almost directly with the size of parts.

FN1, **light drive fits**, are for mating parts requiring light assembly pressures, and produce more or less permanent assemblies. They are suitable for thin sections or long fits, or in cast-iron external members.

FN2, **medium drive fits**, are suitable for ordinary steel parts or for shrink fits in light sections.

FN3, **heavy drive fits**, are suitable for ordinary steel parts or for shrink fits in medium sections.

FN4, **force fits**, are for parts that can be highly stressed or for shrink fits where heavy pressing forces are impractical.

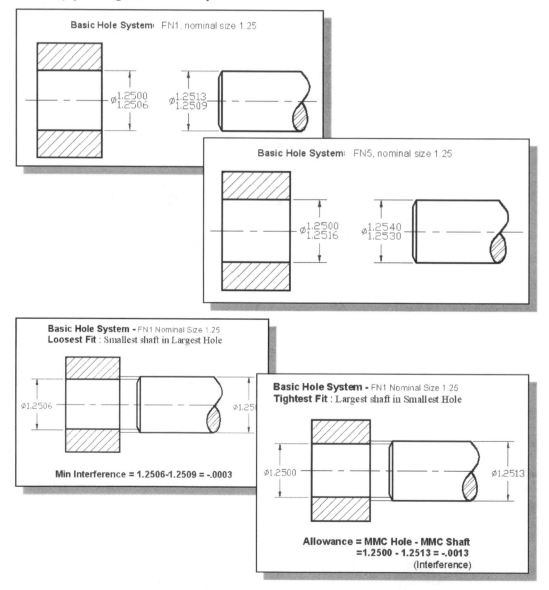

Example 9.2 Basic Hole System

A design requires the use of two mating parts with a nominal size of **0.75** inches. The shaft is to run with moderate speed but with a fairly heavy journal pressure. The fit class chosen is **Basic Hole System RC6**. Determine the size limits of the two mating parts.

From Appendix A, the RC6 fit for 0.75 inch nominal size:

Nominal size Range	Limits of Clearance	RC6 Standard Limits – Basic Hole	
		Hole	**Shaft**
0.71 - 1.19	1.6	+2.0	-1.6
	4.8	-0.0	-2.8

Using the *Basic Hole System*, the *BASIC* size 0.75 is designated as the minimum hole size. And the maximum hole size is 0.75+0.002 = 0.752 inches.

The maximum shaft size is 0.75-0.0016 = 0.7484 inches
The minimum shaft size is 0.75-0.0028 = 0.7472 inches

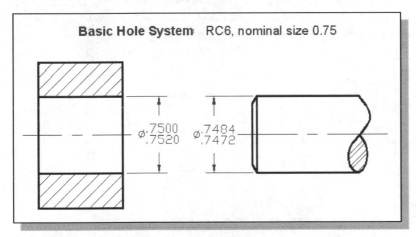

Basic Hole System RC6, nominal size 0.75

> Note the important tolerances and allowance can also be obtained from the table:

Tolerance of the shaft = Max. Shaft – Min. Shaft = 0.7484 – 0.7472 = 0.0012
This can also be calculated from the table above = 2.8 – 2.6 = **1.2 (thousandths)**

Tolerance of the hole = Max. Hole – Min. Hole = 1.2520 – 1.2500 = 0.002
This can also be calculated from the table above = 2.0 – 0.0 = **2.0 (thousandths)**

Allowance = Min. Hole (MMC Hole) – Max. Shaft (MMC Shaft)
           = 0.7500 – 0.7484 = 0.0016
This is listed as the minimum clearance in the table above: **+1.6 (thousandths)**

Max. Clearance = Max. Hole (LMC Hole) – Min. Shaft (LMC Shaft)
              = 0.7520 – 0.7472 = 0.0048
This is listed as the maximum clearance in the table above: **+4.8 (thousandths)**

Example 9.3 Basic Hole System

A design requires the use of two mating parts with a nominal size of **1.25** inches. The shaft and hub is to be fastened permanently using a drive fit. The fit class chosen is **Basic Hole System FN4**. Determine the size limits of the two mating parts.

From ANSI/ASME B4.1 Standard, the FN4 fit for 1.25 inches nominal size:

Nominal size Range	Limits of Interference	FN4 Standard Limits – Basic Hole	
		Hole	Shaft
1.19 - 1.97	1.5	+1.0	+3.1
	3.1	-0.0	+2.5

Using the *Basic Hole System*, the *BASIC* size 1.25 is designated as the minimum hole size. And the maximum hole size is 1.25+0.001 = 1.251 inches.

The maximum shaft size is 1.25+0.0031 = 1.2531 inches
The minimum shaft size is 1.25+0.0025 = 1.2525 inches

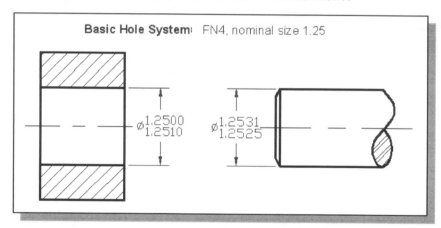

Basic Hole System: FN4, nominal size 1.25

➢   Note the important tolerances and allowance can also be obtained from the table:

Tolerance of the shaft = Max. Shaft – Min. Shaft = 1.2531 – 1.2525 = 0.0006
This can also be calculated from the table above = 3.1 – 2.5 = 0.6 (thousandths)

Tolerance of the hole = Max. Hole – Min. Hole = 1.2510 – 1.2500 = 0.001
This can also be calculated from the table above = 1.0 – 0.0 = 1.0 (thousandths)

Allowance = Min. Hole (MMC Hole) – Max. Shaft (MMC Shaft)
            = 1.2500 – 1.2531 = -0.0031
This is listed as the maximum interference in the table above: **3.1 (thousandths)**

Min. interference = Max. Hole (LMC Hole) – Min. Shaft (LMC Shaft)
            = 1.2510 – 1.2525 = -0.0015
This is listed as the minimum interference in the table above: **1.5 (thousandths)**

Example 9.4 Basic Shaft System

Use the **Basic Shaft System** from Example 9.2. A design requires the use of two mating parts with a nominal size of **0.75** inches. The shaft is to run with moderate speed but with a fairly heavy journal pressure. The fit class chosen is **Basic Shaft System RC6**. Determine the size limits of the two mating parts.

From Appendix A, the Basic Hole System RC6 fit for 0.75 inch nominal size:

Nominal size Range	Limits of Clearance	RC6 Standard Limits – Basic Hole	
		Hole	Shaft
0.71 - 1.19	1.6	+2.0	-1.6
	4.8	-0.0	-2.8

Using the Basic Shaft System, the BASIC size 0.75 is designated as the MMC of the shaft, the maximum shaft size. The conversion of limits to the Basic shaft System is performed by adding 1.6 (thousandths) to the four size limits as shown:

Nominal size Range	Limits of Clearance	RC6 Standard Limits – Basic Shaft	
		Hole	Shaft
0.71 - 1.19	1.6	+3.6	-0.0
	4.8	+1.6	-1.2

Therefore, the maximum shaft size is 0.75
The minimum shaft size is 0.75-0.0012 = 0.7488 inches.
The maximum hole size is 0.75+0.0036 = 0.7536 inches
The minimum hole size is 0.75+0.0016 = 0.7516 inches

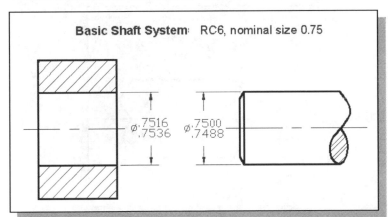

Basic Shaft System: RC6, nominal size 0.75

> Note that all the tolerances and allowance are maintained in the conversion.

Allowance = Min. Hole – Max. Shaft = 0.7517 – 0.7500 = 0.0016
This is listed as the minimum clearance in the table above: **1.6 (thousandths)**

Max. Clearance = Max. Hole – Min. Shaft = 0.7537 – 0.7488 = 0.0048
This is listed as the maximum clearance in the table above: **4.8 (thousandths)**

Example 9.5 Basic Shaft System

Use the **Basic Shaft System** from Example 9.3. A design requires the use of two mating parts with a nominal size of **1.25** inches. The shaft and hub is to be fastened permanently using a drive fit. The fit class chosen is **Basic Shaft System FN4**. Determine the size limits of the two mating parts.

From ANSI/ASME B4.1 Standard, the FN4 fit for 1.25 inches nominal size:

Nominal size Range	Limits of Interference	FN4 Standard Limits – Basic Hole	
		Hole	Shaft
1.19 - 1.97	1.5	+1.0	+3.1
	3.1	-0.0	+2.5

With the *Basic Shaft System*, the *BASIC* size 1.25 is designated as the MMC of the shaft, the maximum shaft size. The conversion of limits to the Basic shaft System is performed by subtracting 3.1 (thousandths) to the four size limits as shown:

Nominal size Range	Limits of Interference	FN4 Standard Limits – Basic Shaft	
		Hole	Shaft
1.19 - 1.97	1.5	-2.1	-0.0
	3.1	-3.1	-0.6

Therefore, the maximum shaft size is 1.25
The minimum shaft size is 1.25-0.0006 = 1.2494 inches.
The maximum hole size is 1.25-0.0021 = 1.2479 inches
The minimum hole size is 1.25-0.0031 = 1.2469 inches

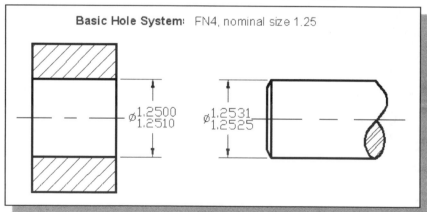

Basic Hole System: FN4, nominal size 1.25

ø1.2500 / 1.2510   ø1.2531 / 1.2525

➢ Note that all the tolerances and allowance are maintained in the conversion.

Allowance = Min. Hole – Max. Shaft = 1.2469 – 1.2500 = -0.0031
This is listed as the maximum interference in the table above: **3.1 (thousandths)**

Min. Interference = Max. Hole – Min. Shaft = 1.2479 – 1.2494 = -0.0015
This is listed as the minimum interference in the table above: **1.5 (thousandths)**

Tolerancing – Metric System

The concepts described in previous sections on the English tolerancing system are also applicable to the Metric System. The commonly used system of preferred metric tolerancing and fits is outlined by the International Organization for Standardization (ISO). The system is specified for holes and shafts, but it is also adaptable to fits between features of parallel surfaces.

- **Basic size** is the theoretical size from which limits or deviations of size are derived.

- **Deviation** is the difference between the basic size and the hole or shaft size. This is equivalent to the word "*Tolerance*" in the *English* system.

- **Upper Deviation** is the difference between the basic size and the permitted maximum size. This is equivalent to "*Maximum Tolerance*" in the *English* system.

- **Lower Deviation** is the difference between the basic size and the permitted minimum size. This is equivalent to "*Minimum Tolerance*" in the *English* system.

- **Fundamental Deviation** is the deviation closest to the basic size. This is equivalent to "*Minimum Allowance*" in the *English* system.

- **Tolerance** is the difference between the permissible variations of a size. The tolerance is the difference between the limits of size.

- **International tolerance grade (IT)** is a set of tolerances that varies according to the basic size and provides a uniform level of accuracy within the grade. There are 18 IT grades- IT01, IT0, and IT1 through IT16.

  ❖ IT01 through IT7 is typically used for measuring tools; IT5 through IT11 are used for Fits and IT12 through IT16 for large manufacturing tolerances.

  ❖ The following is a list of IT grade related to machining processes:

International Tolerance Grade (IT)							
4	5	6	7	8	9	10	11
Lapping or Honing							
	Cylindrical Grinding						
	Surface Grinding						
	Diamond Turning or Boring						
	Broaching						
	Powder Metal-sizes						
		Reaming					
			Turning				
			Powder Metal-sintered				
			Boring				
						Milling, Drilling	
						Planing & Shaping	
						Punching	
							Die Casting

Metric Tolerances and Fits Designation

Metric tolerances and fits are specified by using the fundamental deviation symbol along with the IT grade number. An uppercase **H** is used for the fundamental deviation for the hole tolerance using the **Hole Basis system**, where a lowercase **h** is used to specify the use of the **Shaft Basis System**.

1. **Hole: 30H8**

    Basic size = 30 mm
    Hole size is based on hole basis system using IT8 for tolerance.

2. **Shaft: 30f7**

    Basic size = 30 mm
    Shaft size is based on hole basis system using IT7 for tolerance.

3. **Fit: 30H8/f7**

    Basic size = 30 mm
    Hole basis system with IT8 hole tolerance and IT7 for shaft tolerance.

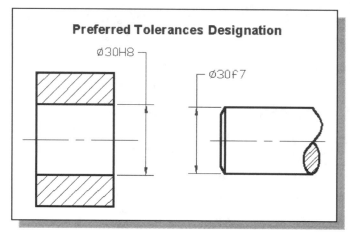

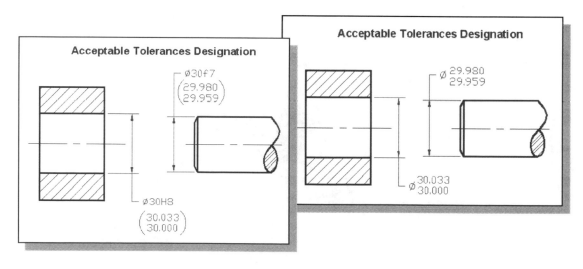

Preferred ISO Metric Fits

The symbols for either the hole-basis or shaft-basis preferred fits are given in the table below. These are the preferred fits; select the fits from the table whenever possible. See **Appendix B** for the complete *ISO Metric Fits* tables.

- The **Hole-Basis** system is a system in which the basic size is the minimum size of the hole. The *hole-basis* system is the preferred method; the fundamental deviation is specified by the uppercase **H**.

- The **Shaft-Basis** system is a system in which the basic size is the maximum size of the shaft. The fundamental deviation of the *shaft-basis* system is specified by the lowercase **h**.

	Hole Basis	Shaft Basis	Description
			ISO Preferred Metric Fits
Clearance Fits	**H11/c11**	**C11/h11**	*Loose running* fit for wide commercial tolerance or allowances.
	H9/d9	**D9/h9**	*Free running* fit not for use when accuracy is essential, but good for large temperature variation, high running speed, or heavy journal pressure.
	H8/f7	**F8/h7**	*Close running* fit for running on accurate machines running at moderate speeds and journal pressure.
	H7/g6	**G7/h6**	*Sliding* fit not intended to run freely, but to move and turn freely and locate accurately.
Transition Fits	**H7/h6**	**H7/h6**	*Locational clearance* fit provides snug fit for locating stationary parts but can be freely assembled and disassembled.
	H7/k6	**K7/h6**	*Locational transition* fit for accurate location, a compromise between clearance and interference.
	H7/n6	**N7/h6**	*Locational transition* fit for accurate location where interference is permissible.
	H7/p6	**P7/h6**	*Locational interference* fit for parts requiring rigidity and alignment with prime accuracy but without special bore pressure.
Interference Fits	**H7/s6**	**S7/h6**	*Medium drive* fit for ordinary steel parts or shrink fits on light sections, the tightest fit usable with cast iron.
	H7/u6	**U7/h6**	*Force* fit suitable for parts which can be highly stressed or for shrink fits where the heavy pressing forces required are impractical.

Example 9.6 Metric Hole Basis System

A design requires the use of two mating parts with a nominal size of **25** millimeters. The shaft and hub are to be fastened permanently using a drive fit. The fit class chosen is **Hole-Basis System H7/s6**. Determine the size limits of the two mating parts.

From Appendix B, the H7/s6 fit for 25 millimeters size:

Basic size	H7/s6 Medium Drive –Hole Basis		
	Hole H7	Shaft s6	Fit
25	25.021	25.048	-0.014
	25.000	25.035	-0.048

Using the *Hole Basis System*, the *BASIC* size 25 is designated as the minimum hole size. And the maximum hole size is 25.021 millimeters.
The maximum shaft size is 25.048 millimeters.
The minimum shaft size is 25.035 millimeters.

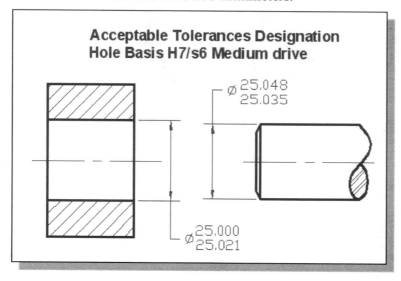

Tolerance of the shaft = Max. Shaft – Min. Shaft = 25.048 – 25.035 = 0.013

Tolerance of the hole = Max. Hole – Min. Hole = 25.021 – 25.000 = 0.021

Allowance = Min. Hole (MMC Hole) – Max. Shaft (MMC Shaft)
= 25.000 – 25.048 = -0.0048
This is listed as the maximum interference in the table above: **-0.048**

Min. interference = Max. Hole (LMC Hole) – Min. Shaft (LMC Shaft)
= 25.021 – 25.035 = -0.0014
This is listed as the minimum interference in the table above: **-0.0014**

Example 9.7 Shaft Basis System

A design requires the use of two mating parts with a nominal size of **25** millimeters. The shaft is to run with moderate speed but with a fairly heavy journal pressure. The fit class chosen is **Shaft Basis System F8/h7**. Determine the size limits of the two mating parts.

From Appendix B, the Shaft Basis System F8/h7 fit for 25 millimeters size:

Basic size	F8/h7 Medium Drive –Shaft Basis		
	Hole F8	Shaft h7	Fit
25	25.053	25.000	0.074
	25.020	24.979	0.020

Therefore, the maximum shaft size is 25
The minimum shaft size is 24.979
The maximum hole size is 25.053
The minimum hole size is 25.020

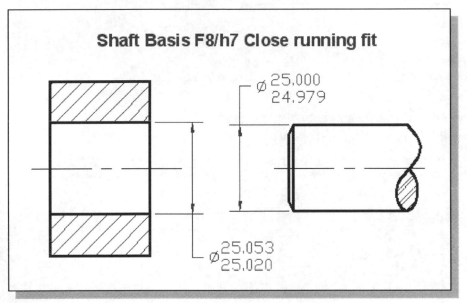

Shaft Basis F8/h7 Close running fit

Tolerance of the shaft = Max. Shaft – Min. Shaft = 25.000 – 24.979 = 0.021

Tolerance of the hole = Max. Hole – Min. Hole = 25.053 – 25.020 = 0.023

Allowance = Min. Hole – Max. Shaft = 25.020 – 25.000 = 0.020
This is listed as the minimum clearance in the table above: **0.020**

Max. Clearance = Max. Hole – Min. Shaft = 25.053 – 24.979 = 0.074
This is listed as the maximum clearance in the table above: **0.074**

Updating the U-Bracket Drawing

1. Select the **Autodesk Inventor 2024** option on the *Program* menu or select the **Autodesk Inventor 2024** icon on the *Desktop*.

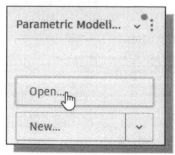

2. In the Autodesk Inventor *Startup* dialog box, select **Open** with a single click of the left-mouse-button.

3. In the *Open* window, select the **U-Bracket.idw** file. Use the *browser* to locate the file if it is not displayed in the *File name* list box.

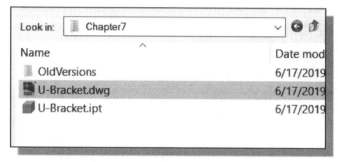

4. Click on the **Open** button in the *Startup* dialog box to accept the selected settings.

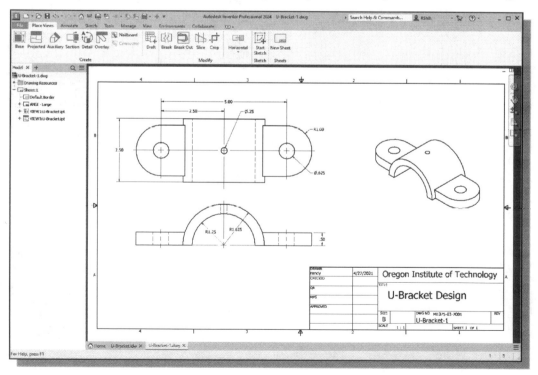

Determining the Tolerances Required

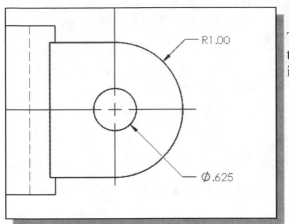

The *Bracket* design requires one set of tolerances: the nominal size of the **0.625** inch holes.

RC2 - Nominal size of 0.625

A pin is to be placed through the diameter 0.625 hole. The fit class chosen is **Basic Hole System RC2**. We will first determine the size limits of the two mating parts.

From Appendix A, the RC2 fit for 0.625 inch nominal size:

Nominal size Range	Limits of Clearance	RC2 Standard Limits – Basic Hole	
		Hole	Shaft
0.40 - 0.71	0.25	+0.4	-0.25
	0.95	-0.0	-0.55

Using the *Basic Hole System*, the *BASIC* size 0.625 is designated as the minimum hole size.

The maximum hole size is 0.625+0.0004 = 0.6254 inches
The minimum hole size is 0.625-0.00 = 0.62500 inches

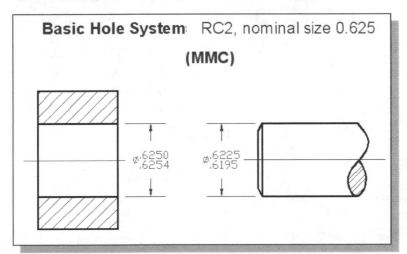

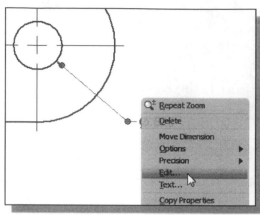

1. Move the cursor on top of the diameter **0.625** dimension.

2. Right-click once on the **0.625** dimension to bring up the option menu and select **Edit** as shown.

3. In the *Edit Dimension* dialog box select **Limit - Stacked** as the *Tolerance Method* as shown.

4. Also set the precision display to 4 digits after the decimal point as shown.

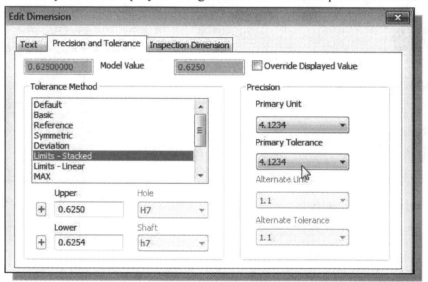

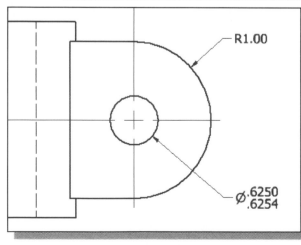

5. Enter +0.6254 in the lower bound box as shown. Note that we are setting the limits using the MMC (Maximum material condition) approach.

➢   The tolerances for the hole are set as shown.

Review Questions:

1. Why are **Tolerances** important to a technical drawing?

2. Using the **Basic Shaft system**, calculate the size limits of a nominal size of 1.5 inches.

   Use the following FN4 fit table for 1.5 inches nominal size: (numbers are in thousandth of an inch)

Nominal size Range	Limits of Interference	FN4 Standard Limits – Basic Hole	
		Hole	Shaft
1.19 - 1.97	1.5	+1.0	+3.1
	3.1	-0.0	+2.5

3. Using the **Shaft Basis D9/h9** fits, calculate the size limits of a nominal size of 30 mm.

4. What is the procedure to set tolerances in Autodesk Inventor?

5. Explain the following terms:

   (a) Limits

   (b) MMC

   (c) Tolerances

   (d) Basic Hole System

   (e) Basic Shaft System

6. Given the dimensions as shown in the below figure, determine the tolerances of the two parts, and the allowance between the parts.

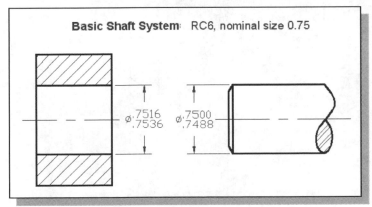

Exercises: (Create the drawings with the following Tolerances.)

1. **Shaft Guide** (Dimensions are in inches.)
   Fits: Diameter 0.25, Basic Hole system RC1, Diameter 1.00, Basic Hole system RC6

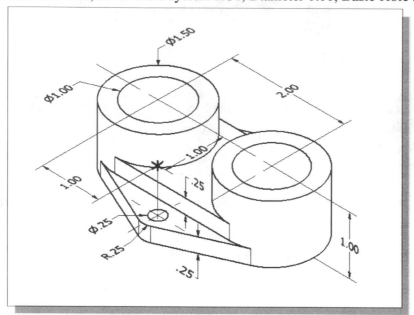

2. **Pivot Lock** (Dimensions are in inches.)
   Fits: Diameter 0.5, Basic Hole system RC2, Diameter 1.00, Basic Shaft system RC5

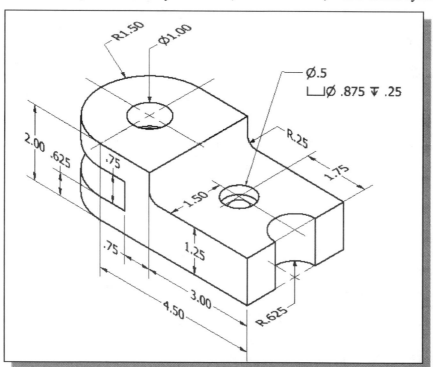

3. **U-Bracket** (Dimensions are in millimeters.) Fits: Diameter 40, H7/g6.

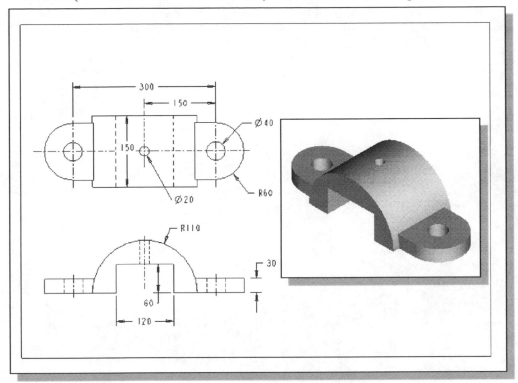

4. **Cylinder Support** (Dimensions are in inches.) Fits: Diameter 1.50, Basic Hole system RC8.

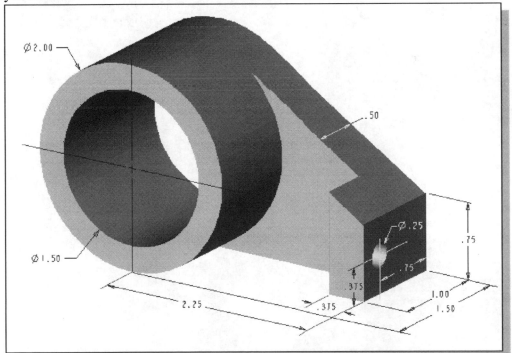

Notes:

Chapter 10
Pictorials and Sketching

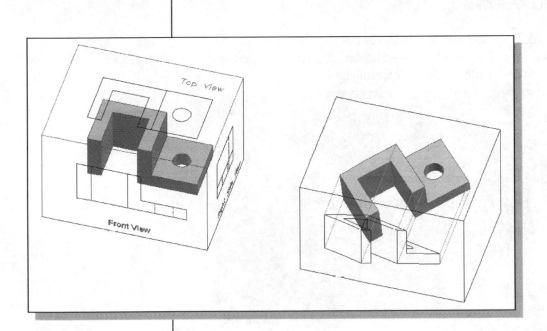

Learning Objectives

- ♦ **Understand the Importance of Freehand Sketching**
- ♦ **Understand the Terminology Used in Pictorial Drawings**
- ♦ **Understand the Basics of the Following Projection Methods: Axonometric, Oblique and Perspective**
- ♦ **Be Able to Create Freehand 3D Pictorials**

Engineering Drawings, Pictorials and Sketching

One of the best ways to communicate one's ideas is through the use of a picture or a drawing. This is especially true for engineers and designers. Without the ability to communicate well, engineers and designers will not be able to function in a team environment and therefore will have only limited value in the profession.

For many centuries, artists and engineers used drawings to express their ideas and inventions. The two figures below are drawings by Leonardo da Vinci (1452-1519) illustrating some of his engineering inventions.

Engineering design is a process to create and transform ideas and concepts into a product definition that meets the desired objective. The engineering design process typically involves three stages: (1) Ideation/conceptual design stage: this is the beginning of an engineering design process, where basic ideas and concepts take shape. (2) Design development stage: the basic ideas are elaborated and further developed. During this stage, prototypes and testing are commonly used to ensure the developed design meets the desired objective. (3) Refine and finalize design stage: This stage of the design process is the last stage of the design process, where the finer details of the design are further refined. Detailed information of the finalized design is documented to assure the design is ready for production.

Two types of drawings are generally associated with the three stages of the engineering process: (1) Freehand Sketches and (2) Detailed Engineering Drawings.

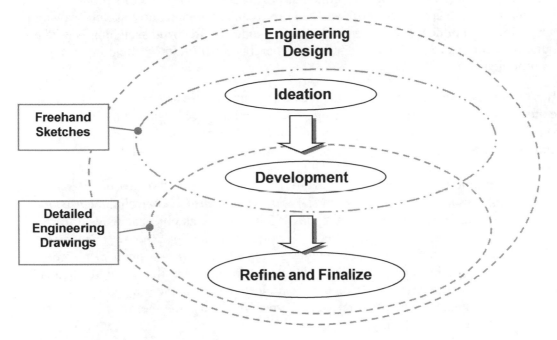

Freehand sketches are generally used in the beginning stages of a design process: (1) to quickly record designer's ideas and help formulate different possibilities, (2) to communicate the designer's basic ideas with others and (3) to develop and elaborate further the designer's ideas/concepts.

During the initial design stage, an engineer will generally picture the ideas in his/her head as three-dimensional images. The ability to think visually, specifically three-dimensional visualization, is one of the most essential skills for an engineer/designer. And freehand sketching is considered as one of the most powerful methods to help develop visualization skills.

Detailed engineering drawings are generally created during the second and third stages of a design process. The detailed engineering drawings are used to help refine and finalize the design and also to document the finalized design for production. Engineering drawings typically require the use of drawing instruments, from compasses to computers, to bring precision to the drawings.

Freehand Sketches and Detailed Engineering Drawings are essential communication tools for engineers. By using the established conventions, such as perspective and isometric drawings, engineers/designers are able to quickly convey their design ideas to others.

The ability to sketch ideas is absolutely essential to engineers. The ability to sketch is helpful, not just to communicate with others, but also to work out details in ideas and to identify any potential problems. Freehand sketching requires only simple tools, a pencil and a piece of paper, and can be accomplished almost anywhere and anytime. Creating freehand sketches does not require any artistic ability. Detailed engineering drawing is employed only for those ideas deserving a permanent record.

Freehand sketches and engineering drawings are generally composed of similar information, but there is a tradeoff between time required to generate a sketch/drawing verses the level of design detail and accuracy. In industry, freehand sketching is used to quickly document rough ideas and identify general needs for improvement in a team environment.

Besides the 2D views, described in the previous chapter, there are three main divisions commonly used in freehand engineering sketches and detailed engineering drawings: (1) **Axonometric**, with its divisions into **isometric**, **dimetric** and **trimetric**; (2) **Oblique**; and (3) **Perspective**.

1. **Axonometric projection**: The word *Axonometric* means "to measure along axes." Axonometric projection is a special *orthographic projection* technique used to generate *pictorials*. **Pictorials** show a 2D image of an object as viewed from a direction that reveals three directions of space. In the figure below, the adjuster model is rotated so that a *pictorial* is generated using *orthographic projection* (projection lines perpendicular to the projection plane) as described in Chapter 7. There are three types of axonometric projections: isometric projection, dimetric projection, and trimetric projection. Typically, in an axonometric drawing, one axis is drawn vertically.

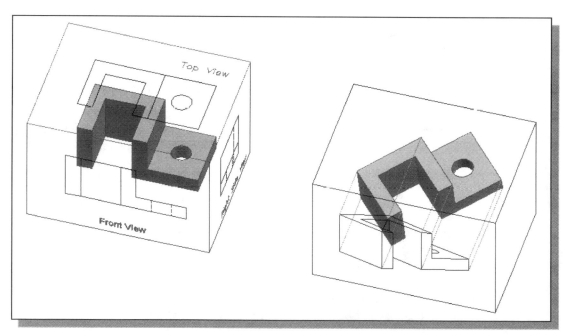

In **isometric projections**, the direction of viewing is such that the three axes of space appear equally foreshortened, and therefore the angles between the axes are equal. In **dimetric projections**, the directions of viewing are such that two of the three axes of space appear equally foreshortened. In **trimetric projections**, the direction of viewing is such that the three axes of space appear unequally foreshortened.

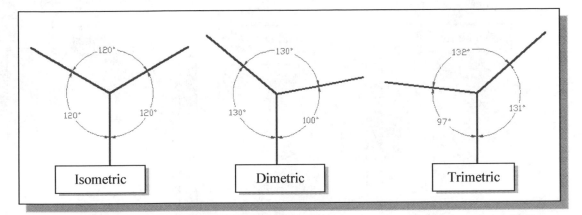

Isometric projection is perhaps the most widely used for pictorials in engineering graphics, mainly because isometric views are the most convenient to draw. Note that the different projection options described here are not particularly critical in freehand sketching as the emphasis is generally placed on the proportions of the design, not the precision measurements. The general procedure to constructing isometric views is illustrated in the following sections.

2. **Oblique Projection** represents a simple technique of keeping the front face of an object parallel to the projection plane and still reveals three directions of space. An **orthographic projection** is a parallel projection in which the projection lines are perpendicular to the plane of projection. An **oblique projection** is one in which the projection lines are other than perpendicular to the plane of projection.

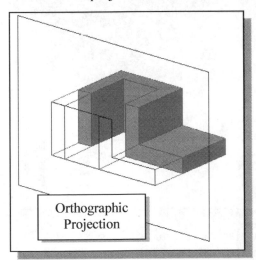

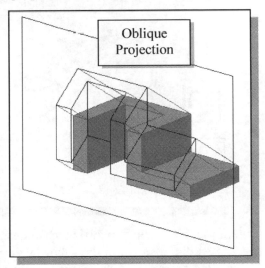

In an oblique drawing, geometry that is parallel to the frontal plane of projection is drawn true size and shape. This is the main advantage of the oblique drawing over the axonometric drawings. The three axes of the oblique sketch are drawn horizontal, vertical, and the third axis can be at any convenient angle (typically between 30 and 60 degrees). The proportional scale along the 3rd axis is typically a scale anywhere between ½ and 1. If the scale is ½, then it is a **Cabinet** oblique. If the scale is 1, then it is a **Cavalier** oblique.

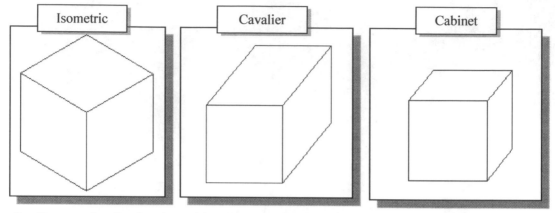

3. **Perspective Projection** adds realism to the three-dimensional pictorial representation; a perspective drawing represents an object as it appears to an observer; objects that are closer to the observer will appear larger to the observer. The key to the perspective projection is that parallel edges converge to a single point, known as the **vanishing point**. If there is just one vanishing point, then it is called a one-point perspective. If two sets of parallel edge lines converge to their respective vanishing points, then it is called a two-point perspective. There is also the case of a three-point perspective in which all three sets of parallel lines converge to their respective vanishing points.

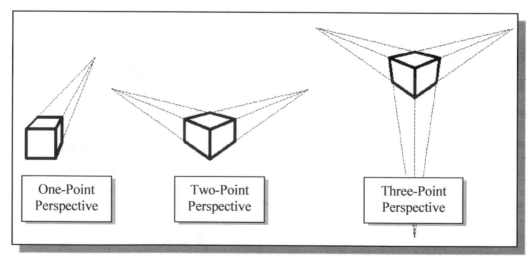

*   Although there are specific techniques available to create precise pictorials with known dimensions, in the following sections, the basic concepts and procedures relating to freehand sketching are illustrated.

Isometric Sketching

Isometric drawings are generally done with one axis aligned to the vertical direction. A **regular isometric** is when the viewpoint is looking down on the top of the object, and a **reversed isometric** is when the viewpoint is looking up on the bottom of the object.

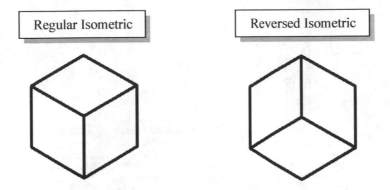

Two commonly used approaches in creating isometric sketches are (1) the **enclosing box** method and (2) the **adjacent surface** method. The enclosing box method begins with the construction of an isometric box showing the overall size of the object. The visible portions of the individual 2D-views are then constructed on the corresponding sides of the box. Adjustments of the locations of surfaces are then made, by moving the edges, to complete the isometric sketch.

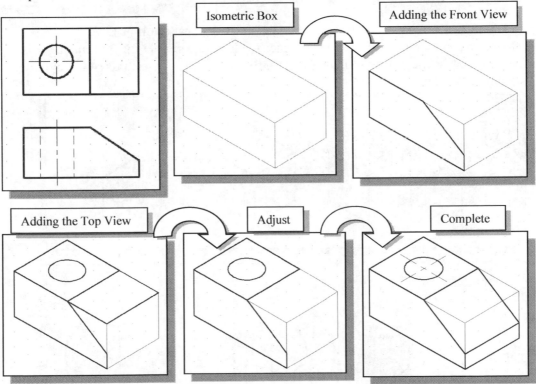

The adjacent surface method begins with one side of the isometric drawing, again with the visible portion of the corresponding 2D-view. The isometric sketch is completed by identifying and adding the adjacent surfaces.

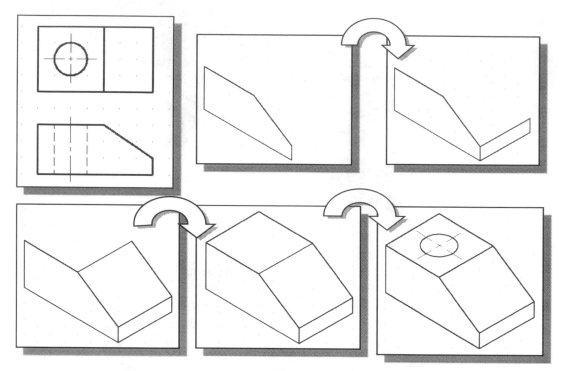

In an isometric drawing, cylindrical or circular shapes appear as ellipses. It can be confusing in drawing the ellipses in an isometric view; one simple rule to remember is the **major axis** of the ellipse is always **perpendicular** to the **center axis** of the cylinder as shown in the figures below.

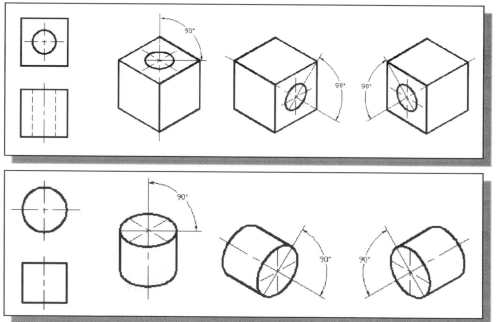

Chapter 10 - Isometric Sketching Exercise 1:

Given the Orthographic Top view and Front view, create the isometric view.

(Note that 3D Models of chapter examples and exercises are available at: http://www.sdcpublications.com/downloads/978-1-63057-583-0)

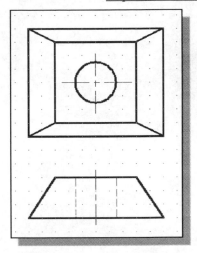

Name: _____   Date: _____

Chapter 10 - Isometric Sketching Exercise 2:

Given the Orthographic Top view and Front view, create the isometric view.
(Note that 3D Models of chapter examples and exercises are available at: http://www.sdcpublications.com/downloads/978-1-63057-583-0)

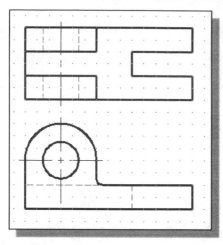

Name: _____   Date: _____

Chapter 10 - Isometric Sketching Exercise 3:

Given the Orthographic Top view and Front view, create the isometric view.

(Note that 3D Models of chapter examples and exercises are available at: http://www.sdcpublications.com/downloads/978-1-63057-583-0)

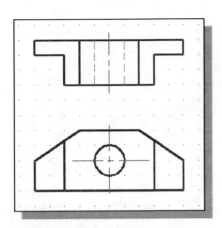

Name: _____   Date: _____

Chapter 10 - Isometric Sketching Exercise 4:

Given the Orthographic Top view and Front view, create the isometric view.

(Note that 3D Models of chapter examples and exercises are available at: http://www.sdcpublications.com/downloads/978-1-63057-583-0)

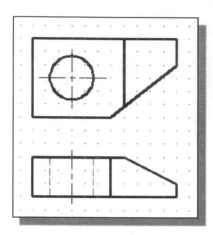

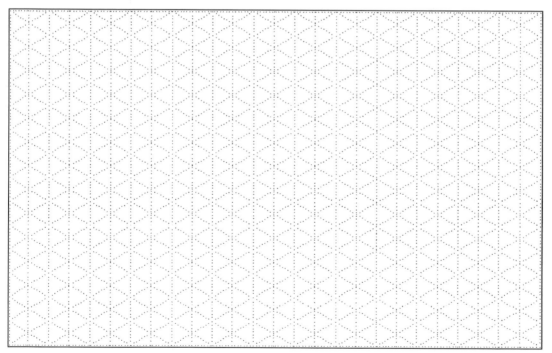

Name: _____   Date: _____

Chapter 10 - Isometric Sketching Exercise 5:

Given the Orthographic Top view, Front view, and Side view create the isometric view.
(Note that 3D Models of chapter examples and exercises are available at: http://www.sdcpublications.com/downloads/978-1-63057-583-0)

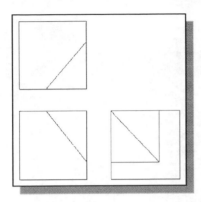

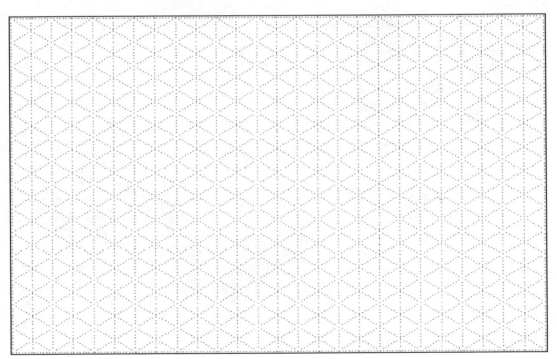

Name: _____   Date: _____

Chapter 10 - Isometric Sketching Exercise 6:

Given the Orthographic Top view, Front view, and Side view create the isometric view.
(Note that 3D Models of chapter examples and exercises are available at: http://www.sdcpublications.com/downloads/978-1-63057-583-0)

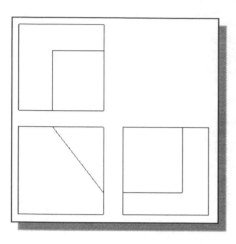

Name: _____   Date: _____

Chapter 10 - Isometric Sketching Exercise 7:

Given the Orthographic Top view, Front view, and Side view create the isometric view.
(Note that 3D Models of chapter examples and exercises are available at: http://www.sdcpublications.com/downloads/978-1-63057-583-0)

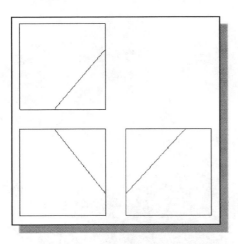

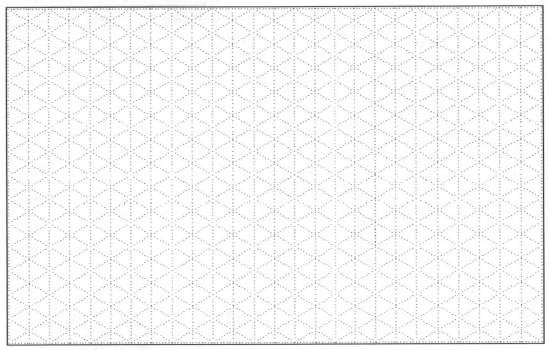

Name: _____   Date: _____

Chapter 10 - Isometric Sketching Exercise 8:

Given the Orthographic Top view, Front view, and Side view create the isometric view.

(Note that 3D Models of chapter examples and exercises are available at: http://www.sdcpublications.com/downloads/978-1-63057-583-0)

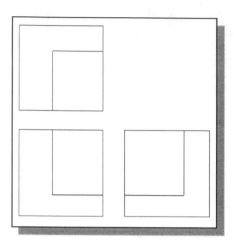

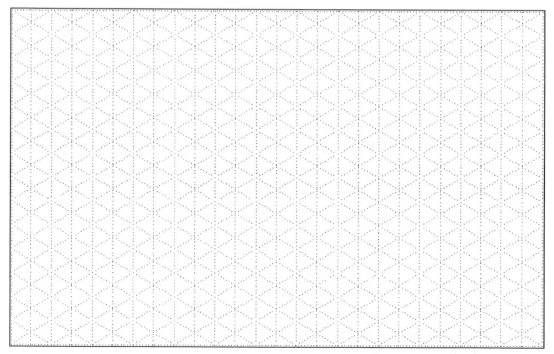

Name: _____   Date: _____

Chapter 10 - Isometric Sketching Exercise 9:

Given the Orthographic Top view, Front view, and Side view create the isometric view.
(Note that 3D Models of chapter examples and exercises are available at: http://www.sdcpublications.com/downloads/978-1-63057-583-0)

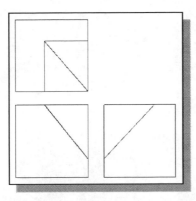

Name: _____   Date: _____

Chapter 10 - Isometric Sketching Exercise 10:

Given the Orthographic Top view, Front view, and Side view create the isometric view.
(Note that 3D Models of chapter examples and exercises are available at: http://www.sdcpublications.com/downloads/978-1-63057-583-0)

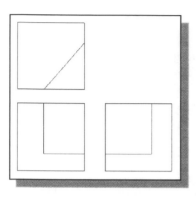

Name: _____   Date: _____

Oblique Sketching

Keeping the geometry that is parallel to the frontal plane true size and shape is the main advantage of the oblique drawing over the axonometric drawings. Unlike isometric drawings, circular shapes that are paralleled to the frontal view will remain as circles in oblique drawings. Generally speaking, an oblique drawing can be created very quickly by using a 2D view as the starting point. For designs with most of the circular shapes in one direction, an oblique sketch is the ideal choice over the other pictorial methods.

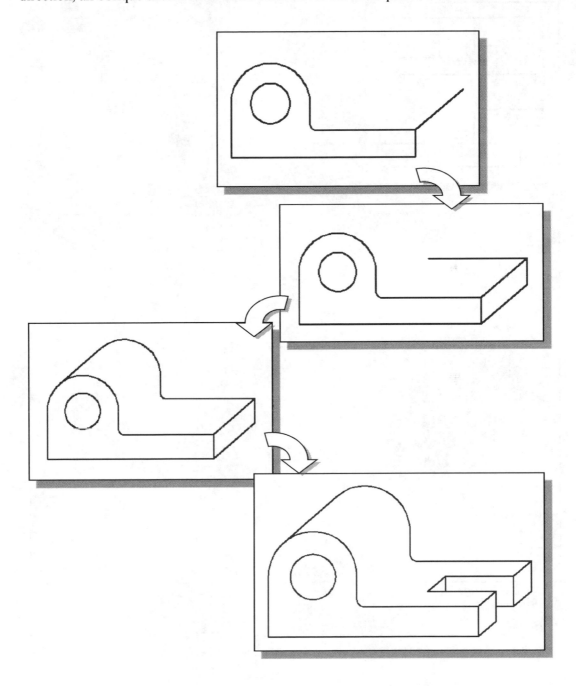

Chapter 10 - Oblique Sketching Exercise 1:

Given the Orthographic Top view and Front view, create the oblique view.
(Note that 3D Models of chapter examples and exercises are available at: http://www.sdcpublications.com/downloads/978-1-63057-583-0)

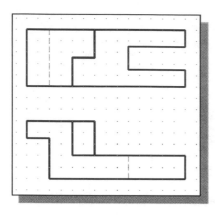

Name: _____   Date: _____

Chapter 10 - Oblique Sketching Exercise 2:

Given the Orthographic Top view and Front view, create the oblique view.
(Note that 3D Models of chapter examples and exercises are available at:
http://www.sdcpublications.com/downloads/978-1-63057-583-0)

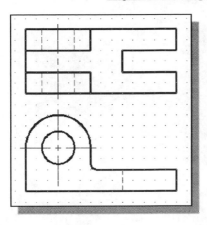

Name: _____   Date: _____

Chapter 10 - Oblique Sketching Exercise 3:

Given the Orthographic Top view and Front view, create the oblique view.

(Note that 3D Models of chapter examples and exercises are available at: http://www.sdcpublications.com/downloads/978-1-63057-583-0)

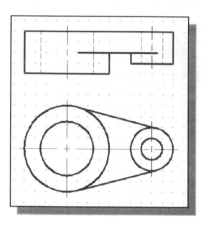

Name: _____   Date: _____

Chapter 10 - Oblique Sketching Exercise 4:

Given the Orthographic Top view and Front view, create the oblique view.
(Note that 3D Models of chapter examples and exercises are available at:
http://www.sdcpublications.com/downloads/978-1-63057-583-0)

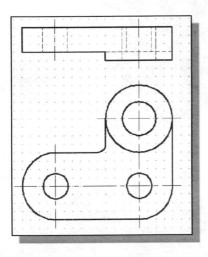

Name: _____   Date: _____

Chapter 10 - Oblique Sketching Exercise 5:

Given the Orthographic Top view and Front view, create the oblique view.
(Note that 3D Models of chapter examples and exercises are available at: http://www.sdcpublications.com/downloads/978-1-63057-583-0)

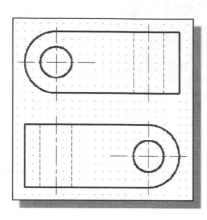

Name: _____   Date: _____

Chapter 10 - Oblique Sketching Exercise 6:

Given the Orthographic Top view and Front view, create the oblique view.
(Note that 3D Models of chapter examples and exercises are available at:
http://www.sdcpublications.com/downloads/978-1-63057-583-0)

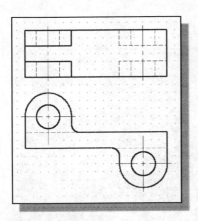

Name: _____   Date: _____

Perspective Sketching

A perspective drawing represents an object as it appears to an observer; objects that are closer to the observer will appear larger to the observer. The key to the perspective projection is that parallel edges converge to a single point, known as the **vanishing point**. The vanishing point represents the position where projection lines converge.

The selection of the locations of the vanishing points, which is the first step in creating a perspective sketch, will affect the looks of the resulting images.

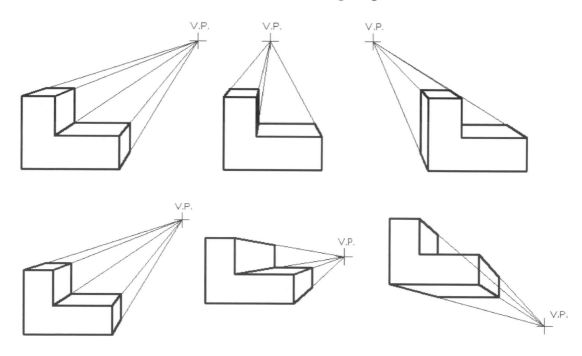

One-Point Perspective

One-point perspective is commonly used because of its simplicity. The first step in creating a one-point perspective is to sketch the front face of the object just as in oblique sketching, followed by selecting the position for the vanishing point. For Mechanical designs, the vanishing point is usually placed above and to the right of the picture. The use of construction lines can be helpful in locating the edges of the object and to complete the sketch.

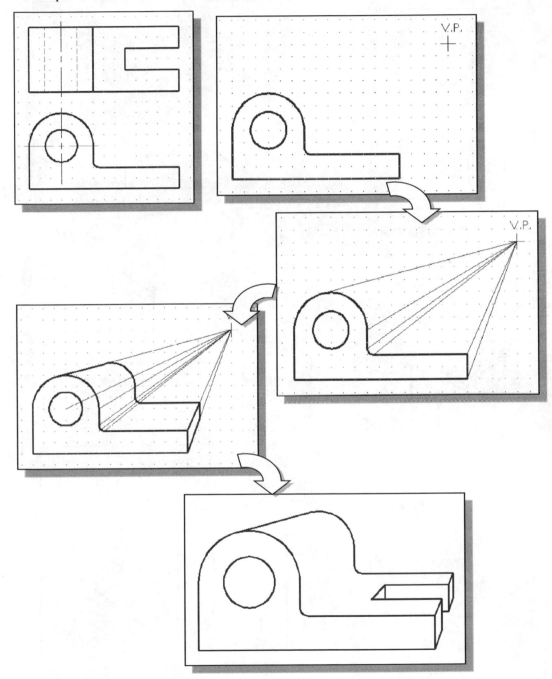

Two-Point Perspective

Two-point perspective is perhaps the most popular of all perspective methods. The use of the two vanishing points gives very true to life images. The first step in creating a two-point perspective is to select the locations for the two vanishing points, followed by sketching an enclosing box to show the outline of the object. The use of construction lines can be very helpful in locating the edges of the object and to complete the sketch.

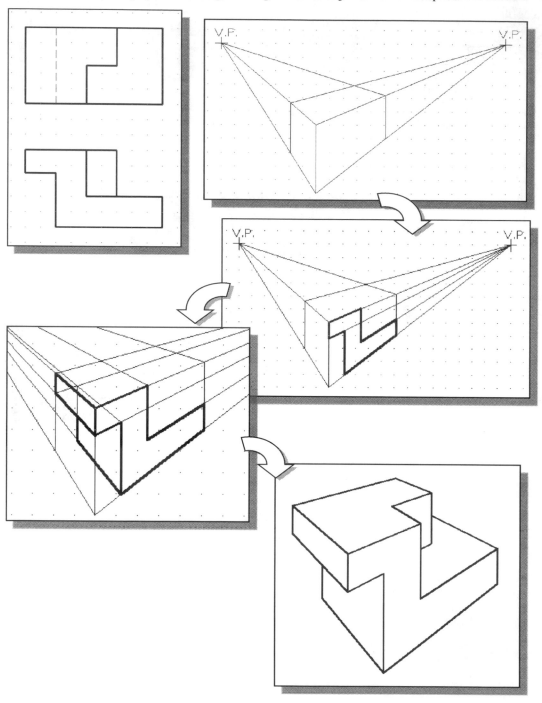

Chapter 10 - Perspective Sketching Exercise 1:

Given the Orthographic Top view and Front view, create one-point or two-point perspective views.
(Note that 3D Models of chapter examples and exercises are available at: http://www.sdcpublications.com/downloads/978-1-63057-583-0)

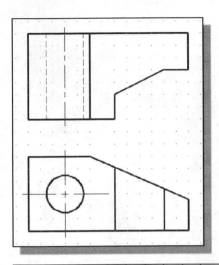

Name: _____   Date: _____

Chapter 10 - Perspective Sketching Exercise 2:

Given the Orthographic Top view and Front view, create one-point or two-point perspective views.
(Note that 3D Models of chapter examples and exercises are available at: http://www.sdcpublications.com/downloads/978-1-63057-583-0)

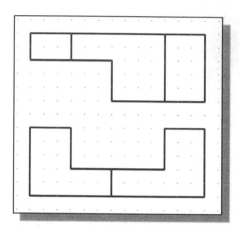

Name: _____   Date: _____

Chapter 10 - Perspective Sketching Exercise 3:

Given the Orthographic Top view and Front view, create one-point or two-point perspective views.
(Note that 3D Models of chapter examples and exercises are available at: http://www.sdcpublications.com/downloads/978-1-63057-583-0)

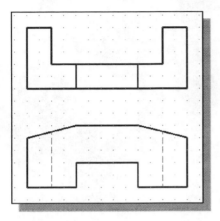

Name: _____   Date: _____

Chapter 10 - Perspective Sketching Exercise 4:

Given the Orthographic Top view and Front view, create one-point or two-point perspective views.
(Note that 3D Models of chapter examples and exercises are available at: http://www.sdcpublications.com/downloads/978-1-63057-583-0)

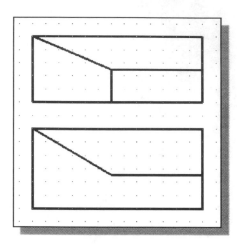

Name: _____   Date: _____

Chapter 10 - Perspective Sketching Exercise 5:

Given the Orthographic Top view and Front view, create one-point or two-point perspective views.
(Note that 3D Models of chapter examples and exercises are available at:
http://www.sdcpublications.com/downloads/978-1-63057-583-0)

Name: _____   Date: _____

Chapter 10 - Perspective Sketching Exercise 6:

Given the Orthographic Top view and Front view, create one-point or two-point perspective views.
(Note that 3D Models of chapter examples and exercises are available at: http://www.sdcpublications.com/downloads/978-1-63057-583-0)

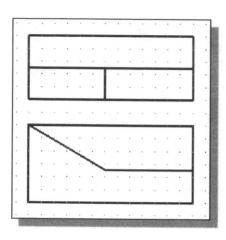

Name: _____   Date: _____

Review Questions:

1. What are the three types of Axonometric projection?

2. Describe the differences between an Isometric drawing and a Trimetric drawing.

3. What is the main advantage of Oblique projection over the isometric projection?

4. Describe the differences between a one-point perspective and a two-point perspective.

5. Which pictorial methods maintain true size and shape of geometry on the frontal plane?

6. What is a vanishing point in a perspective drawing?

7. What is a Cabinet Oblique?

8. What is the angle between the three axes in an isometric drawing?

9. In an Axonometric drawing, are the projection lines perpendicular to the projection plane?

10. A cylindrical feature, in a frontal plane, will remain a circle in which pictorial methods?

11. In an Oblique drawing, are the projection lines perpendicular to the projection plane?

12. Create freehand pictorial sketches of:
    *   Your desk
    *   Your computer
    *   One corner of your room
    *   The tallest building in your area

Exercises:

Complete the missing views. (Hint: Create a pictorial sketch as an aid in reading the views.)

1.

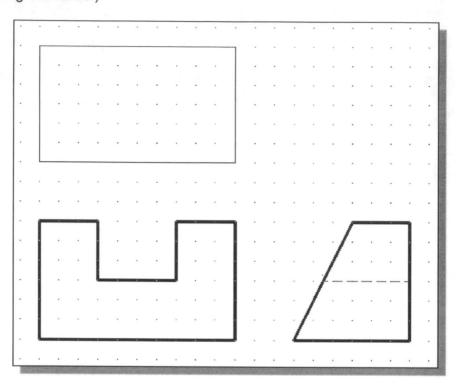

2.

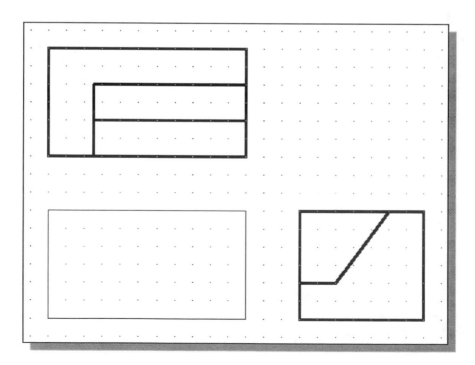

3.

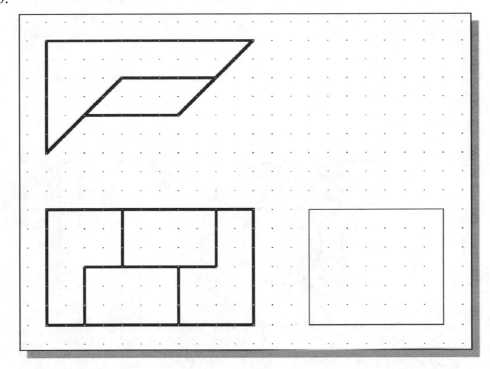

4.

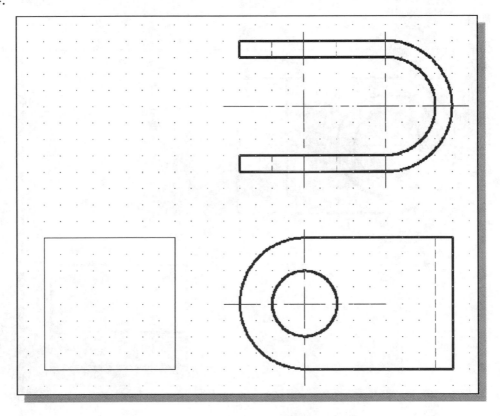

5.

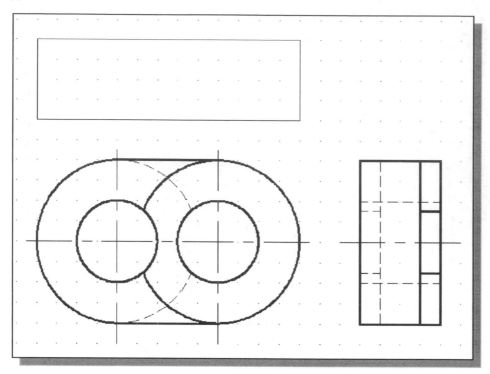

6.

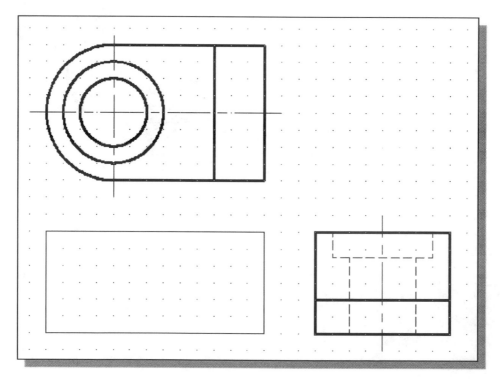

Chapter 11
Auxiliary Views and Reference Geometry

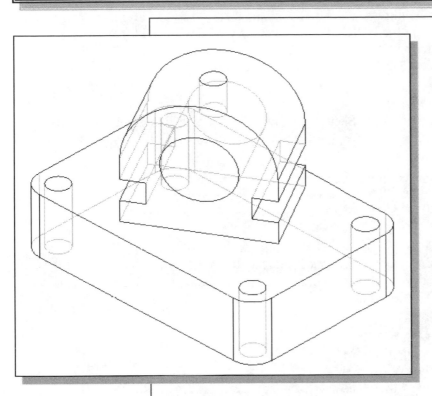

Learning Objectives

♦ **Understand the Principles of Creating Auxiliary Views**

♦ **Understand the Concepts and the Use of Reference Geometry**

♦ **Use the Different Options to Create Reference Geometry**

♦ **Create Auxiliary Views in 2D Drawing Mode**

♦ **Create and Adjust Centerlines**

♦ **Create Shaded Images in the 2D Drawing Mode**

Autodesk Inventor Certified User Exam Objectives Coverage

Parametric Modeling Basics

Section 3: Sketches

Objectives: Creating 2D Sketches, Draw Tools, Sketch Constraints, Pattern Sketches, Modify Sketches, Format Sketches, Sketch Doctor, Shared Sketches, Sketch Parameters

Section 4: Parts

Objectives: Creating parts, Work Features, Pattern Features, Part Properties

Section 6: Drawings

Objectives: Create drawings

Introduction

An important rule concerning multiview drawings is to draw enough views to accurately describe the design. This usually requires two or three of the regular views, such as a front view, a top view and/or a side view.

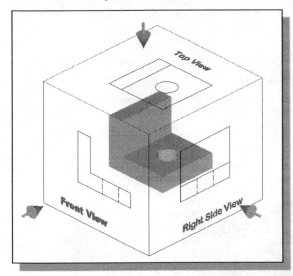

In the left figure, the L-shaped object is placed with the surfaces parallel to the principal planes of projection.
The top, front and right side views show the true shape of the different surfaces of the object. Note especially that planes of projection are parallel to the top, front and right side of the object.

Based on the principle of orthographic projection, it is clear that a plane surface is shown in true shape when the direction of view is perpendicular to the surface.

Many designs have features located on inclined surfaces that are not parallel to the regular planes of projection. To truly describe the feature, the true shape of the feature must be shown using an **auxiliary view**. An *auxiliary view* has a line of sight that is perpendicular to the inclined surface, as viewed looking directly at the inclined surface. An *auxiliary view* is a supplementary view that can be constructed from any of the regular views. A primary *auxiliary view* is projected onto a plane that is perpendicular to one of the principal planes of projection and is inclined to the other two. A secondary *auxiliary view* is projected onto a plane that is inclined to all three principal planes of projection. In the figures below, the use of the standard views does not show the true shape of the upper feature of the design; the use of an auxiliary view provided the true shape of the feature and also eliminated the need of front and right side views.

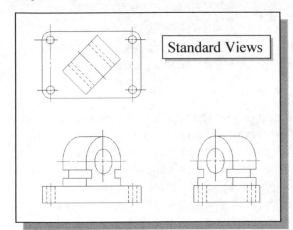

Standard Views

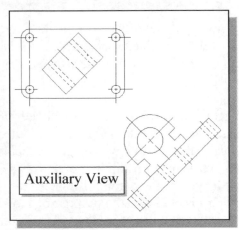

Auxiliary View

Normal View of an Inclined Surface

No matter what the position of a surface may be, the fundamentals of projecting a normal view of the surface remain the same: **The projection plane is placed parallel to the surface to be projected. The line of sight is set to be perpendicular to the projection plane and therefore perpendicular to the surface to be projected.** This type of view is known as **normal view**. In geometry, the word "*normal*" means "*perpendicular.*"

In the figure below, the design has an inclined face that is inclined to the horizontal and profile planes and **perpendicular to the frontal plane**.

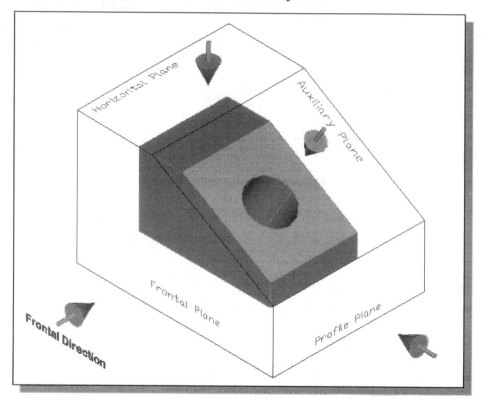

The principal views (Top, Front and Right side views) do not show the **true size and shape** of the inclined surface. Note the inclined surface does appear as an edge in the front view. To show the true size and shape of the inclined surface, a *normal* view is needed.

To get the normal view of the inclined surface, a projection is made perpendicular to the surface. This projection is made from the view where the surface shows as an edge, in this case, the front view. The perpendicularity between the surface and the line of sight is seen in true relationship in the 2D views. These types of extra normal views are known as **auxiliary views** to distinguish them from the principal views. However, since an auxiliary is made for the purpose of showing the true shape of a surface, the terms *normal view* and *edge view* are also used to describe the relations of the views.

An auxiliary is constructed following the rules of orthographic projection. An auxiliary view is aligned to the associated views, in this case, the front view. The line of sight is perpendicular to the edge view of the surface, as shown in the below figures.

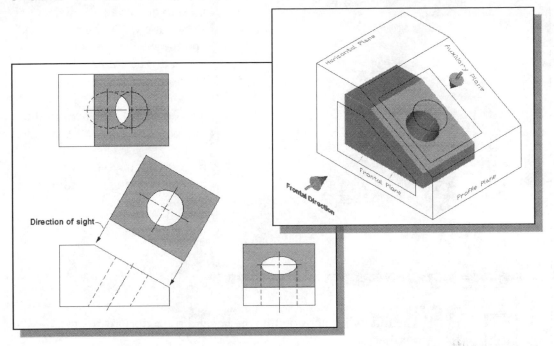

The orientation of the inclined surface may be different, but the direction of the normal view remains the same as shown in the below figures.

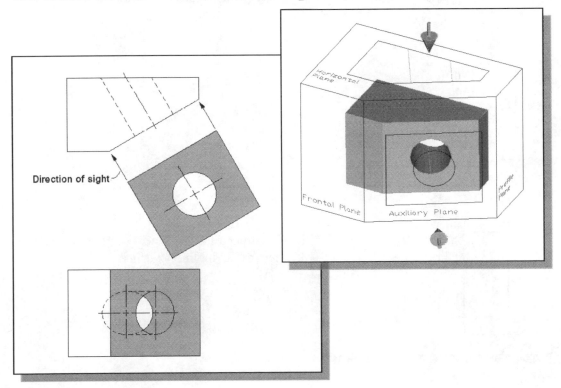

Construction Method I – Folding Line Method

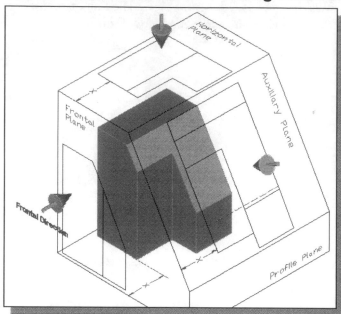

Two methods are commonly used to construct auxiliary views: the **folding-line** method and the **reference plane** method. The folding-line method uses the concept of placing the object inside a glass-box; the distances of the object to the different projection planes are used as measurements to construct the necessary views, including the auxiliary views.

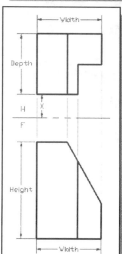

The following steps outline the general procedure to create an auxiliary view:

1. Construct the necessary principal views, in this case, the front and top views.

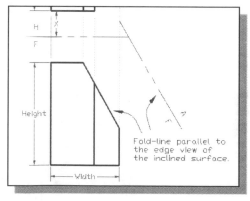

2. Construct a folding line parallel to the edge view of the inclined surface.

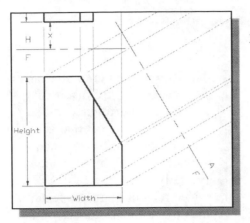

3. Construct the projection lines perpendicular to the edge view of the inclined surface and also perpendicular to the folding line.

4. Use the corresponding distances, X, and the depth of the object from the principal views to construct the inclined surface.

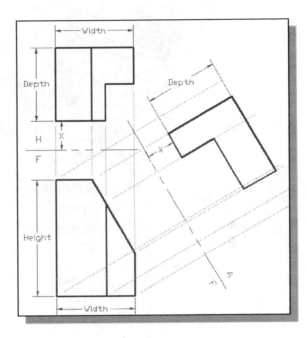

5. Complete the auxiliary view by following the principles of orthographic projection.

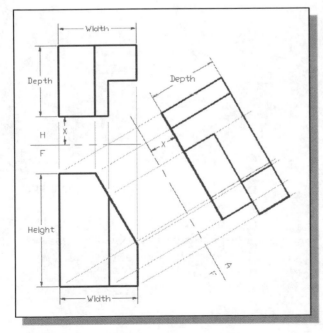

Construction Method II – Reference Plane Method

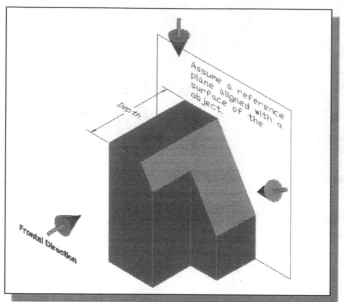

The **Reference plane method** uses the concept of placing a reference plane aligned with a surface of the object that is perpendicular to the inclined surface. The reference plane is typically a flat surface or a plane which runs through the center of the object. The distances of the individual corner to the reference plane are then used as measurements to construct the auxiliary view.

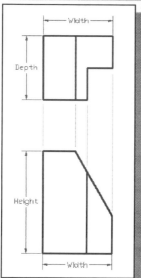

The following steps outline the general procedure to create an auxiliary view:

1. Construct the necessary principal views, in this case, the front and top views.

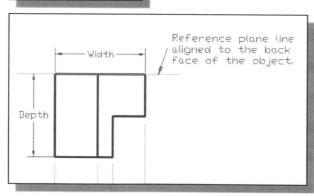

2. Construct a reference plane line aligned to a flat face and perpendicular to the inclined surface of the object, in this case, the back face of the object.

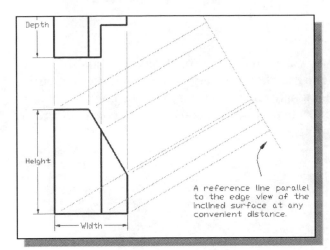

3. Construct a reference line parallel to the edge view of the inclined surface and construct the projection lines perpendicular to the edge view of the inclined surface.

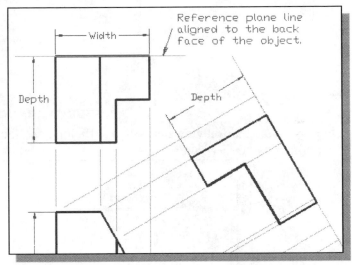

4. Using the corresponding distances, such as the depth of the object, construct the inclined surface.

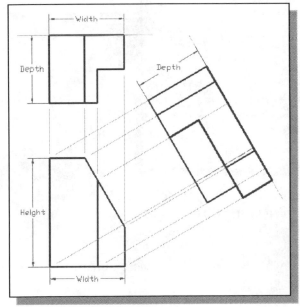

5. Complete the auxiliary view by following the principles of orthographic projection.

Partial Views

The primary purpose of using auxiliary views is to provide detailed descriptions of features that are on inclined surfaces. The use of an auxiliary view often makes it possible to omit one or more standard views. But it may not be necessary to draw complete auxiliary views, as the completeness of detail may be time consuming to construct and add nothing to the clearness of the drawing.

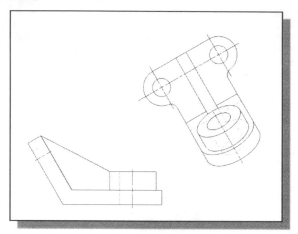

In these cases, partial views are often sufficient, and the resulting drawings are much simplified and easier to read.

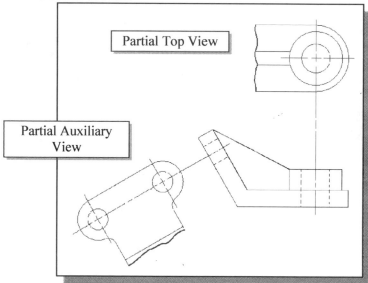

Partial Top View

Partial Auxiliary View

- To clarify the relationship of partial auxiliary views and the standard views, a center line or a few projection lines can be used. This is especially important when the partial views are small.

- In practice, hidden lines are generally omitted in auxiliary views, unless they provide more clearness to the drawings.

Work Features

Feature-based parametric modeling is a cumulative process. The relationships that we define between features determine how a feature reacts when other features are changed. Because of this interaction, certain features must, by necessity, precede others. A new feature can use previously defined features to define information such as size, shape, location and orientation. Autodesk Inventor provides several tools to automate this process. Work features can be thought of as user-definable datum, which are updated with the part geometry. We can create work planes, axes, or points that do not already exist. Work features can also be used to align features or to orient parts in an assembly. In this chapter, the use of the **Offset** option and the **Angled** option to create new work planes, surfaces that do not already exist, is illustrated. By creating parametric work features, the established feature interactions in the CAD database assure the capturing of the design intent. The default work features, which are aligned to the origin of the coordinate system, can be used to assist the construction of the more complex geometric features.

Auxiliary Views in 2D Drawings

An important rule concerning multiview drawings is to draw enough views to accurately describe the design. This usually requires two or three of the regular views, such as a front view, a top view and/or a side view. However, many designs have features located on inclined surfaces that are not parallel to the regular planes of projection. To truly describe the feature, the true shape of the feature must be shown using an **auxiliary view**. An *auxiliary view* has a line of sight that is perpendicular to the inclined surface, as viewed looking directly at the inclined surface. An *auxiliary view* is a supplementary view that can be constructed from any of the regular views. Using the solid model as the starting point for a design, auxiliary views can be easily created in 2D drawings. In this chapter, the general procedure of creating auxiliary views in 2D drawings from solid models is illustrated.

The Rod-Guide Design

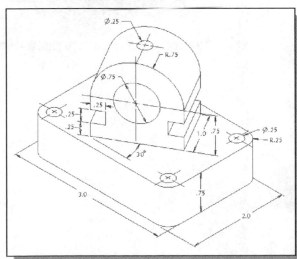

❖ Based on your knowledge of Autodesk Inventor so far, how would you create this design? What are the most difficult features involved in the design? Take a few minutes to consider a modeling strategy and do preliminary planning by sketching on a piece of paper. You are also encouraged to create the design on your own prior to following through the tutorial.

Modeling Strategy

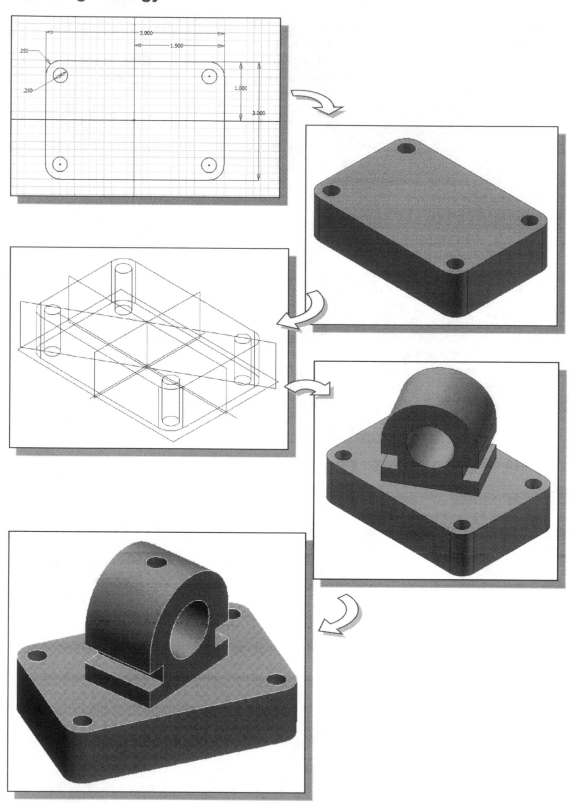

Starting Autodesk Inventor

1. Select the **Autodesk Inventor** option on the *Start* menu or select the **Autodesk Inventor** icon on the desktop to start Autodesk Inventor. The Autodesk Inventor main window will appear on the screen.

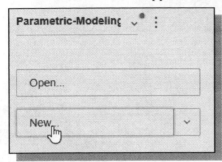

2. Select the **New File** icon with a single click of the left-mouse-button as shown.

3. On your own, confirm the project is set to the *Parametric-Modeling* project.

4. In the *New File* dialog box, select the **English** units set and then select **Standard(in).ipt**.

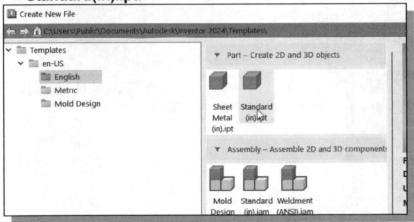

5. Click **Create** in the *New File* dialog box to accept the selected settings.

Apply the BORN Technique

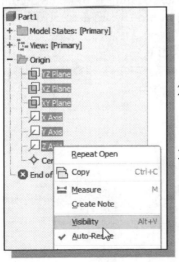

1. In the *Part Browser* window, click on the [**+**] symbol in front of the **Origin** feature to display more information on the feature.

2. Inside the *browser* window, select all of the work features by holding down the **[Control]** or **[Shift]** keys and click with the left-mouse-button.

3. Move the right-mouse-button on any of the work features to display the option menu. Click on **Visibility** to toggle *ON* the display of the selected work features.

4. On your own, use the dynamic viewing options (3D Rotate, Zoom and Pan) to view the work features established.

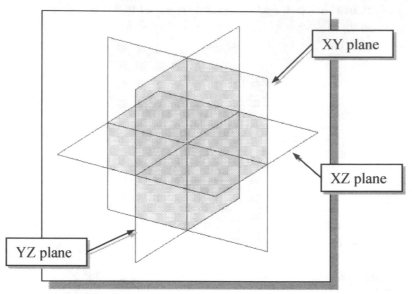

5. In the *Sketch* toolbar select the **Start 2D Sketch** command by left-clicking once on the icon.

6. In the *Status Bar* area, the message "*Select plane to create sketch or an existing sketch to edit*" is displayed. *Autodesk Inventor* expects us to identify a planar surface where the 2D sketch of the next feature is to be created. Move the graphics cursor on top of **XZ Plane**, inside the *browser* window as shown, and notice that Autodesk Inventor automatically highlights the corresponding plane in the graphics window. Left-click once to select the XZ Plane as the sketching plane.

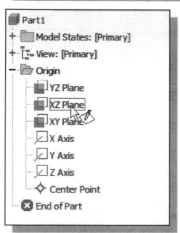

7. Single left-click to activate the **Home View** option as shown. The view will be adjusted back to the default *isometric view*.

Creating the Base Feature

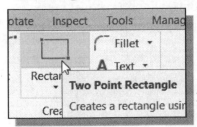

1. Select the **Two point rectangle** command by clicking once with the left-mouse-button on the icon in the *Sketch* toolbar.

2. Create a rectangle of arbitrary size with the center point near the center of the rectangle as shown.

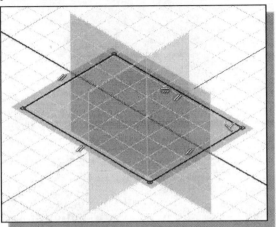

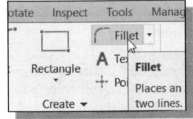

3. Click on the **Fillet** icon in the *Sketch* panel.

4. The *2D Fillet* radius dialog box appears on the screen. Use the default radius value and create four rounded corners of the rectangle.

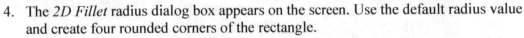

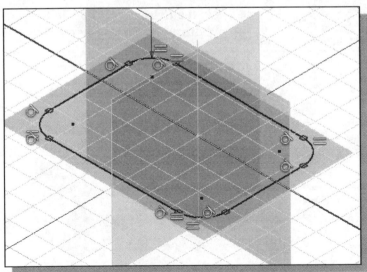

5. On your own, create four circles of the same diameter and with the centers aligned to the centers of the arcs. Also create and modify the six dimensions as shown in the figure.

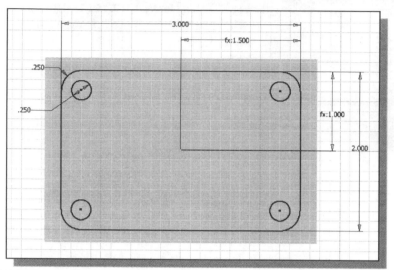

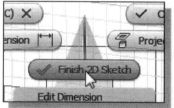

6. Inside the graphics window, click once with the **right-mouse-button** and select **Finish 2D Sketch** in the pop-up menu to end the Sketch option.

7. In the *3D Model* tab, select the **Extrude** command by clicking the left-mouse-button on the icon.

8. Select the inside regions of the 2D sketch to create a profile as shown.

9. In the *Extrude* pop-up window, enter **0.75** as the extrusion distance and create the feature.

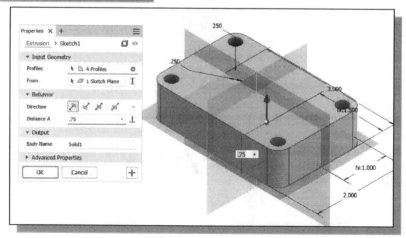

Create an Angled Work Plane

1. Activate the **View tab** in the *Ribbon* panel as shown.

2. Select **Wireframe** display mode under the *Visual Style* icon as shown.

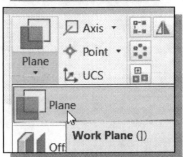

3. Activate the *3D Model* tab and select the **Work Plane** command by left-clicking the icon as shown.

4. In the *Status Bar* area, the message "*Define work plane by highlighting and selecting geometry*" is displayed. Autodesk Inventor expects us to select any existing geometry to be used as a reference to create the new work plane.

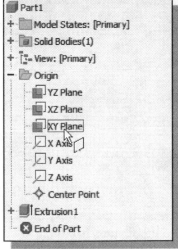

5. Inside the *browser* window, left-click once to select the **XY Plane** as the first reference of the new work plane.

6. Inside the *browser* window, left-click once to select the **Y Axis** as the second reference of the new work plane.

7. In the *Angle* pop-up window, enter **30** as the rotation angle for the new work plane.

❖ Note that the *angle* is measured relative to the selected reference plane, XY Plane.

8. Click on the **check mark** button to accept the setting.

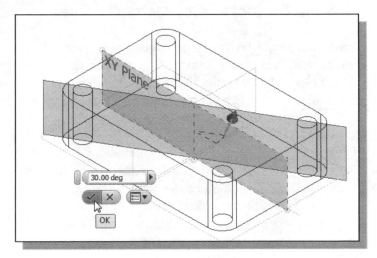

Create a 2D Sketch on the Work Plane

1. In the *Sketch* toolbar select the **Start 2D Sketch** command by left-clicking once on the icon.

2. In the *Status Bar* area, the message *"Select face, work plane, sketch or sketch geometry"* is displayed. Pick the **work plane** by clicking the work plane name inside the *browser* as shown below.

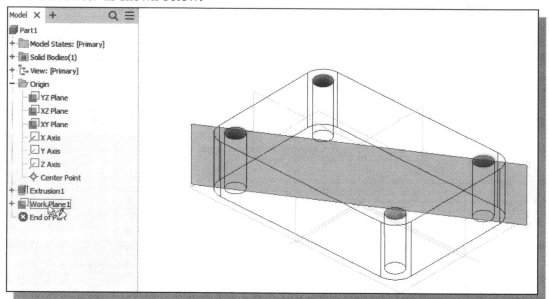

Use the Projected Geometry Option

Projected geometry is another type of *reference geometry*. The **Project Geometry** tool can be used to project geometry from previously defined sketches or features onto the sketch plane. The position of the projected geometry is fixed to the feature from which it was projected. We can use the **Project Geometry** tool to project geometry from a sketch or feature onto the active sketch plane.

Typical uses of projected geometry include:
- Project a silhouette of a 3D feature onto the sketch plane for use in a 2D profile.
- Project the default center point onto the sketch plane to constrain a sketch to the origin of the coordinate system.
- Project a sketch from a feature onto the sketch plane so that the projected sketch can be used to constrain a new sketch.

1. Choose **View** in the *Ribbon* tabs.

2. Select **Object Visibility → Origin Axes** in the options list to toggle *OFF* the axes.

3. Note that quick-key option, [CTRL]+[/], is also available to toggle *ON/OFF* the different reference geometry.

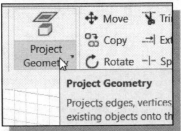

4. Select the **Project Geometry** command in the *Sketch* panel. The *Project Geometry* command allows us to project existing features to the active sketching plane.

5. Select the top back edge of the base feature to create a projected line on the sketching plane.

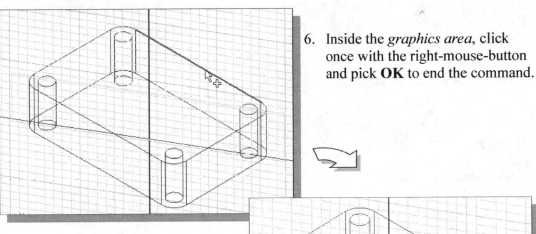

6. Inside the *graphics area*, click once with the right-mouse-button and pick **OK** to end the command.

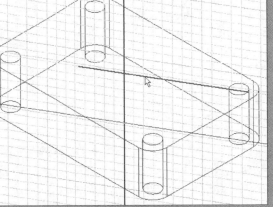

➢ The projected line will be used in the 2D profile of the next solid feature.

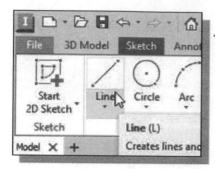

7. In the *Sketch* toolbar, click on the **Line** icon with the left-mouse-button to activate the Line command.

8. Create a rough sketch using the projected edge as the bottom line as shown in the figure. (Note that all edges are either horizontal or vertical.)

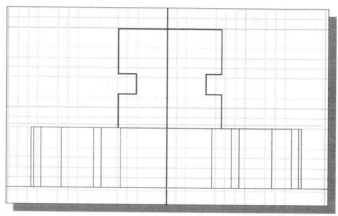

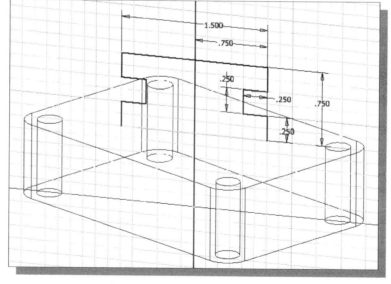

9. Left-click once on the **General Dimension** icon to activate the General Dimension command.

10. On your own, create and modify the dimensions as shown; note that the sketch is symmetrical vertically.

➢ Note that the projected line is used only as a reference; the gap at the bottom of the 2D sketch indicates the 2D sketch does not form a closed region profile.

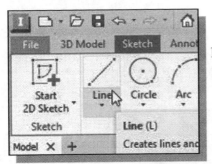

11. Select the **Line** command in the *2D Sketch* panel.

12. On your own, create an additional **bottom line** connecting the bottom of the 2D sketch as shown in the figure.

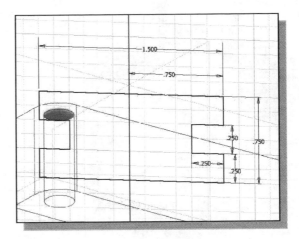

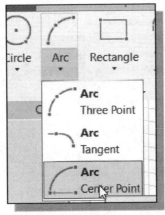

13. Select the **Center point Arc** option in the *2D Sketch* panel as shown.

14. On your own, create the arc so that the center point is aligned to the mid-point of the top edge.

15. On your own, add a **0.75** circle, and complete the 2D sketch as shown in the figure.

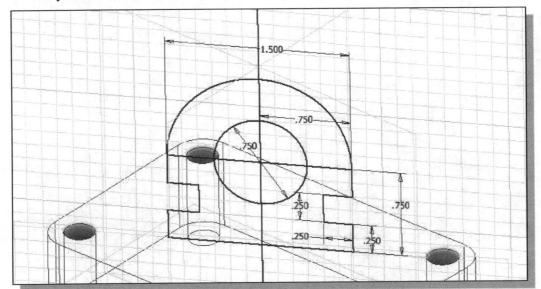

Complete the Solid Feature

1. Inside the graphics window, click once with the right-mouse-button and select **Finish Sketch** in the pop-up menu to end the Sketch option.

2. In the *3D Model* toolbar, select the **Extrude** command by left clicking on the icon.

3. Select the inside regions of the 2D sketch to create a profile as shown.

4. In the *Extrude* pop-up window, enter **1.0** as the extrusion distance.

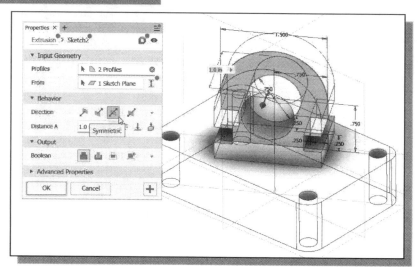

5. Set the extrusion direction to **Symmetric** as shown.

6. Click on the **OK** button to proceed with creating the feature.

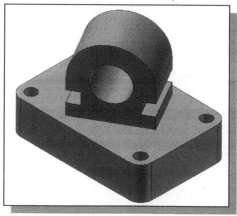

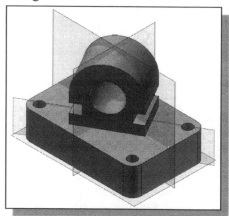

Create an Offset Work Plane

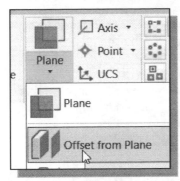

1. In the *3D Model* toolbar, select the **Offset from Plane** command by left-clicking the icon.

❖ In the *Status Bar* area, the message "*Define work plane by highlighting and selecting geometry*" is displayed. Autodesk Inventor expects us to select any existing geometry, which will be used as a reference to create the new work plane.

2. Inside the graphics window, select the **top plane** of the base feature as the reference of the new work plane.

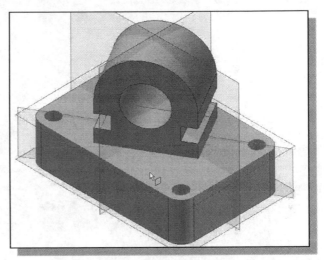

3. Set the value in the *Offset* pop-up window to **0.75** as the offset distance for the new work plane.

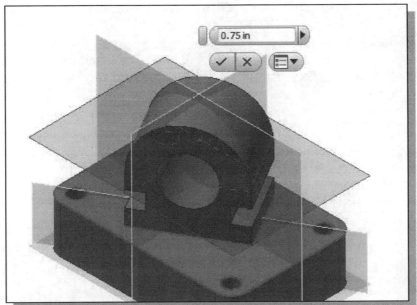

4. Click on the **check mark** button to accept the setting and create the reference plane.

Create another Cut Feature Using the Work Plane

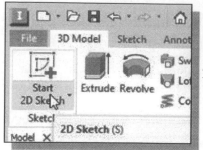

1. In the *Sketch* toolbar select the **Start 2D Sketch** command by left-clicking once on the icon.

2. In the *Status Bar* area, the message *"Select face, work plane, sketch or sketch geometry"* is displayed. Pick the work plane we just created by clicking on one of the edges of the work plane as shown below.

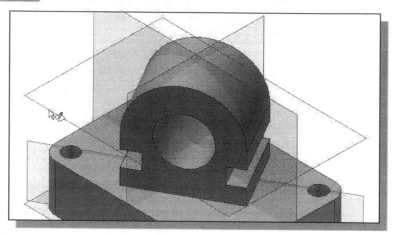

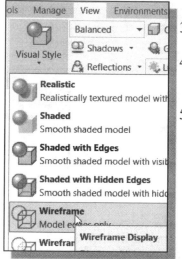

3. Activate the **View tab** in the *Ribbon* panel as shown.

4. Select **Wireframe** display mode under the *Visual Style* icon as shown.

5. Use the **[Ctrl+]]** quick-key combination to toggle off the display of the *Origin Planes*.

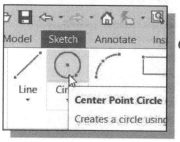

6. Select the **Center Point Circle** command by clicking once with the left-mouse-button on the icon in the *Sketch* tab on the Ribbon.

7. On your own, create a **circle** with the center point aligned to the projected **Center Point** as shown in the figure below.

8. Using the Dimension command, set the circle to **Ø0.25**.

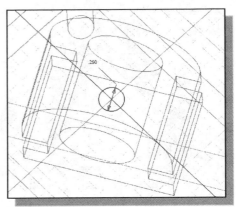

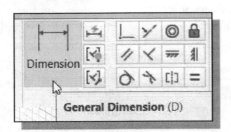

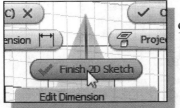

9. Inside the graphics window, click once with the right-mouse-button and select **Finish 2D Sketch** in the pop-up menu to end the Sketch option.

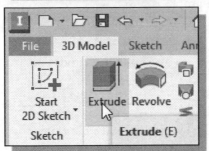

10. In the *3D Model* toolbar, select the **Extrude** command by left clicking on the icon.

11. On your own, complete the **cut** feature as shown.

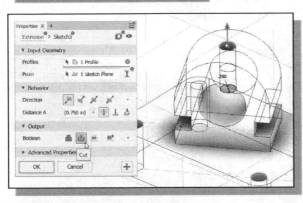

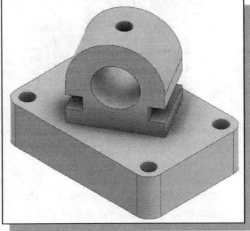

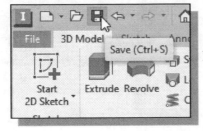

12. Save the design as *Rod-Guide.ipt* in the **Chapter 9** folder.

Start a New 2D Drawing

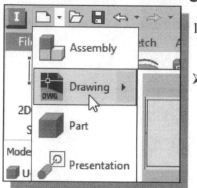

1. Select **New File → Drawing**, without choosing any drawing option, in the *Quick Access* toolbar as shown.

➢ Note that a new graphics window appears on the screen. We can switch between the solid model and the drawing by clicking the corresponding graphics windows.

❖ In the graphics window, Autodesk Inventor displays a default drawing sheet that includes a title block. The drawing sheet is placed on the 2D paper space, and the title block also indicates the paper size being used.

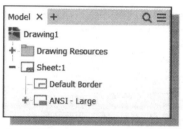

➢ In the *browser* area, the Drawing1 icon is displayed at the top, which indicates that we have switched to *Drawing Mode*. **Sheet:1** is the current drawing sheet that is displayed in the graphics window.

❖ Different types of pre-defined borders and title blocks are available in Autodesk Inventor.

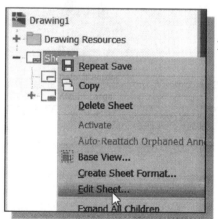

2. In the *browser* area, click once with the right-mouse-button on **Sheet:1** and select **Edit Sheet** in the pop-up menu.

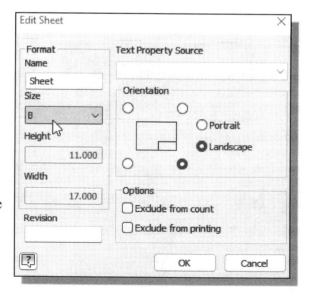

3. In the *Edit Sheet* window, set the size option to **B** size as shown.

4. Click on the **OK** button to accept the settings.

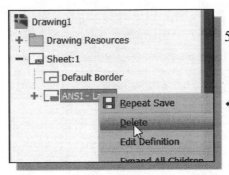

5. In the *browser* area, click once with the right-mouse-button on **ANSI-Large** and select **Delete** in the pop-up menu.

❖ Before applying a different title block, the existing title block must be removed.

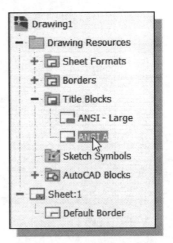

6. In the *browser* area, click on the [>] symbol in front of the **Drawing Resources** and **Title Blocks** options to expand the lists.

7. **Double-click** on the **ANSI-A** title block to place a copy of this title block into the current sheet.

Add a Base View

In Autodesk Inventor *Drawing Mode*, the first drawing view we create is called a **base view**. A *base view* is the primary view in the drawing; other views can be derived from this view. When creating a *base view*, Autodesk Inventor allows us to specify the view to be shown. By default, Autodesk Inventor will treat the *world XY plane* as the front view of the solid model. Note that there can be more than one *base view* in a drawing.

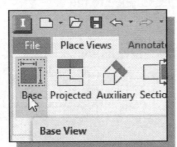

1. Click on the **Base View** in the *Drawing Views* panel to create a base view.

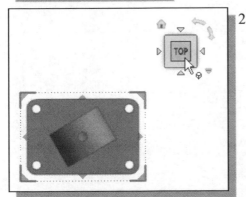

2. In the *Drawing View* dialog box, confirm that the settings are set to **Top** view with **Hidden Line** displayed and **Scale to 1 : 1** as shown.

3. Inside the graphics window, place the **base view** near the upper left corner of the graphics window as shown below. (If necessary, drag the *Create View* dialog box to another location on the screen.)

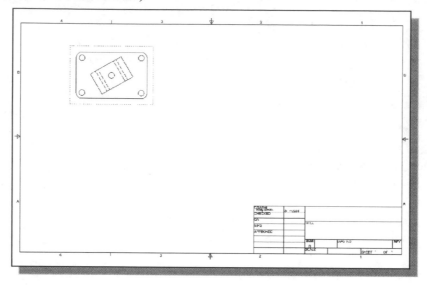

4. Click the [**OK**] key once to end the *Base View* command.

Create an Auxiliary View

In Autodesk Inventor *Drawing Mode*, the **Projected View** command is used to create standard views such as the *top* view, *front* view or *isometric* view. For non-standard views, the **Auxiliary View** command is used. *Auxiliary views* are created using orthographic projections. Generally, orthographic projections are aligned to the base view and inherit the base view's scale and display settings.

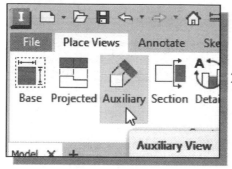

1. Click on the **Auxiliary View** button in the *Drawing Views* panel.

2. Click on the **base view** to select the view as the referenced view for projection.

3. Confirm the settings in the *Auxiliary View* window are set as shown. (**DO NOT** click on the **OK** button yet.)

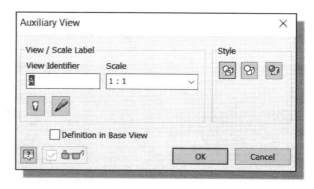

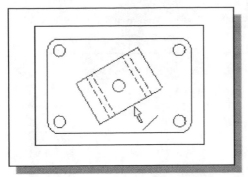

4. Pick the front edge of the upper section of the model as shown.

❖ The orthographic projection direction will be perpendicular to the selected edge.

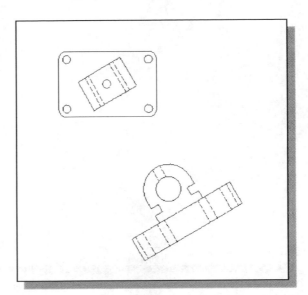

5. Move the cursor below the base view and click once with the **left mouse button** to select a location to position the auxiliary view of the model as shown.

6. Click on **Base View** in the *Drawing Views* panel to create a base view.

7. In the *Drawing View* dialog box, confirm that the settings are set to **Iso Top Right**, **Scale 1 : 1** and **Hidden Line Removed**. (**DO NOT** click on the **OK** button at this point.)

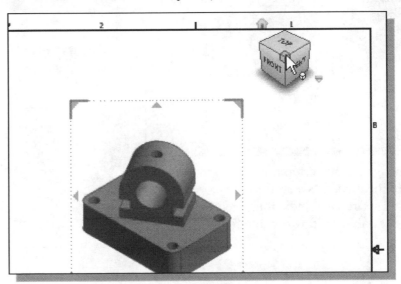

8. Drag and drop with the mouse and position the *isometric* view toward the right side of the title block as shown below.

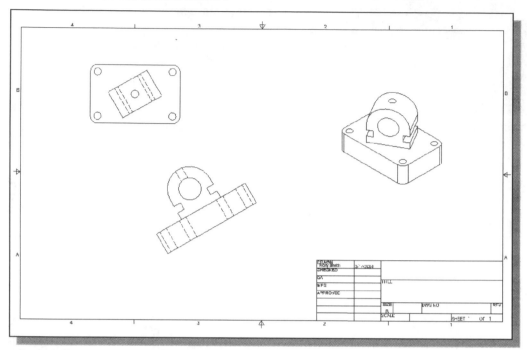

 9. Right-click and choose **OK** to end the command.

Display Feature Dimensions

By default, feature dimensions are not displayed in 2D views in Autodesk Inventor. We can change the default settings while creating the views or switch on the display of the parametric dimensions using the option menu.

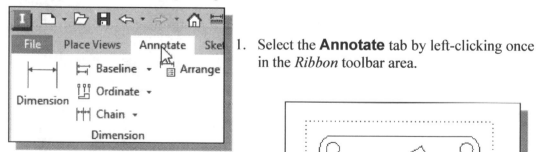

1. Select the **Annotate** tab by left-clicking once in the *Ribbon* toolbar area.

2. Move the cursor on top of the *top* view of the model. Watch for the box around the entire view indicating the view is selectable as shown in the figure.

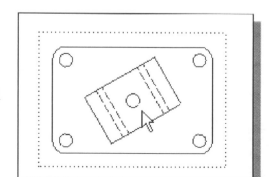

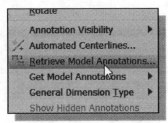

3. Inside the *graphics window*, **right-click** once to bring up the option menu.

4. Select **Retrieve Model Annotations** to display the parametric dimensions used to create the model.

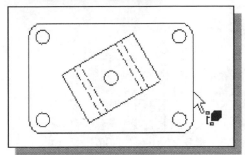

5. Move the cursor to the *top* view and select anywhere on the *Rod-Guide* part as shown.

6. In the *Sketch and Feature Dimensions* tab, set the *Select Source* option to **Select Parts** as shown.

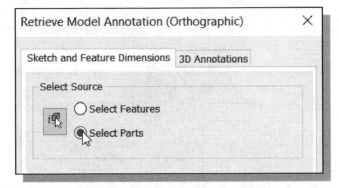

➢ Note that all the dimensions used to create the part are now displayed in the selected view.

➢ The system now expects us to select the dimensions to be retrieved.

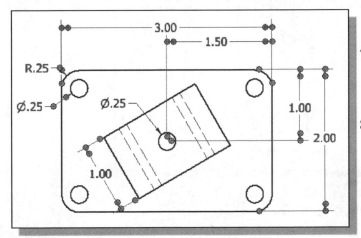

7. On your own, select the dimensions in the top view as shown.

8. Click on the **OK** button to end the Retrieve Dimensions command.

Adjust the View Scale

1. Move the cursor on top of the isometric view and watch for the box around the entire view indicating the view is selectable as shown in the figure. Right-click once to bring up the option menu. Select **Edit View** in the option menu as shown.

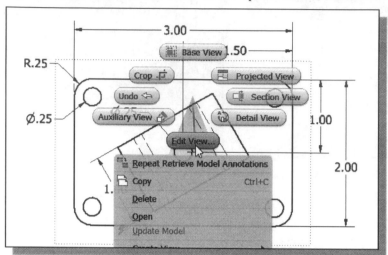

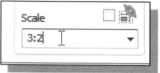

2. Inside the *Drawing View* dialog box set the *Scale* to **3:2** as shown in the figure.

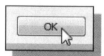

3. Click on the **OK** button to accept the settings.

4. On your own, reposition the views and dimensions by clicking and dragging the individual entities.

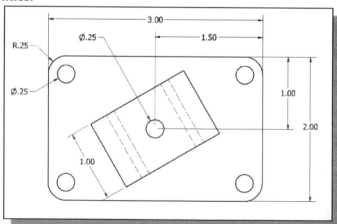

❖   Note that in parametric modeling software, dimensions are always associated with the geometry, even in the *2D Drawing Mode*.

Retrieving Dimensions in the Auxiliary View

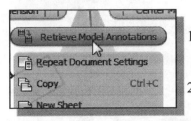

1. Inside the *graphics window*, **right-click** once in a blank area to bring up the option menu.

2. Select **Retrieve Model Annotations** to display the parametric dimensions used to create the model.

3. Select the **auxiliary view** to retrieve the associated dimension.

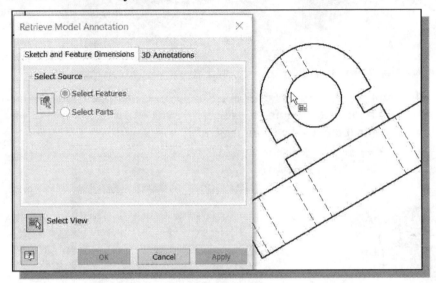

4. Select the six dimensions as shown in the below figure.

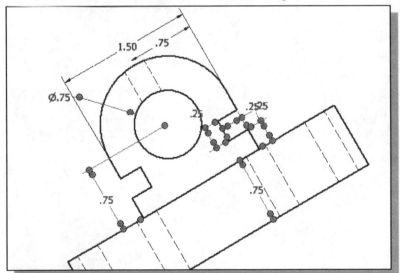

5. Click **OK** to accept the selection.

6. Reposition the dimensions by clicking and dragging the individual entities.

➢ Note that additional options are available through the **Options** menu.

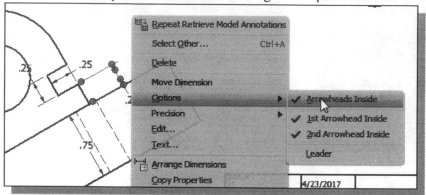

➢ Note that the different grip points can also be used to adjust the dimensions. You are encouraged to experiment with dragging the different parts of the dimensions and understand how to control the displayed dimensions.

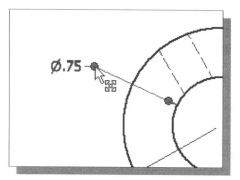

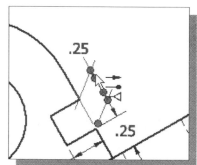

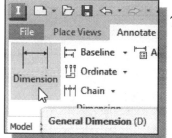

7. On your own, create and position the angle dimension for the design as shown in the figure below.

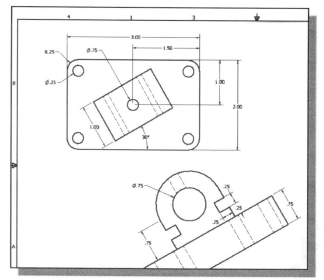

Add Center Marks and Center Lines

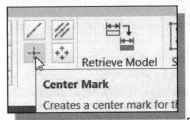

1. Click on the **Center Mark** button in the *Drawing Annotation* window.

2. Click on the five circles in the top view to add the center marks as shown.

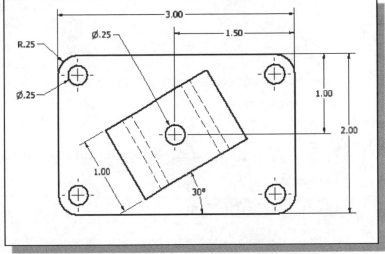

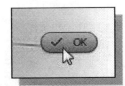

3. Inside the graphics window, click once with the right-mouse-button to display the option menu. Select **OK** in the pop-up menu to end the **Center Mark** command.

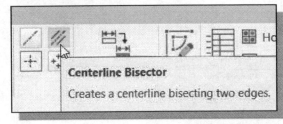

4. Select **Centerline Bisector** in the Ribbon toolbar area.

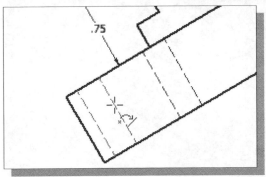

5. Inside the graphics window, click on the two hidden edges of one of the *drill* features and create a center line in the auxiliary view as shown in the figure.

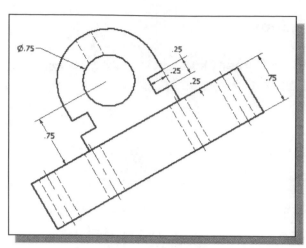

6. On your own, repeat the above step and create additional centerlines as shown.

7. Inside the graphics window, click once with the right-mouse-button to display the option menu. Select **Cancel** in the pop-up menu to end the Centerline Bisector command.

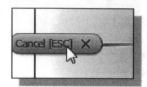

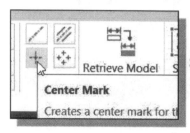

8. Select **Center Mark** in the *Drawing Annotation* window.

9. Click on the arc in the auxiliary view to create the centerlines as shown.

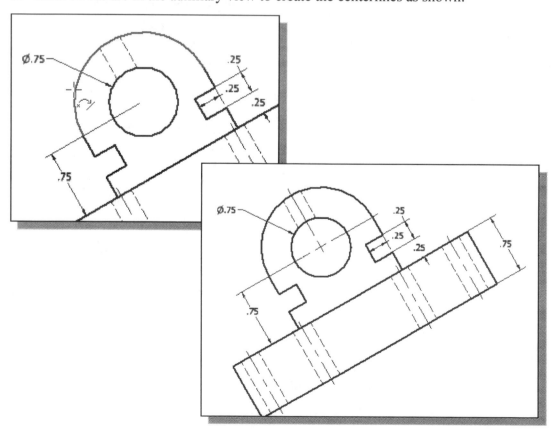

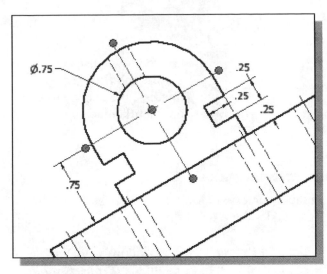

10. Hit the **[ESC]** key once to end the Center Mark command.

11. Click on the **centerlines** in the auxiliary view as shown.

12. Adjust the length of the horizontal centerline by dragging on one of the grip points as shown.

13. On your own, repeat the above steps and adjust the dimensions/centerlines as shown below.

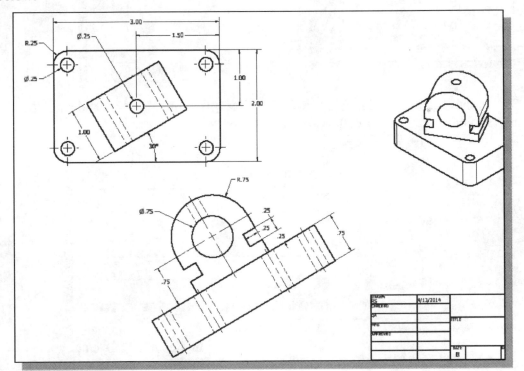

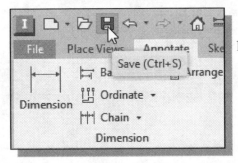

14. Click on the **Save** icon in the *Standard* toolbar and save the drawing as ***Rod-Guide.dwg*** in the **Chapter 9** folder.

Complete the Title Block with iProperties

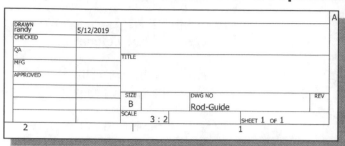

1. On your own, use the **Zoom** and **Pan** commands to adjust the display to work on the title block area.

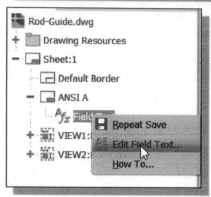

2. In the *Browser* window, right-click on the **Field Text** under the ANSI title block.

3. In the *Edit Property Fields* window, click the **iProperties** icon to bring up the *iProperty dialog box*.

4. In the *Summary* tab, enter **Rod-Guide Design** in the *Title box*, and enter the name of your organization in the *Company box* as shown.

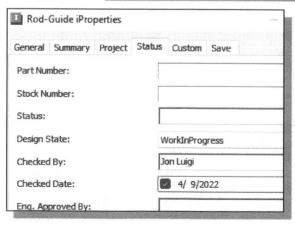

5. In the *Status* tab, enter your instructor's name in the *Checked By box* and set the *Checked Date* as shown.

6. Click **OK** to accept the settings.

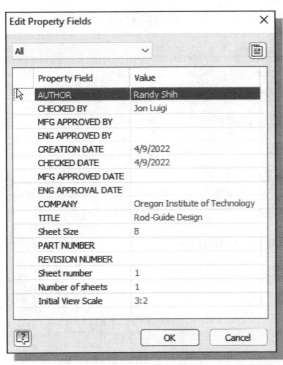

7. Note that the items in the Property fields list have been updated to reflect the changes.

8. On your own, open up the *iProperty dialog box*, and experiment with filling in the other items listed in the Property fields list.

9. Click **OK** to accept the settings.

• Note that any information entered in the iProperty dialog box will be automatically placed in the *Title block* as shown.

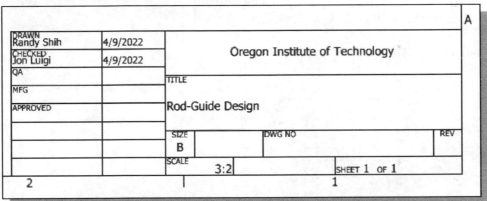

10. On your own, also create a general note at the lower left corner of the border as shown.

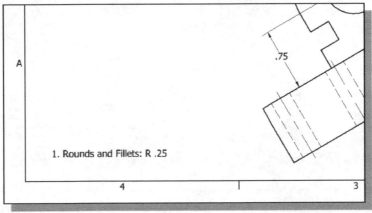

Edit the Isometric View

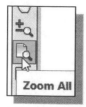

Zoom All

1. Click on the **Zoom All** button in the *Standard* toolbar.

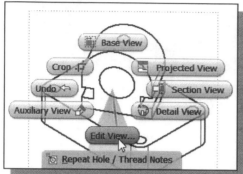

2. Right-click inside the *isometric* view to bring up the option menu as shown.

3. Select **Edit View** in the option menu.

4. In the *Drawing View* dialog box, set the scale to **3:2** and the display *Style* to **Shaded**.

5. Click on the **OK** button to accept the settings.

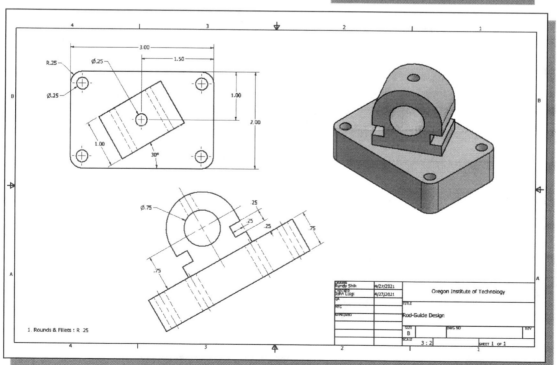

Review Questions: (Time: 25 minutes)

1. What are the different types of work features available in Autodesk Inventor?

2. Why are work features important in parametric modeling?

3. Describe the purpose of auxiliary views in 2D drawings?

4. What are the required elements in order to generate an auxiliary view?

5. Can we use a different Title Block in the Drawing Mode? How?

6. Describe the different methods used to create centerlines in the chapter.

7. Can we change the *View Scale* of existing views? How?

8. What is the main difference between an auxiliary view and a projected view in Autodesk Inventor?

9. Describe the steps to change the *display style* of a drawing view.

10. Describe the difference between the centerlines created with the **Centerline Bisector** and the **Center Mark** commands.

Exercises: Create the Solid models and the associated 2D drawings.

1. **Rod Slide** (Dimensions are in inches. The overall height of the design is 2.50.)

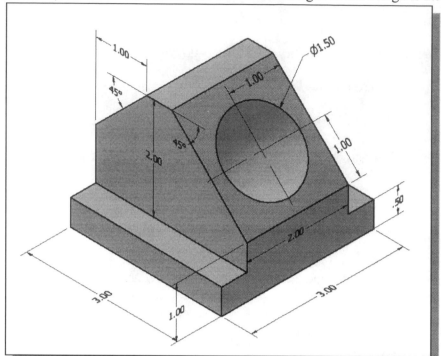

2. **Anchor Base** (Dimensions are in inches.)

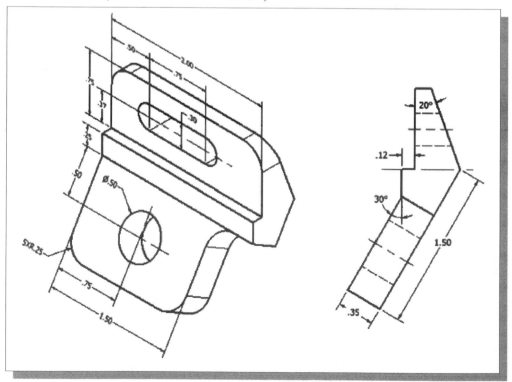

3. **Bevel Washer** (Dimensions are in inches.)

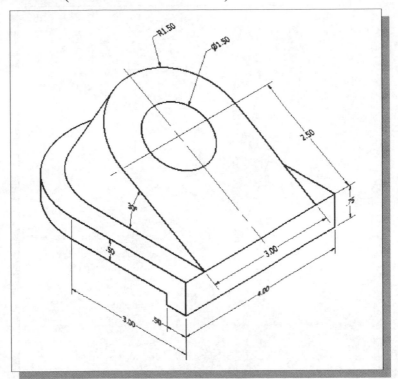

4. **Angle V-Block** (Dimensions are in inches.)

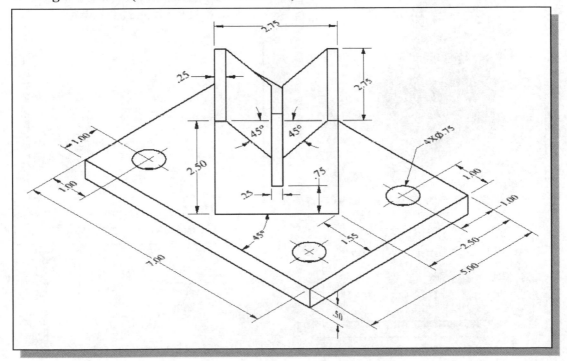

5. **Angle Support** (Dimensions are in millimeters.)

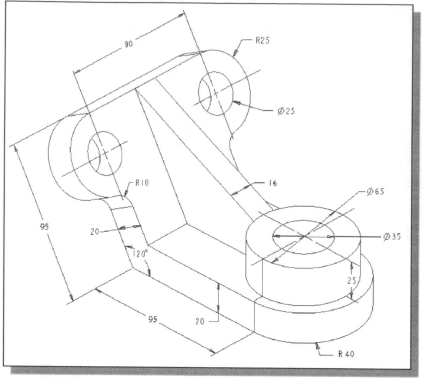

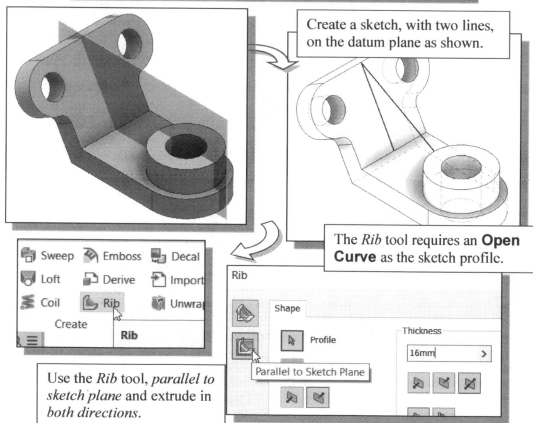

Create a sketch, with two lines, on the datum plane as shown.

The *Rib* tool requires an **Open Curve** as the sketch profile.

Use the *Rib* tool, *parallel to sketch plane* and extrude in *both directions*.

6. **Jig Base** (Dimensions are in millimeters.)

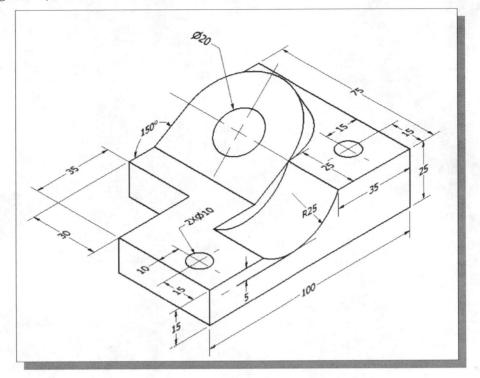

Notes:

Chapter 12
Section Views & Symmetrical Features in Designs

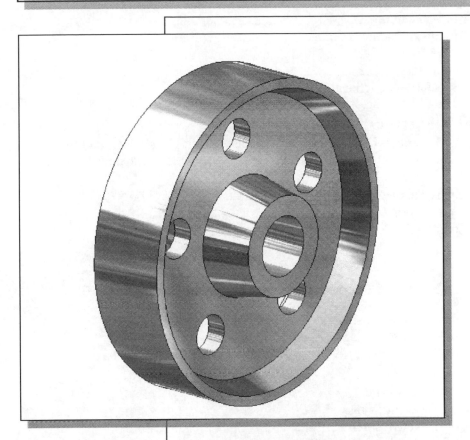

Learning Objectives

♦ **Create Revolved Features**
♦ **Use the Mirror Feature Command**
♦ **Create New Borders and Title Blocks**
♦ **Create Circular Patterns**
♦ **Create and Modify Linear Dimensions**
♦ **Use Autodesk Inventor's Associative Functionality**
♦ **Identify Symmetrical Features in Designs**
♦ **Create a Section View in a Drawing**

Autodesk Inventor Certified User Reference Guide

Autodesk Inventor Certified User Exam Objectives Coverage

Parametric Modeling Basics

Section 3: Sketches

Objectives: Creating 2D Sketches, Draw Tools, Sketch Constraints, Pattern Sketches, Modify Sketches, Format Sketches, Sketch Doctor, Shared Sketches, Sketch Parameters

Section 4: Parts

Objectives: Creating parts, Work Features, Pattern Features, Part Properties

Section 6: Drawings

Objectives: Create drawings

Introduction

In the previous chapters, we have explored the basic CAD methods of creating orthographic views. By carefully selecting a limited number of views, the external features of most complicated designs can be fully described. However, we are frequently confronted with the necessity of showing the interiors of parts, or when parts are assembled, that cannot be shown clearly by means of hidden lines.

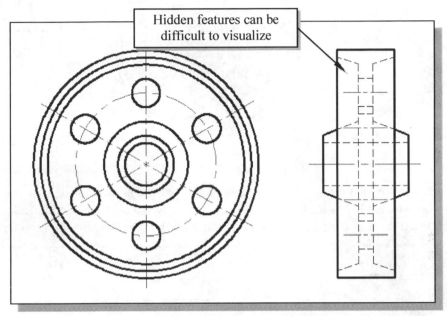

Hidden features can be
difficult to visualize

In cases of this kind, to aid in describing the object, one or more views are drawn to show the object as if a portion of the object had been cut away to reveal the interior.

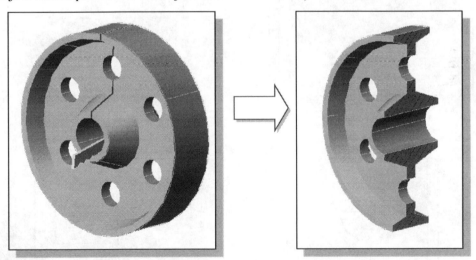

This type of convention is called a **section**, which is defined as an imaginary cut made through an object to expose the interior of a part. Such kind of cutaway view is known as a **section view**.

In a *section view*, the place from which the section is taken must be identifiable on the drawing. If the place from which the section is taken is obvious, as it is for the below figure, no further description is needed.

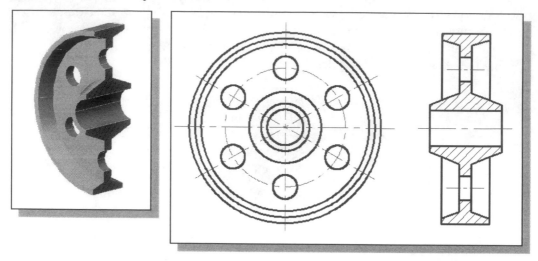

If the place from which the section is taken is not obvious, as it is for the below figure, a **cutting plane** is needed to identify the section. Two arrows are also used to indicate the viewing direction. A *cutting plane line* is drawn with the *phantom* or *hidden* line.

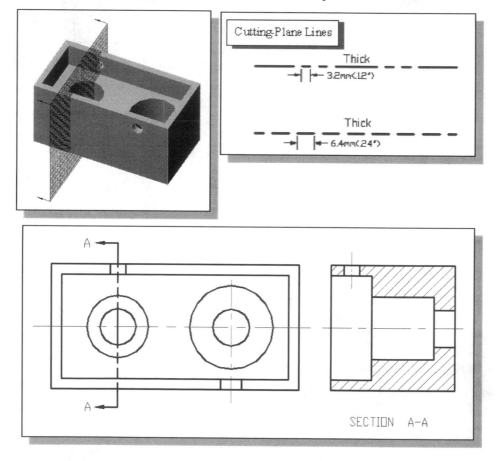

General Rules of Section Views

Section views are used to make a part drawing more understandable, showing the internal details of the part. Since the sectioned drawing is showing the internal features there is generally no need to show hidden lines.

A *section view* still follows the general rules of any view in a multiview drawing. Cutting planes may be labeled at their endpoints to clarify the association of the cut and the section views, especially when multiple cutting plane lines are used. When using multiple cutting planes, each sectioned drawing is drawn as if the other cutting plane lines do not exist. The *cutting plane line* takes precedence over center lines and, remember, cutting plane lines may be omitted when their location is obvious.

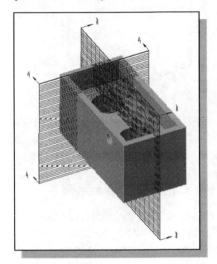
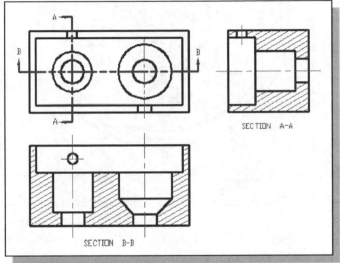

Section lines or **cross-hatch lines** are drawn where the cutting plane passes through the object. This is done as if a saw was used to cut the part and then section lines are used to represent the cutting marks left by the saw blade. Different materials are represented by the use of different section line types. The line type for iron may be used as the general section line type for any material.

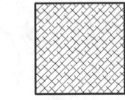

| Cast iron, Malleable iron and general use for all materials | Steel | Bronze, Brass and composition Materials | Magnesium, aluminum and aluminum alloys |

Section lines should not be parallel or perpendicular to object lines. Section lines are generally drawn at *45°*, *30°* or *60°*, unless there are conflicts with other rules. Section lines should be oriented at different angles for different parts.

Section Drawing Types

- **Full Section:** The cutting plane passes completely through the part as a single flat plane.

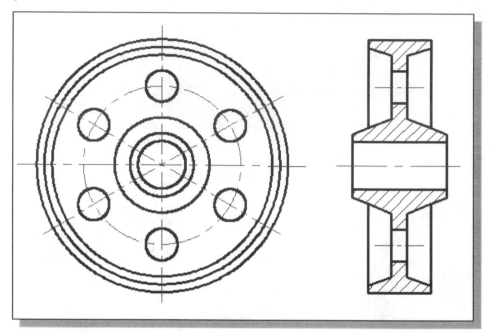

- **Half Section:** The cutting plane only passes half-way through the part and the other half is drawn as usual. Hidden lines are generally not shown on either half of the part. However, hidden lines may be used in the un-sectioned half if necessary for more clarity of the design. Half section is mostly used on cylindrical parts, and a center line is used to separate the two halves.

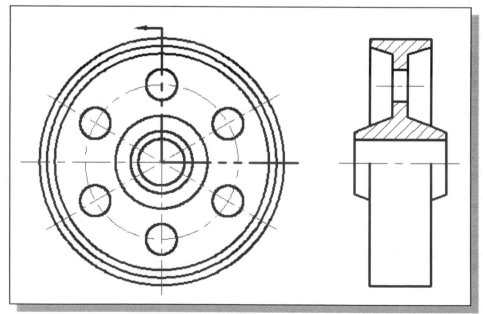

- **Offset Section:** It is often difficult to interpret a multiview drawing when there are multiple features on the object. An **offset section** allows the cutting plane to pass through several places by **"offsetting"** or **bending** the cutting plane. The offsets or bends in the cutting plane are all 90° and are *not* shown in the section view.

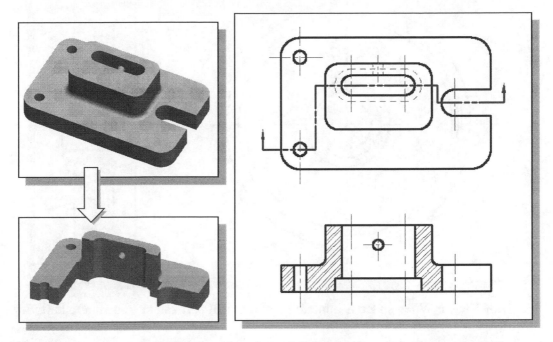

- **Broken-Out Section:** It is quite often that a full section is not necessary, but only a partial section is needed to show the interior details of a part. In a **broken-out section**, only a portion of the view is sectioned and a jagged break-line is used to divide the sectioned and un-sectioned portion of the view.

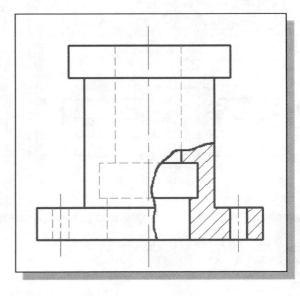

- **Aligned Section:** To include certain angled features in a section view, the cutting plane may be bent to pass through those features to their true radial position.

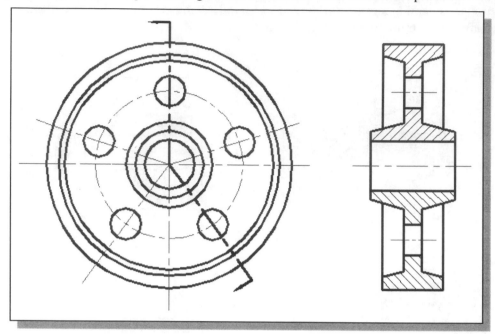

- **Half Views:** When space is limited, a symmetrical part can be shown as a half view.

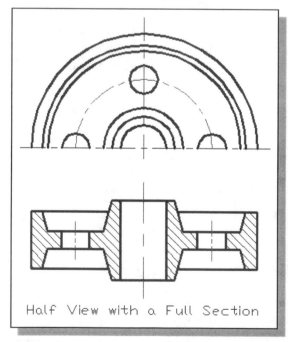

Half View with a Full Section

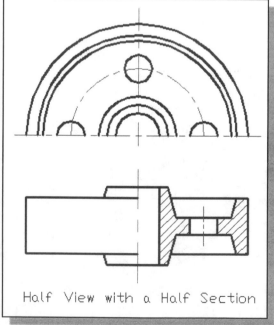

Half View with a Half Section

- **Thin Sections:** For thin parts, typically parts that are less than 3mm thickness, such as sheet metal parts and gaskets, section lines are ineffective and therefore should be omitted. Thin sections are generally shown in solid black without section lines.

- **Revolved Section:** A cross section of the part is revolved 90° and superimposed on the drawing.

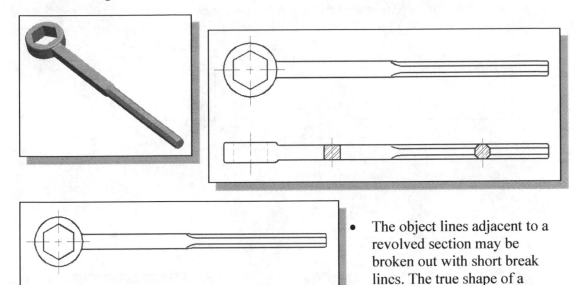

- The object lines adjacent to a revolved section may be broken out with short break lines. The true shape of a revolved section is retained regardless of the object lines in the view.

- **Removed Section:** Similar to the revolved section except that the sectioned drawing is not superimposed on the drawing but placed adjacent to it. The removed section may also be drawn at a different scale. Very useful for detailing small parts or features.

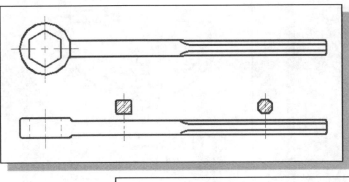

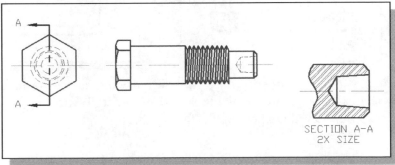

- **Conventional Breaks:** In making the details of parts that are long and with a uniform cross section, it is rarely necessary to draw its whole length. Using the Conventional Breaks to shorten such an object will allow the part to be drawn with a larger scale. S-Breaks are preferred for cylindrical objects and jagged lines are used to break non-circular objects.

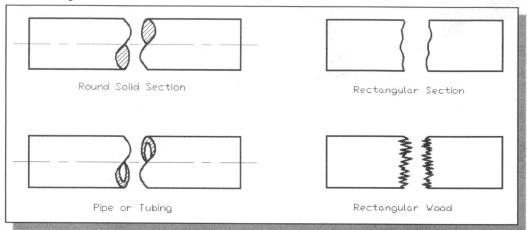

- **Ribs and Webs in Sections:** To avoid a misleading effect of thickness and solidity, ribs, webs, spokes, gear teeth and other similar parts are drawn without crosshatching when the cutting plane passes through them lengthwise. Ribs are sectioned if the cutting plane passes through them at other orientations.

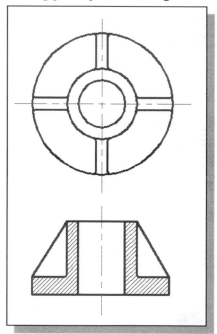 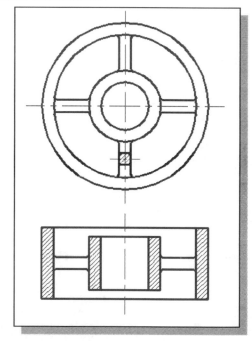

- **Parts Not Sectioned:** Many machine elements, such as fasteners, pins, bearings and shafts, have no internal features, and are generally more easily recognized by their exterior views. Do not section these parts.

Section Views in Autodesk Inventor

In parametric modeling, it is important to identify and determine the features that exist in the design. *Feature-based parametric modeling* enables us to build complex designs by working on smaller and simpler units. This approach simplifies the modeling process and allows us to concentrate on the characteristics of the design. Symmetry is an important characteristic that is often seen in designs. Symmetrical features can be easily accomplished by the assortment of tools that are available in feature-based modeling systems, such as Autodesk Inventor.

The modeling technique of extruding two-dimensional sketches along a straight line to form three-dimensional features, as illustrated in the previous chapters, is an effective way to construct solid models. For designs that involve cylindrical shapes, shapes that are symmetrical about an axis, revolving two-dimensional sketches about an axis can form the needed three-dimensional features. In solid modeling, this type of feature is called a *revolved feature*.

In Autodesk Inventor, besides using the **Revolve** command to create revolved features, several options are also available to handle symmetrical features. For example, we can create multiple identical copies of symmetrical features with the **Feature Pattern** command or create mirror images of models using the **Mirror Feature** command. We can also use *construction geometry* to assist the construction of more complex features. In this lesson, the construction and modeling techniques of these more advanced options are illustrated.

A Revolved Design: Pulley

❖ Based on your knowledge of Autodesk Inventor, how many features would you use to create the design? Which feature would you choose as the **base feature** of the model? Identify the symmetrical features in the design and consider other possibilities in creating the design. You are encouraged to create the model on your own prior to following through the tutorial.

Modeling Strategy – A Revolved Design

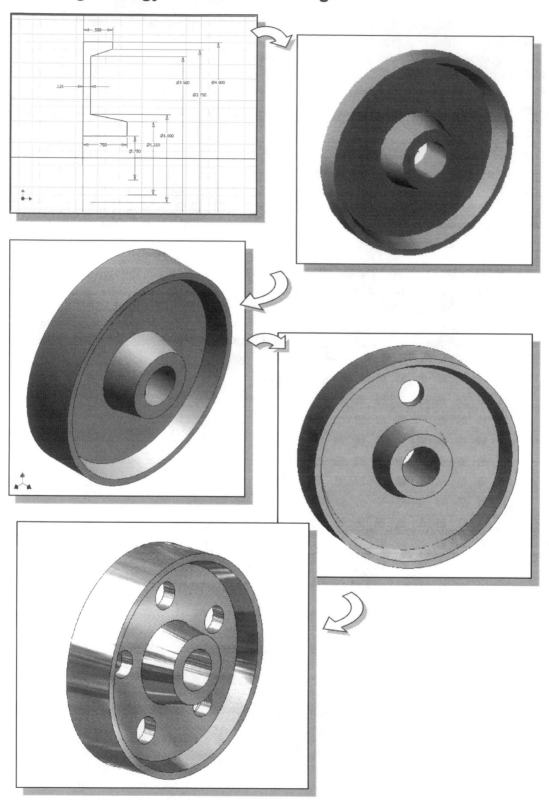

Starting Autodesk Inventor

1. Select the **Autodesk Inventor** option on the *Start* menu or select the **Autodesk Inventor** icon on the desktop to start Autodesk Inventor. The Autodesk Inventor main window will appear on the screen.

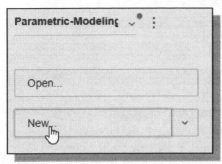

2. Select the **New File** icon with a single click of the left-mouse-button.

3. Select the **English** units set and in the *Part File* area, select **Standard(in).ipt**.

4. Click **Create** in the *New File* dialog box to start a new model.

Set Up the Display of the Sketch Plane

1. In the *Part Browser* window, click on the [**+**] symbol in front of the **Origin** feature to display more information on the feature.

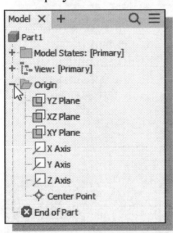

❖ In the *browser* window, notice a new part name appeared with seven work features established. The seven work features include three *work planes*, three *work axes* and a *work point*. By default, the three work planes and work axes are aligned to the **world coordinate system** and the work point is aligned to the *origin* of the **world coordinate system**.

2. Inside the *browser* window, select the three work planes by holding down the **[Ctrl]** key and clicking with the left-mouse-button.

3. Click the right-mouse-button on any of the work features to display the option menu. Click on **Visibility** to toggle *ON* the display of the plane.

Creating the 2D Sketch for the Base feature

1. In the *Sketch* toolbar select the **Start 2D Sketch** command by left-clicking once on the icon.

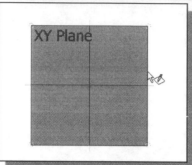

2. In the *Status Bar* area, the message "*Select plane to create sketch or an existing sketch to edit.*" is displayed. Select the **XY Plane** by clicking on the plane inside the graphics window, as shown.

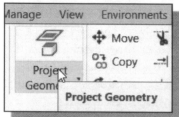

3. Select the **Project Geometry** command in the *2D Sketch* panel. The Project Geometry command allows us to project existing features onto the active sketching plane. Left-click once on the icon to activate the Project Geometry command.

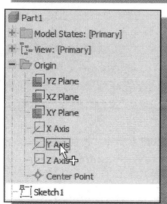

4. In the *Status Bar* area, the message "*Select edge, vertex, work geometry or sketch geometry to project.*" is displayed. Inside the *browser* window, select the **X-axis** and **Y-axis** to project these entities onto the sketching plane.

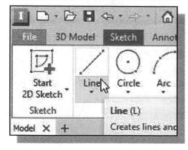

5. Select the **Line** option in the *2D Sketch* panel. A *Help-tip box* appears next to the cursor and a brief description of the command is displayed at the bottom of the drawing screen: "*Creates Straight lines and arcs.*"

6. Create a closed-region sketch with the starting point aligned to the projected Y-axis as shown below. (Note that the *Pulley* design is symmetrical about a horizontal axis as well as a vertical axis, which allows us to simplify the 2D sketch as shown below.)

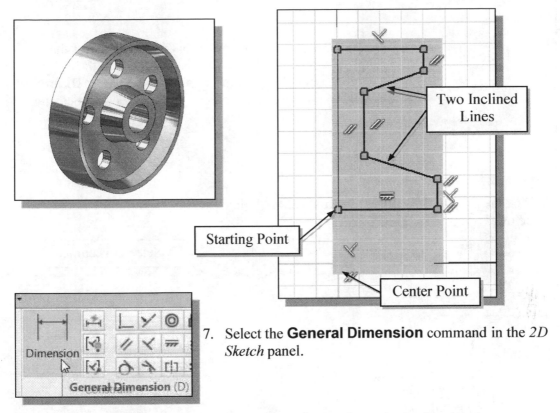

Two Inclined Lines

Starting Point

Center Point

7. Select the **General Dimension** command in the *2D Sketch* panel.

8. Pick the **X-axis** as the first entity to dimension as shown in the figure below.

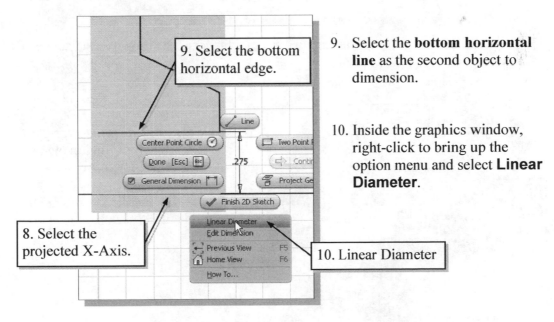

9. Select the bottom horizontal edge.

8. Select the projected X-Axis.

9. Select the **bottom horizontal line** as the second object to dimension.

10. Inside the graphics window, right-click to bring up the option menu and select **Linear Diameter**.

10. Linear Diameter

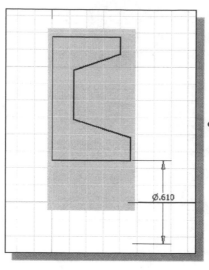

11. Place the dimension text to the right side of the sketch.

• To create a dimension that will account for the symmetrical nature of the design, pick the axis of symmetry, pick the entity, select **Linear Diameter** in the option menu, and then place the dimension.

12. Pick the projected **X-axis** as the first entity to dimension as shown in the figure below.

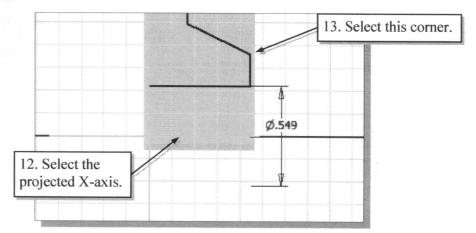

13. Select this corner.

12. Select the projected X-axis.

13. Select the **corner point** as the second object to dimension as shown in the above figure.

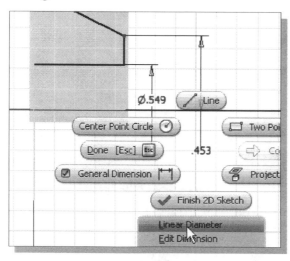

14. Inside the graphics window, right-click to bring up the option menu and select **Linear Diameter**.

15. Place the dimension text toward the right side of the sketch.

16. On your own, create and adjust the vertical size/location dimensions as shown below. (Hint: Modify the larger dimensions first.)

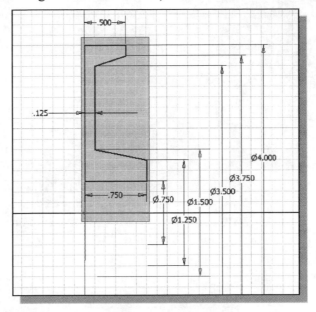

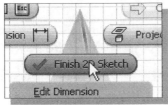

17. Inside the graphics window, click once with the right-mouse-button to display the option menu. Select **Finish 2D Sketch** in the pop-up menu to end the Sketch option.

➢ On your own, use the **3D-Rotate** command to confirm the completed sketch and dimensions are on a 2D plane.

18. Select **Home View** in the ViewCube to adjust the display of the 2D sketch to the isometric view.

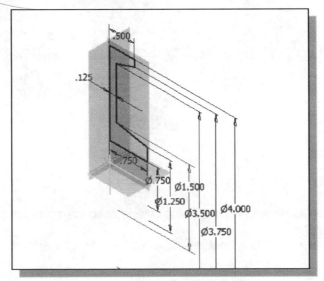

Create the Revolved Feature

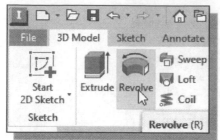

1. In the *Create Feature* panel select the **Revolve** command by clicking the left-mouse-button on the icon.

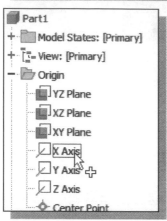

2. In the *Revolve* dialog box, the Axis button is activated indicating Autodesk Inventor expects us to select the revolution axis for the revolved feature. Select **X-Axis** as the axis of rotation in the *browser* window as shown.

3. In the *Revolve* dialog box, confirm the termination *Extents* option is set to **Full** as shown.

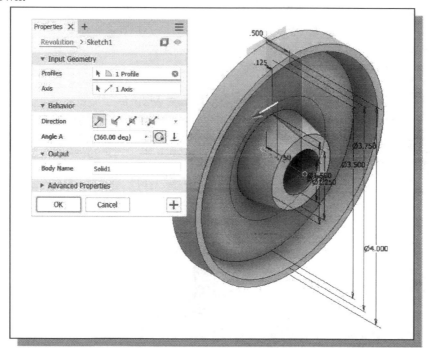

4. Click on the **OK** button to accept the settings and create the revolved feature.

Mirroring Features

In Autodesk Inventor, features can be mirrored to create and maintain complex symmetrical features. We can mirror a feature about a work plane or a specified surface. We can create a mirrored feature while maintaining the original parametric definitions, which can be quite useful in creating symmetrical features. For example, we can create one quadrant of a feature, and then mirror it twice to create a solid with four identical quadrants.

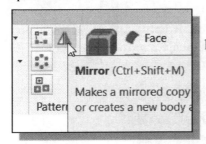

1. In the *Pattern* panel, select the **Mirror Feature** command by left-clicking on the icon.

2. In the *Mirror Pattern* dialog box, the **Features** button is activated. Autodesk Inventor expects us to select features to be mirrored. In the prompt area, the message *"Select feature to pattern"* is displayed. Select **any edge** of the 3D base feature.

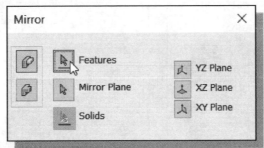

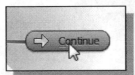

3. Inside the graphics window, **right-click** to bring up the **option menu**.

4. Select **Continue** in the option list to proceed with the Mirror Feature command.

5. In the *Mirror Pattern* dialog box, we can also activate the **Mirror Plane** option by clicking on the icon. Autodesk Inventor expects us to select a planar surface about which to mirror.

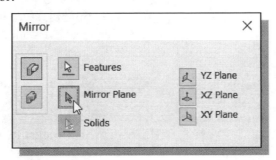

6. On your own, use the ViewCube or the 3D-Rotate function key [F4] to dynamically rotate the solid model so that we are viewing the back surface as shown.

7. Select the surface as shown to the left as the planar surface about which to mirror.

8. Click on the **OK** button to accept the settings and create a mirrored feature.

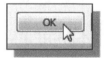

9. On your own, use the ViewCube or the 3D-Rotate function key [F4] to dynamically rotate the solid model and view the resulting solid.

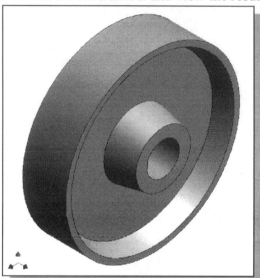   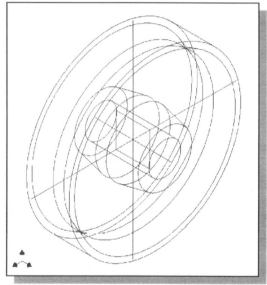

➢ Now is a good time to save the model (quick-key: [Ctrl] + [S]). It is a good habit to save your model periodically, just in case something might go wrong while you are working on it. You should also save the model after you have completed any major constructions.

10. Inside the graphics window, right-click to bring up the option menu.

11. Select **Home View** in the option list to adjust the display of the 2D sketch on the screen.

Create a Pattern Leader Using Construction Geometry

In Autodesk Inventor, we can also use **construction geometry** to help define, constrain, and dimension the required geometry. **Construction geometry** can be lines, arcs, and circles that are used to line up or define other geometry but are not themselves used as the shape geometry of the model. When profiling the rough sketch, Autodesk Inventor will separate the construction geometry from the other entities and treat them as construction entities. Construction geometry can be dimensioned and constrained just like any other profile geometry. When the profile is turned into a 3D feature, the construction geometry remains in the sketch definition but does not show in the 3D model. Using construction geometry in profiles may mean fewer constraints and dimensions are needed to control the size and shape of geometric sketches. We will illustrate the use of the construction geometry to create a cut feature.

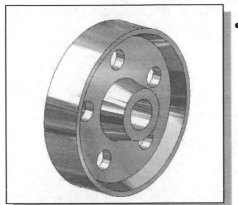

- The *Pulley* design requires the placement of five identical holes on the base solid. Instead of creating the five holes one at a time, we can simplify the creation of these holes by using the Pattern command, which allows us to create duplicate features. Prior to using the Pattern command, we will first create a *pattern leader*, which is a regular extruded feature.

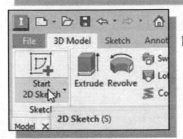

1. In the *Sketch* toolbar select the **Start 2D Sketch** command by left-clicking once on the icon.

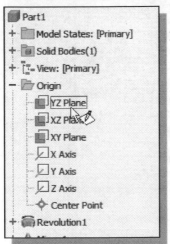

2. In the *Status Bar* area, the message "*Select face, work plane, sketch or sketch geometry.*" is displayed. Pick the YZ Plane, inside the *browser* window, as shown.

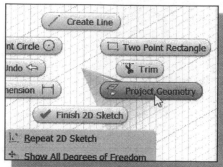

3. Inside the graphics window, **right-click** to bring up the option menu.

4. Select **Project Geometry** in the option menu. The **Project Geometry** command allows us to project existing geometry to the active sketching plane.

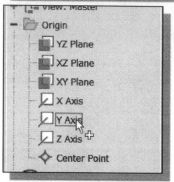

5. In the *Status Bar* area, the message "*Select edge, vertex, work geometry or sketch geometry to project.*" is displayed. Inside the *browser* window, select the **Y-axis** and **Z-axis** to project these entities onto the sketching plane.

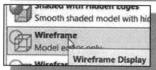

6. On your own, set the display to *wireframe* by clicking on the **Wireframe Display** icon in the *View* panel.

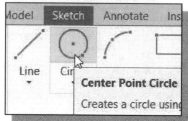

7. Select the **Center Point Circle** command by clicking once with the left-mouse-button on the icon in the *2D Sketch* panel.

8. Create a circle of arbitrary size as shown below.

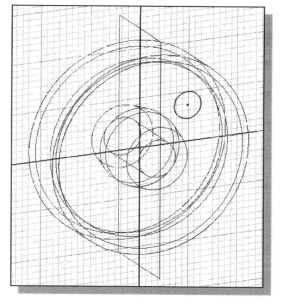

9. In the *Standard* toolbar area, click on the **Look At** button.

10. Select the circle we just created to orient the display of the sketching plane on the screen.

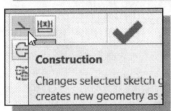

11. Select the **Line** command in the *2D Sketch* panel. A brief description of the command is displayed at the bottom of the drawing screen: *"Creates Straight lines and arcs."*

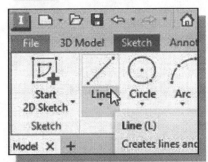

12. Set the *Style* option to **Construction** as shown.

13. Create a *construction line* by connecting from the center of the circle we just created to the projected center point (***origin***) at the center of the 3D model as shown below.

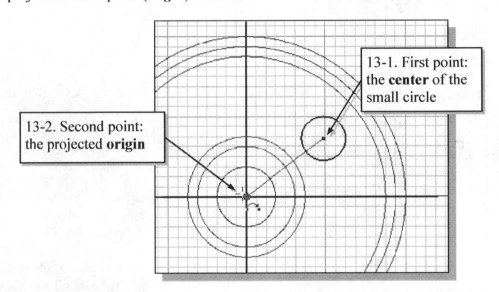

13-1. First point: the **center** of the small circle

13-2. Second point: the projected **origin**

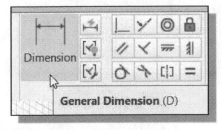

14. Select the **General Dimension** command in the *Constrain* panel.

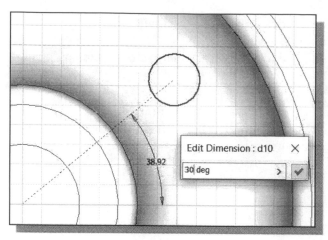

15. Pick the **horizontal axis** as the first entity to dimension as shown in the figure.

16. Select the **construction line** as the second object to dimension.

17. Place the dimension text to the right of the model as shown.

18. On your own, set the angle dimension to **30** as shown in the above figure.

➢ Note that the small circle moves as the location of the construction line is adjusted by the *angle dimension* we created.

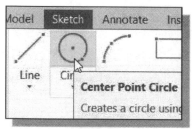

19. Select the **Center Point Circle** command by clicking once with the left-mouse-button on the icon in the *Draw* panel.

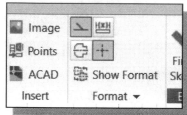

20. Confirm the *Style* option is still set to **Construction**.

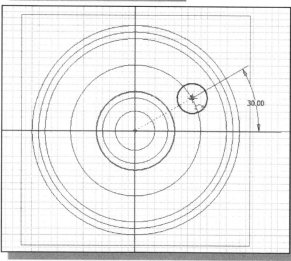

21. Create a *construction circle* by placing the center at the projected center point (*origin*).

22. Pick the **center** of the small circle to set the size of the construction circle as shown.

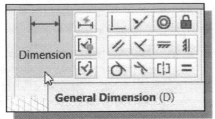

23. Select the **General Dimension** command in the *Constrain* panel.

24. On your own, create the two diameter dimensions, **.5** and **2.5**, as shown below.

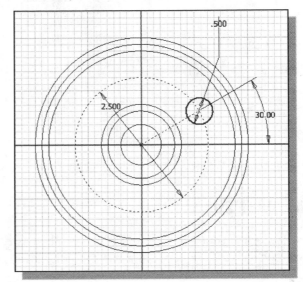

25. Inside the graphics window, click once with the right-mouse-button to display the option menu. Select **Finish Sketch** in the pop-up menu to end the Sketch option.

26. In the *Create Features* panel select the **Extrude** command by left-clicking on the icon.

27. Pick the inside region of the circle to set up the profile of the extrusion.

28. Inside the *Extrude* dialog box, select the **Cut** operation for **Symmetric (both directions)**, and set the *Extents* to **All** as shown.

29. Click on the **OK** button to accept the settings and create the cut feature.

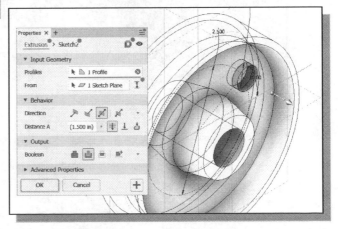

30. On your own, adjust the angle dimension applied to the construction line of the cut feature to **90** and observe the effect of the adjustment.

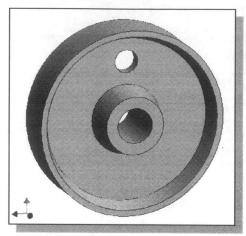
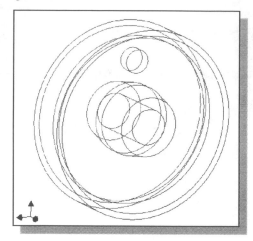

Circular Pattern

In Autodesk Inventor, existing features can be easily duplicated. The **Pattern** command allows us to create both rectangular and polar arrays of features. The patterned features are parametrically linked to the original feature; any modifications to the original feature are also reflected in the arrayed features.

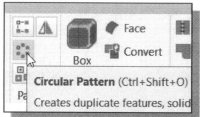

1. In the *Pattern Features* panel, select the **Circular Pattern** command by left-clicking once on the icon.

2. The message "*Select Feature to be arrayed:*" is displayed in the command prompt window. Select the **circular cut feature** when it is highlighted as shown.

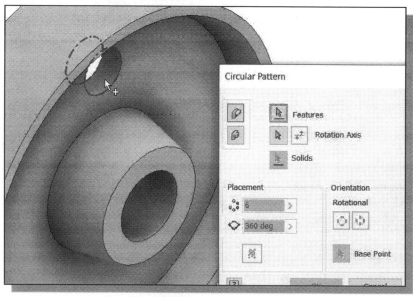

3. Inside the graphics window, **right-click** to bring up the **option menu**.

4. Select **Continue** in the option list to proceed with the Circular Pattern command.

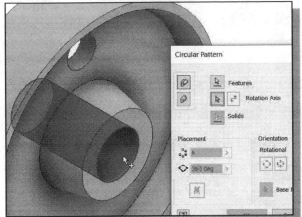

5. Autodesk Inventor expects us to select an axis to pattern about. In the prompt area, the message *"Define the axis of revolution"* is displayed. Select the **X-Axis** in the *browser* window or the inside cylindrical surface in the graphics window.

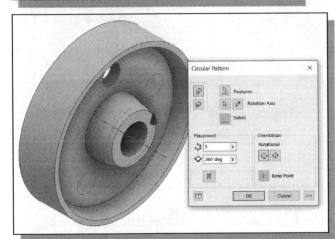

6. In the *Circular Pattern* dialog box, enter **5** in the *Count* box and **360** in the *Angle* box as shown. (Note the different options available for the Circular Pattern command.)

7. Click on the **OK** button to accept the settings and create the *circular pattern*.

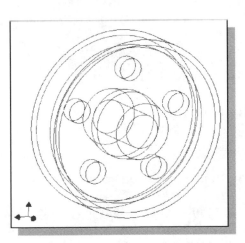

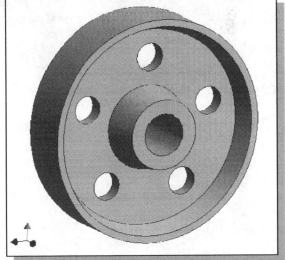

Examine the Design Parameters

1. In the *Manage* tab select the **Parameters** command by left-clicking once on the icon. The *Parameters* pop-up window appears.

2. Scroll down to the bottom of the list and locate the two dimensions used to create the circular pattern (the d15 & d16 dimensions in the below figure).

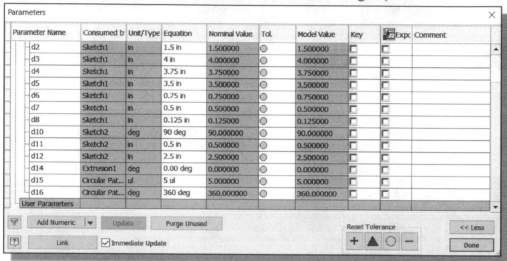

Parameter Name	Consumed b	Unit/Type	Equation	Nominal Value	Tol.	Model Value	Key	Expc	Comment
d2	Sketch1	in	1.5 in	1.500000	○	1.500000	☐	☐	
d3	Sketch1	in	4 in	4.000000	○	4.000000	☐	☐	
d4	Sketch1	in	3.75 in	3.750000	○	3.750000	☐	☐	
d5	Sketch1	in	3.5 in	3.500000	○	3.500000	☐	☐	
d6	Sketch1	in	0.75 in	0.750000	○	0.750000	☐	☐	
d7	Sketch1	in	0.5 in	0.500000	○	0.500000	☐	☐	
d8	Sketch1	in	0.125 in	0.125000	○	0.125000	☐	☐	
d10	Sketch2	deg	90 deg	90.000000	○	90.000000	☐	☐	
d11	Sketch2	in	0.5 in	0.500000	○	0.500000	☐	☐	
d12	Sketch2	in	2.5 in	2.500000	○	2.500000	☐	☐	
d14	Extrusion1	deg	0.00 deg	0.000000	○	0.000000	☐	☐	
d15	Circular Pat...	ul	5 ul	5.000000	○	5.000000	☐	☐	
d16	Circular Pat...	deg	360 deg	360.000000	○	360.000000	☐	☐	
User Parameters									

3. Click on the **Done** button to accept the settings.

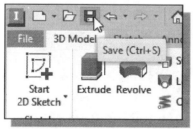

4. Select **Save** in the *Standard* toolbar; we can also use the "**Ctrl-S**" combination (press down the [Ctrl] key and hit the [S] key once) to save the part as ***Pulley*** in the **Chapter11** folder.

Drawing Mode – Defining a New Border and Title Block

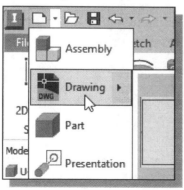

1. Click on the **drop-down arrow** next to the **New File** button in the *Quick Access* toolbar area to display the available new file options.

2. Select **Drawing** from the drop-down list.

➢   Note that a new graphics window appears on the screen. We can switch between the solid model and the drawing by clicking the corresponding graphics window.

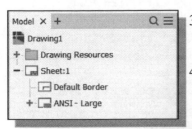

3. Inside the drawing *browser* window, click on the [**+**] symbol in front of **Sheet1** to display the settings.

4. On your own, **delete** the **Default Border** and the **ANSI-Large** *title block*. (Hint: right-click on the names to bring up the option menu.)

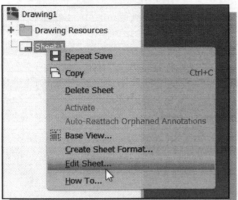

5. Inside the drawing *browser* window, **right-click** on **Sheet1** to bring up the option menu.

6. Select **Edit Sheet** in the option list.

7. In the *Edit Sheet* dialog box, set the sheet size to **A** (8.5 × 11).

8. Confirm the page *Orientation* is set to **Landscape** and click on the **OK** button to accept the settings.

9. In the *Ribbon* toolbar panel, select **Manage → Define New Border**.

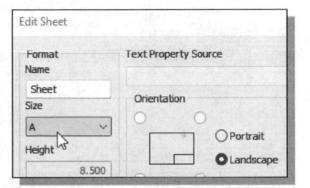

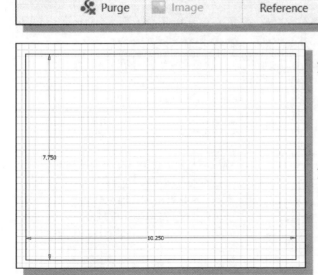

10. On your own, create a rectangle (**7.75 × 10.25**) using the **Two Point Rectangle** command and the **General Dimension** command.

11. Drag and drop the edges of the rectangle and position the rectangle to the center of the sheet.

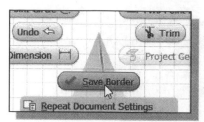

12. Inside the graphics window, **right-click** to bring up the option menu.

13. Select **Save Border** in the option list.

14. In the *Border* dialog box, enter **A-Size** as the new border name.

15. Click on the **Save** button to end the Define New Border command.

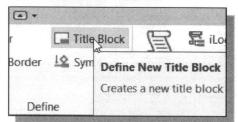

16. Inside the *browser* window, expand the **Drawing Resources** list by clicking on the [**+**] symbol.

17. **Double-click** on the **A-size** border, the border we just created, to place the border in the current drawing. Note that none of the dimensions used to construct the border are displayed.

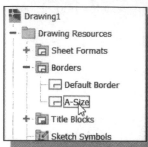

18. In the Define Toolbar, select **[Define]** → **[Title Block]**.

19. On your own, create a title block using the **Two Point Rectangle**, **Line**, **General Dimension**, and **Text** commands.

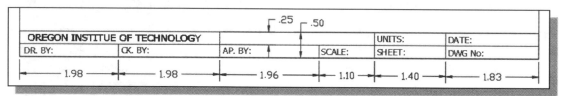

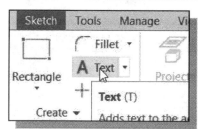

20. In the *Create toolbar*, activate the **Text** command as shown.

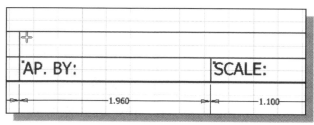

21. Click near the top left corner of the *Title Box*, the second box of the top section in the title block we just created, as shown.

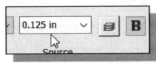

22. In the *Format Text dialog* box, adjust the text size to **0.125** as shown.

23. In the *Format Text dialog* box, select **Standard iProperties** and **Title** as the *Type* and *Property* parameter as shown.

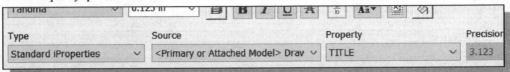

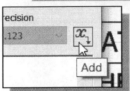

24. Click on the **Add Text Parameter** icon to insert the associated iProperty information to the title block.

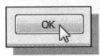

25. Click on the **OK** button to accept the settings.

26. On your own, repeat the above process to add the **Standard iProperty - Creation Date** to the *Date Box* as shown.

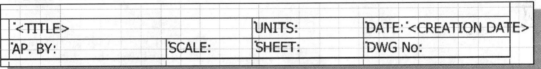

27. On your own, repeat the above steps and add in the associated Standard iProperties for the bottom line of the title block as shown.

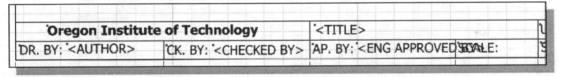

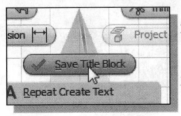

28. Inside the *graphics window*, right-click to bring up the option menu.

29. Select **Save Title Block** in the option list.

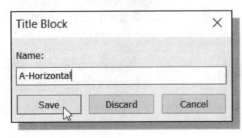

30. In the *Title Block* dialog box, enter **A-Horizontal** as the new title block name.

31. Click on the **Save** button to end the Define New Title Block command.

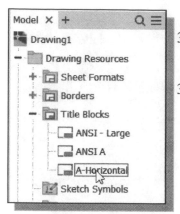

32. Inside the *browser* window, expand the **Title Blocks** list by clicking on the [**+**] symbol.

33. Double-click on the **A-Horizontal** title block, the title block we just created, to place it into the current drawing.

Create a Drawing Template

In Autodesk Inventor, each new drawing is created from a template. During the installation of Autodesk Inventor, a default drafting standard was selected which sets the default template used to create drawings. We can use this template or another predefined template, modify one of the predefined templates, or create our own templates to enforce drafting standards and other conventions. Any drawing file can be used as a template; a drawing file becomes a template when it is saved in the *Templates* folder. Once the template is saved, we can create a new drawing file using the new template.

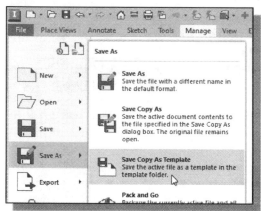

1. Select **Save Copy As Template** in the *File Menu*; we will save the current drawing as a template file.

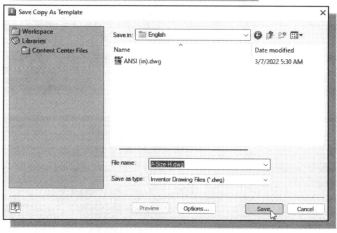

2. In the *Save As* dialog box, switch to the *Public User - Inventor* folder: **Inventor 2024 → Templates → en-US→ English** directory.

3. Enter **A-Size-H.dwg** as the template filename.

4. Click on the **Save** button to create a drawing template file.

Create the Necessary Views

1. Click on **Base View** in the *Place Views* panel to create a base view.

2. In the *Drawing View* dialog box, set the scale to **3 : 4** and select the *left* view as shown in the figure below.

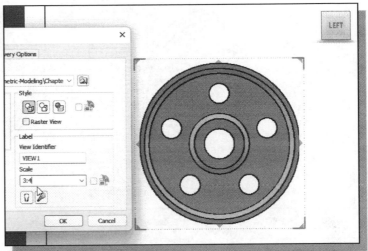

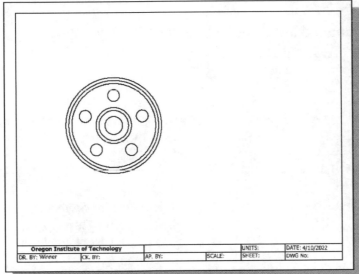

3. Move the cursor inside the graphics window and place the **base view** toward the left side of the border as shown. (If necessary, drag the *Drawing View* dialog box to another location on the screen.)

4. Click on the **OK** button to accept the setting.

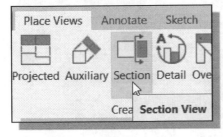

5. Click on the **Section View** button in the *Place Views* panel to create a section view.

6. Click on the **base view** to select the view where the section line is to be created.

7. Inside the graphics window, align the cursor to the center of the base view and create the vertical cutting plane line as shown.

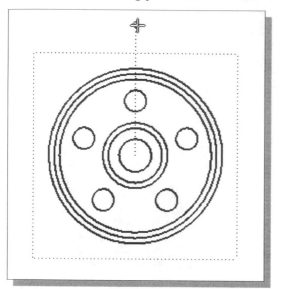
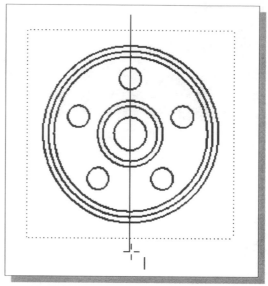

8. Inside the *graphics window*, right-click once to bring up the **option menu**.

9. Select **Continue** to proceed with the Section View command.

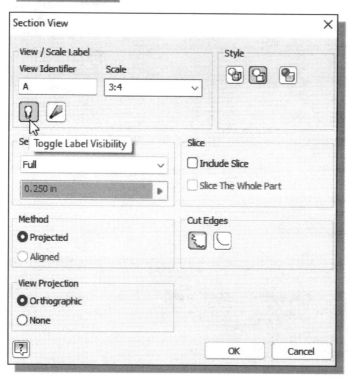

10. Inside the *Section View* dialog box, toggle *OFF* the **Label Visibility** option as shown.

11. Set the *Display Style* to **Hidden Line Removed**, as shown. (**DO NOT** click on the **OK** button.)

12. Next, Autodesk Inventor expects us to place the projected section. Select a location that is toward the right side of the base view as shown in the figure.

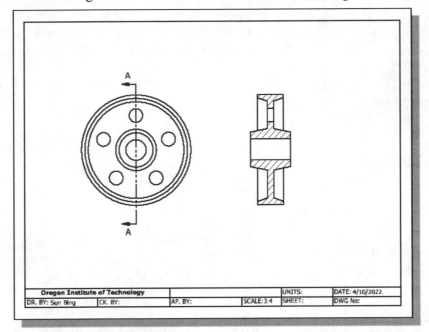

13. On your own, use the **Projected View** option and create an *isometric* **view** (Iso Top Left – ¾ scale) of the design and place the view toward the right side of the section view as shown below.

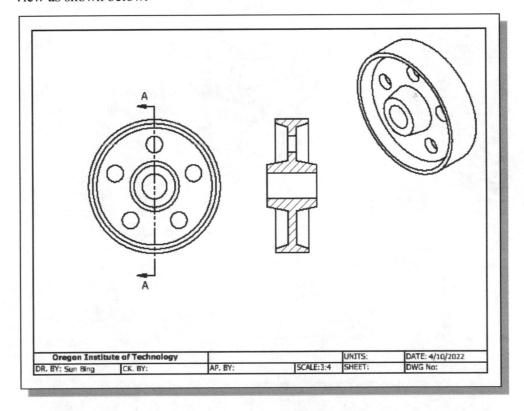

Retrieve Model Annotations – Features Option

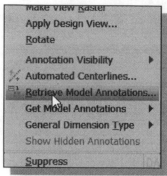

1. Move the cursor on top of the *section* **view** of the model and watch for the box around the entire view indicating the view is selectable as shown in the figure.

2. Inside the *graphics window*, **right-click** once to bring up the option menu.

3. Select **Retrieve Model Annotations** to display the parametric dimensions used to create the model.

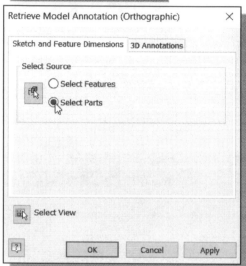

4. In the *Retrieve Model Annotations* dialog box, set the *Select Source* option to **Select Parts** as shown.

➤ The dimensions used to create the revolved feature are now displayed in the section view.

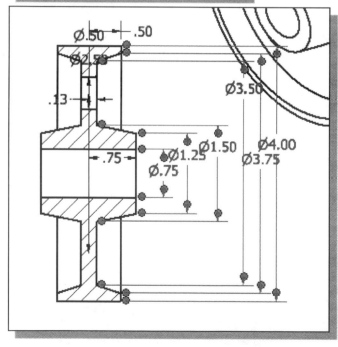

5. On your own, select the desired **model dimensions** to be displayed.

6. Click on the **Apply** button and notice the selected dimensions changed color indicating they have been retrieved.

7. Note that the **Select View** button is activated; click on the Base view to continue with the Retrieve Model Annotations command.

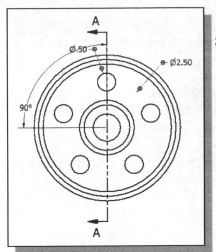

8. On your own, retrieve the model dimensions on the base view as shown.

9. On your own, reposition the views and dimensions as shown in the figure.

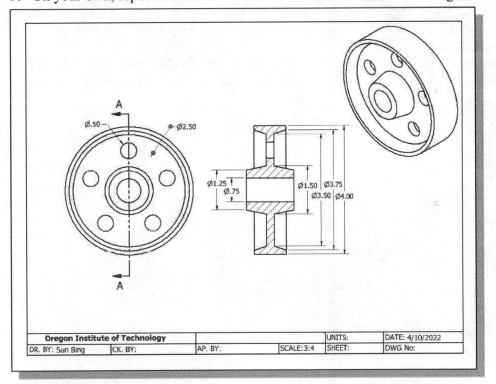

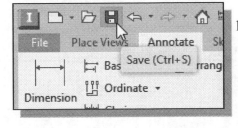

10. On your own, use the **Save** command and save the drawing as ***pulley.dwg***.

Associative Functionality – A Design Change

Autodesk Inventor's *associative functionality* allows us to change the design at any level, and the system reflects the changes at all levels automatically. We will illustrate the associative functionality by changing the circular pattern from five holes to six holes.

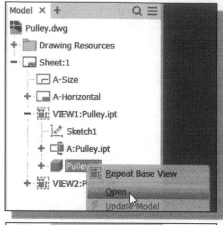

1. Inside the *Model Tree* window, below the VIEW1 list, right-click on the **Pulley.ipt** part name to bring up the option menu.

2. Select **Open** in the pop-up menu to switch to the associated solid model.

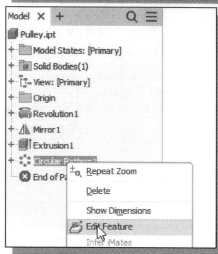

3. Inside the *Model Tree* window, right-click on the CircularPattern1 feature to bring up the option menu.

4. Select **Edit Feature** in the pop-up menu to bring up the associated feature option.

5. In the *Circular Pattern* dialog box, change the number to **6** as shown.

6. Click on the **OK** button to accept the setting.

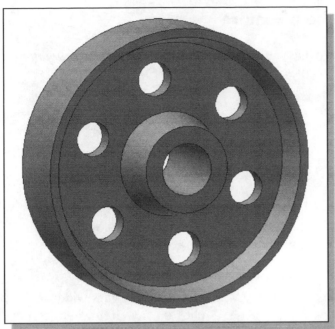

❖ The solid model is updated showing the 6 equally spaced holes as shown.

7. Switch back to the *Pulley* drawing and notice the drawing is also updated.

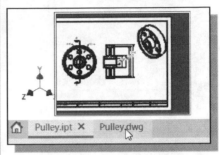

❖ Notice, in the *Pulley* drawing, the circular pattern is also updated automatically in all views.

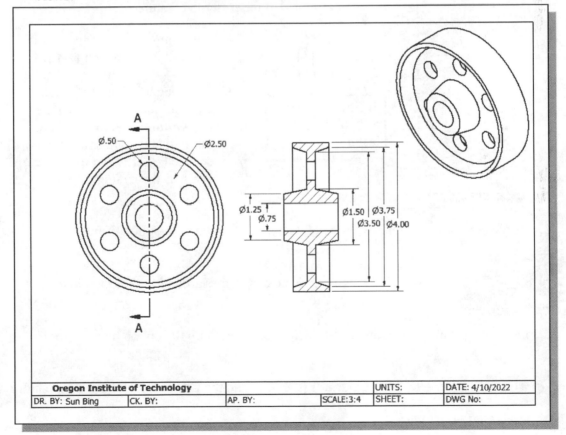

Add Centerlines to the Pattern Feature

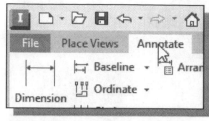

1. On your own, switch to the *Drawing Annotation* panel.

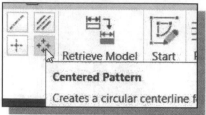

2. Select **Centered Pattern** from the symbol toolbar.

➤ The **Centered Pattern** option allows us to add centerlines to a patterned feature.

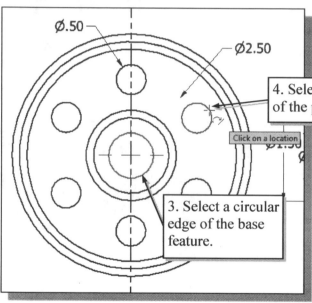

Ø.50

Ø2.50

4. Select any circular edge of the patterned feature.

Click on a location

3. Select a circular edge of the base feature.

3. Inside the graphics window, click on any **circular edge** of the **base feature**.

4. Select any circular edge that is part of the **patterned feature**.

5. Continue to select the circular edges of the patterned features, in a **counterclockwise** manner, until all patterned items are selected. (Select the **first circle** again as the ending item.)

6. Inside the graphics window, **right-click** once to bring up the option menu.

7. Select **Create** to create the centerlines around the selected items.

8. Inside the graphics window, **right-click** once to bring up the option menu.

9. Select **Cancel [ESC]** to end the Centered Pattern command.

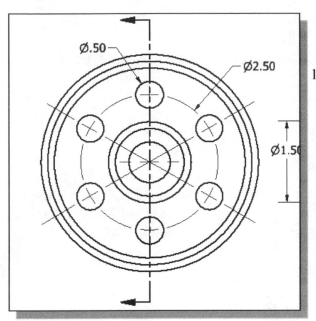

10. On your own, extend the segments of the centerlines so that they pass through the center of the base feature.

Complete the Drawing

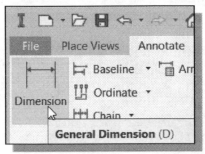

1. On your own, use the **General Dimension** command to create the angle dimension as shown in the figure below.

2. For the diameter of the circular pattern, set the options so that the **Arrowheads Inside** option is switched *ON* and the **Single Dimension Line** option is switched *OFF*.

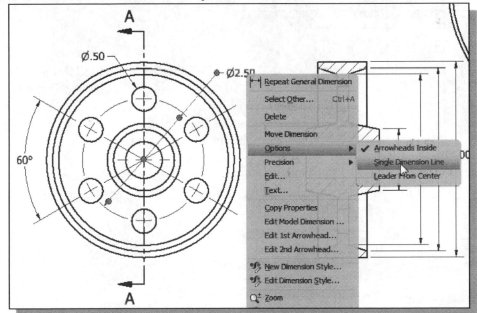

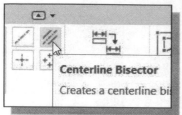

3. Select **Centerline Bisector** from the *Symbols* toolbar.

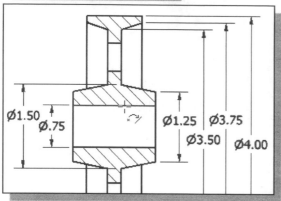

4. Inside the graphics window, click on the **top edge** of the 0.75 diameter hole as shown in the figure.

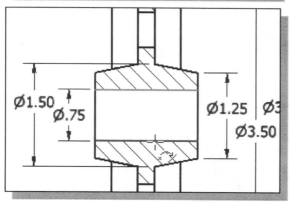

5. Inside the graphics window, click on the **bottom edge** of the 0.75 diameter hole as shown.

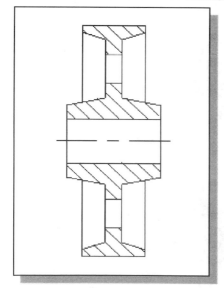

➢ Notice that a centerline is created through the center of the section view as shown in the figure.

6. Repeat the above step and create the centerlines through the other two holes.

7. On your own, complete the drawing and the title block so that the drawing appears as shown on the next page.

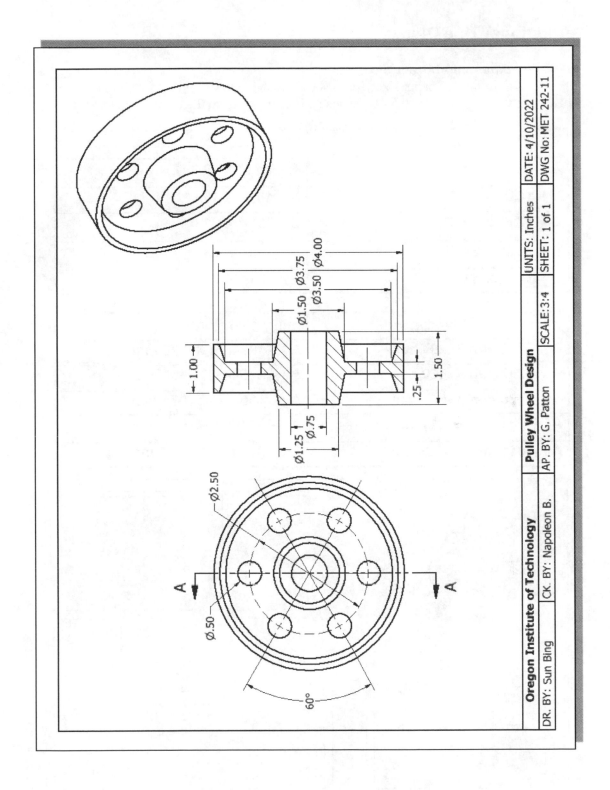

Oregon Institute of Technology | Pulley Wheel Design | UNITS: Inches | DATE: 4/10/2022

DR. BY: Sun Bing | CK. BY: Napoleon B. | AP. BY: G. Patton | SCALE: 3:4 | SHEET: 1 of 1 | DWG No: MET 242-11

Additional Title Blocks

Drawing Paper and Border Sizes

The standard drawing paper sizes are as shown in the below tables. The edges of the title block border are generally 0.5 ~ 1 inches or 10~20 mm from the edges of the paper.

American National Standard	Suggested Border Size
A – 8.5″ X 11.0″	A – 7.75″ X 10.25″
B – 11.0″ X 17.0″	B – 10.0″ X 16.0″
C – 17.0″ X 22.0″	C – 16.0″ X 21.0″
D – 22.0″ X 34.0″	D – 21.0″ X 33.0″
E – 34.0″ X 44.0″	E – 33.0″ X 43.0″

International Standard	Suggested Border Size
A4 – 210 mm X 297 mm	A4 – 190 mm X 276 mm
A3 – 297 mm X 420 mm	A3 – 275 mm X 400 mm
A2 – 420 mm X 594 mm	A2 – 400 mm X 574 mm
A1 – 594 mm X 841 mm	A1 – 574 mm X 820 mm
A0 – 841 mm X 1189 mm	A0 – 820 mm X 1168 mm

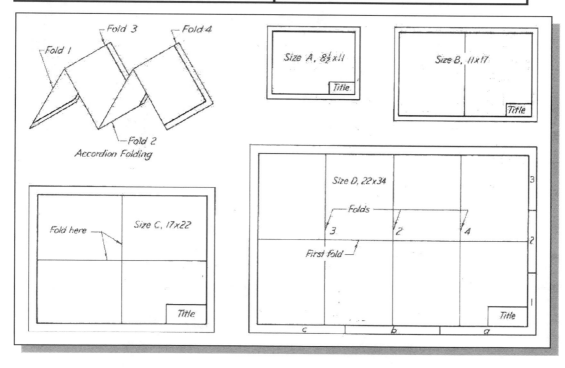

- **English Title Block** (For A size paper, dimensions are in inches.)

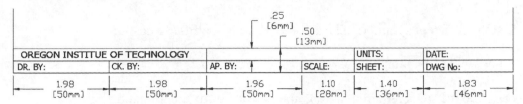

- **Metric Title Block** (For A4 size paper, dimensions are in mm.)

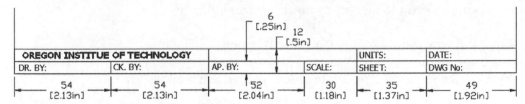

- **English Title Block** (Dimensions are in inches.)

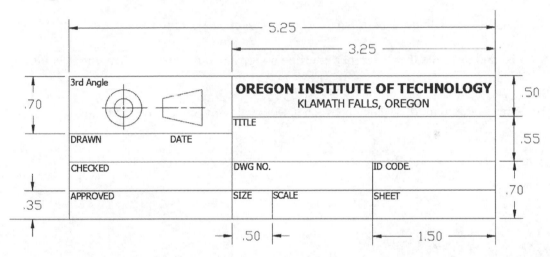

Metric Title Block (Dimensions are in mm.)

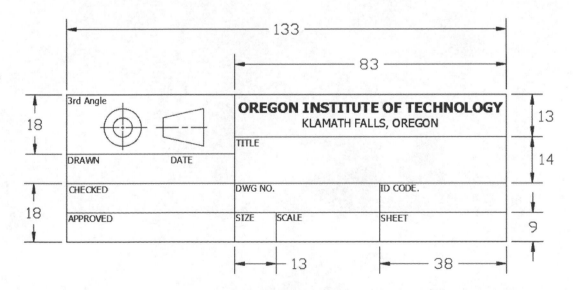

Review Questions: (Time: 25 minutes)

1. List the different symmetrical features created in the *Pulley* design.

2. What are the advantages of using a *drawing template*?

3. Describe the steps required in using the Mirror Feature command.

4. Why is it important to identify symmetrical features in designs?

5. When and why should we use the Pattern option?

6. What are the required elements in order to generate a sectional view?

7. How do we create a *Linear Diameter dimension* for a revolved feature?

8. What is the difference between *construction geometry* and *normal geometry*?

9. List and describe the different centerline options available in the *Drawing Annotation* panel.

10. What is the main difference between a sectional view and a projected view?

Exercises: Create the Solid models and the associated 2D drawings and save the exercises in the Chapter12 folder. (Time: 180 minutes.)

1. **Shaft Support** (Dimensions are in inches.)

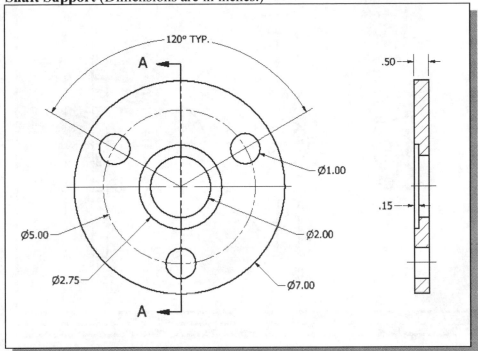

2. **Ratchet Plate** (Dimensions are in inches. Thickness: 0.125 inch.)

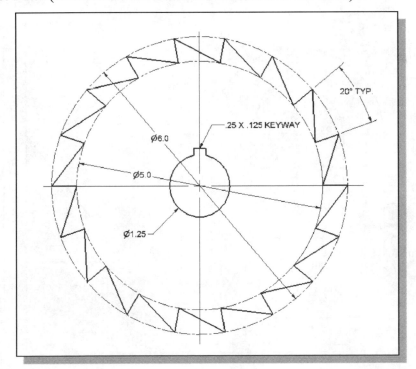

3. **Geneva Wheel** (Dimensions are in inches.)

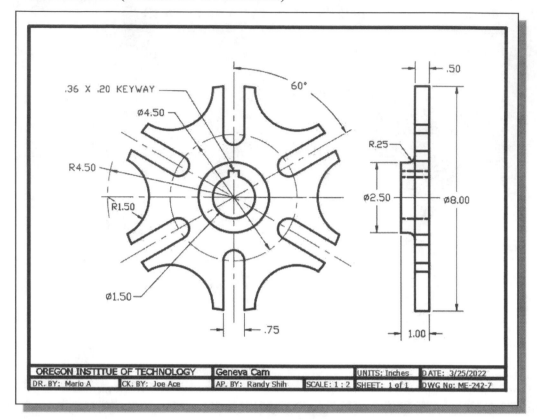

4. **Support Mount** (Dimensions are in inches.)

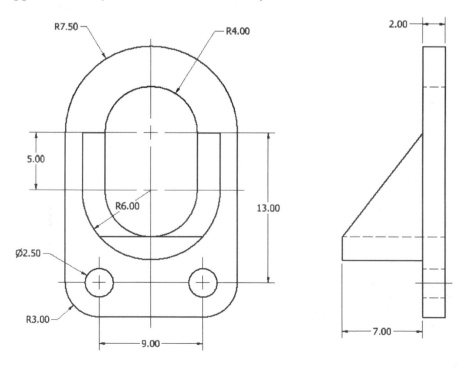

5. **Hub** (Dimensions are in inches.)

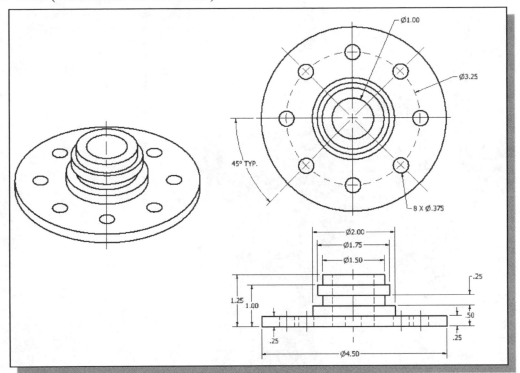

6. **Bearing Seat** (Dimensions are in inches.)

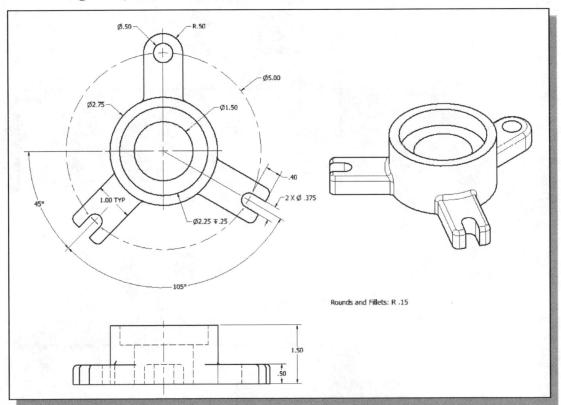

7. **Pulley Wheel** (Dimensions are in inches.)

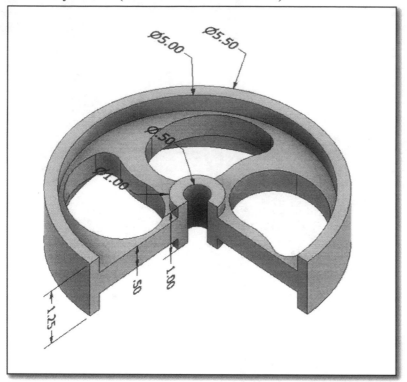

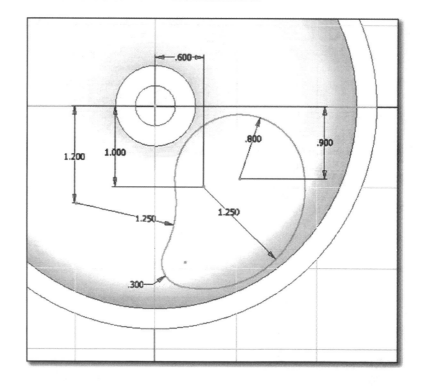

Chapter 13
Threads and Fasteners

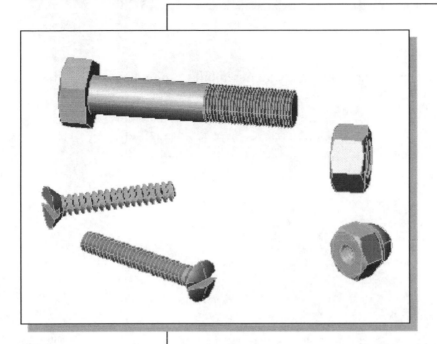

Learning Objectives

- ♦ **Understand the Concepts and the Use of Work Features**
- ♦ **Use the Different Options to Create Work Features**
- ♦ **Create Auxiliary Views in 2D Drawing Mode**
- ♦ **Create and Adjust Centerlines**
- ♦ **Create Shaded Images in the 2D Drawing Mode**

Autodesk Inventor Certified User Exam Objectives Coverage

Parametric Modeling Basics

Section 5: Assemblies

Objectives: Creating Assemblies, Viewing Assemblies, Animation Assemblies, Adaptive Features, Parts, and Subassemblies

Autodesk Inventor Certified User Reference Guide

Introduction

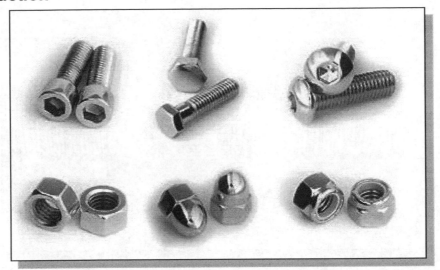

Threads and fasteners are the principal means of assembling parts. Screw threads occur in one form or another on practically all engineering products. Screw threads are designed for many different purposes; the three basic functionalities are to hold parts together, to transmit power and for use as adjustment of locations of parts.

The earliest records of the screw threads are found in the writings of Archimedes (278 to 212 B.C.), the mathematician who described several designs applying the screw principle. Screw thread was a commonly used element in the first century B.C. but was crudely made during that period of time. Machine production of screws started in England during the Industrial Revolution and the standardization of the screw threads was first proposed by Sir Joseph Whitworth in 1841. His system was generally adopted in England but not in the United States during the 1800s. The initial attempt to standardize screw threads in the United States came in 1864 with the adoption of the thread system designed by William Sellers. The "Sellers thread" fulfilled the need for a general-purpose thread; but it became inadequate with the coming of modern devices, such as automobiles and airplanes. Through the efforts of various engineering societies, the National Screw Thread Commission was authorized in 1918. The Unified Screw Thread, a compromise between the American and British systems, was established on November 18, 1948.

The Metric fastener standard was established in 1946, through the cooperative efforts of several organizations: The International Organization for Standardization (ISO), the Industrial Fasteners Institute (IFI) and the American National Standards Institute.

The following sections describe the general thread definitions; for a more complete description refer to the ANSI/ASME standards: B1.1, B1.7M, B1.13M, Y14.6 and Y14.6aM. (The letter M is for metric system.)

Screw-Thread Terminology

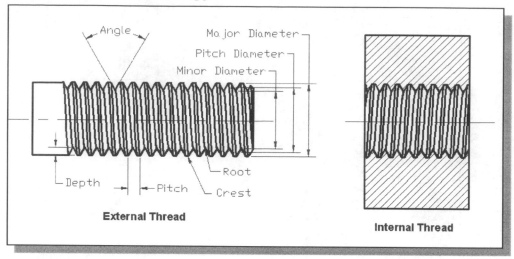

Screw Thread (Thread): A ridge of a uniform section in the form of a helix on the external or internal surface of a cylinder or cone.

External Thread: A thread on the external surface of a cylinder or cone.

Internal Thread: A thread on the internal surface of a cylinder or cone.

Major Diameter: The largest diameter of a screw thread.

Minor Diameter: The smallest diameter of a screw thread.

Pitch Diameter: The diameter of an imaginary cylinder, the surface of which cuts the thread where the width of the thread and groove are equal.

Crest: The outer edge or surface that joins the two sides of a thread.

Root: The bottom edge or surface that joins the sides of two adjacent threads.

Depth of Thread: The distance between crest and root measured normal to the axis.

Angle of Thread: The angle included between the two adjacent sides of the threads.

Pitch: The distance between corresponding points on adjacent thread forms measured parallel to the axis. This distance is a measure of the size of the thread form used, which is equal to 1 divided by the number of threads per inch.

Threads per Inch: The reciprocal of the pitch and the value specified to govern the size of the thread form.

Form of Thread: The profile (cross section) of a thread. The next section shows various forms.

Right-hand Thread (RH): A thread which advances into a nut when turned in a clockwise direction. Threads are always considered to be right-handed unless otherwise specified.

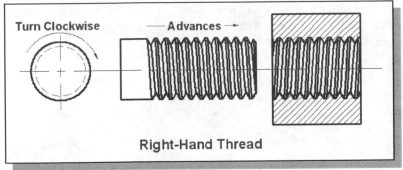

Left-hand Thread (LH): A thread which advances into a nut when turned in a counterclockwise and receding direction. All left-hand threads are labeled **LH**.

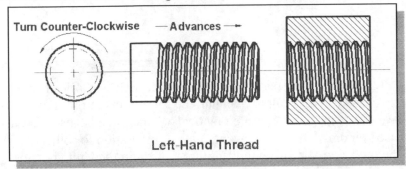

Lead: The distance a threaded part moves axially, with respect to a fixed mating part, in one complete revolution. See the definition and examples of multiple threads in the below figures.

Multiple Threads: A thread having two or more helical curves of the cylinder running side by side. For a single thread, lead and pitch are identical values; for a double thread, lead is twice the pitch; and for a triple thread, lead is three times the pitch. A multiple thread permits a more rapid advance without a larger thread form. The *slope line* can be used to aid the construction of multiple threads; it is the hypotenuse of a right triangle where the short side equals .5P for single threads, P for double threads and so on.

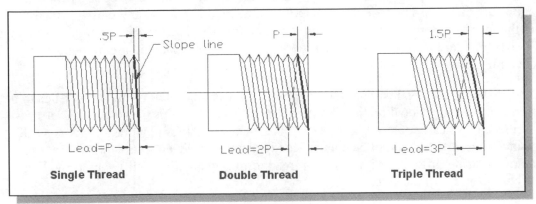

Thread Forms

Screw threads are used on fasteners to fasten parts together, on devices for making adjustments, and for the transmission of power and motion. For these different purposes, a number of thread forms are in use. In practical usage, clearance must be provided between the external and internal threads.

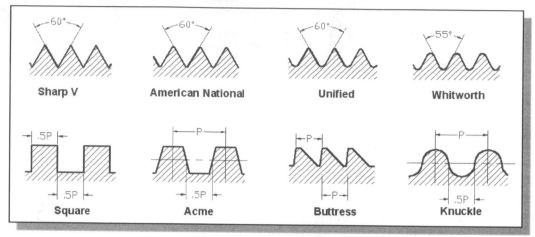

The most common screw thread form is the one with a symmetrical V-Profile. The **Sharp V** is rarely used now, because it is difficult to maintain the sharp roots in production. The form is of interest, however, as the basis of more practical V-type threads; also, because of its simplicity, it is used on drawings as a conventional representation for other (V-profile) threads. The modified V-profile standard thread form in the United States is the **American National**, which is commonly used in fasteners. The V-profile form is prevalent in the **Unified** Screw Thread (UN, UNC, UNF, UNEF) form as well as the ISO/Metric thread. The advantage of symmetrical threads is that they are easier to manufacture and inspect compared to non-symmetrical threads.

The **Unified Screw Thread** is the standard of the United States, Canada, and Great Britain and as such is known as the **Unified** thread. Note that while the crest may be flat or rounded, the root is rounded by design. These are typically used in general purpose fasteners.

The former British standard was the **Whitworth**, which has an angle of 55° with crests and roots rounded. The British Association Standard that uses an angle of 47°, measured in the metric system, is generally used for small threads. The French and the International Metric Standards have a form similar to the American National but are in the metric system.

The V shapes are not desirable for transmitting power since part of the thrust tends to be transmitted to the side direction. **Square thread** is used for this purpose as it transmits all the forces nearly parallel to the axis. The square thread form, while strong, is harder to manufacture. It also cannot be compensated for wear unlike an **Acme** thread. Because of manufacturing difficulties, the square thread form is generally modified by providing a slight taper (5°) to the sides.

The **Acme** is generally used in place of the square thread. It is stronger, more easily produced, and permits the use of a disengaging or split nut that cannot be used on a square thread.

The **buttress**, for transmitting power in one direction, has the efficiency of the square and the strength of the V thread.

The **knuckle** thread is especially suitable when threads are to be molded or rolled in sheet metal. It is commonly used on glass jars and in a shallow form on bases of ordinary light bulbs.

Internal threads are produced by cutting, while external threads are made by cutting or rolling. For internal threads, a hole is first drilled and then the threads are cut using a tap. The tap drill hole is a little bigger than the minor diameter of the mating external thread. The depth of the tap drill is generally deeper than the length of the threads. There are a few useless threads at the end of a normal tap. For external threads, a shaft that is the same size of the major diameter is cut using a die or on a lathe. Chamfers are generally cut to allow easy assembly.

Thread Representations

The true representation of a screw thread is almost never used in making working drawings. In true representation, the crest and root lines appear as the projections of helical curves, which are extremely tedious to draw.

On practical working drawings, three representation methods are generally used: **detailed**, **schematic** and **simplified**.

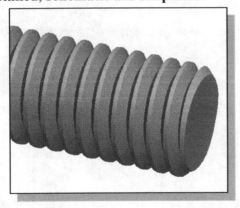

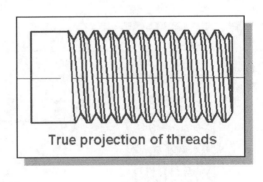

True projection of threads

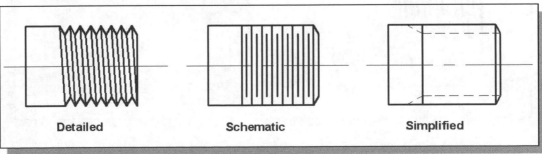

Detailed   Schematic   Simplified

Detailed Representation

The detailed representation simplifies the drawing of the thread principally by constructing the projections of the helical curves into straight lines. And where applicable, further simplifications are made. For example, the 29° angle of the Acme is generally drawn as 30°, and the *American National* and *Unified* threads are represented by the sharp V. In general, true pitch should be shown, although a small increase or decrease in pitch is permissible so as to have even measurements in making the drawing. For example, seven threads per inch may be decreased to six. Remember this is only to simplify the drawing; the actual threads per inch must be specified in the drawing.

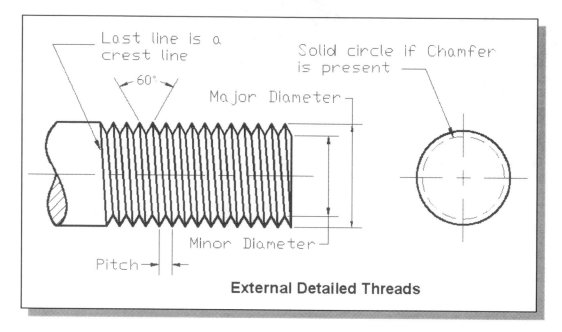

External Detailed Threads

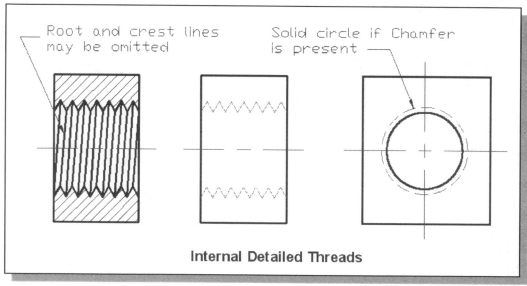

Internal Detailed Threads

Schematic Representation

The schematic representation simplifies the drawing of the thread even further. This method omits the true forms and indicates the crests and roots by lines perpendicular to the main axis. And the schematic representation is nearly as effective as the detailed representation but is much easier to construct. This representation should not be used for section views of external threads or when internal threads are shown as hidden lines; the simplified representation is used instead. Also, to avoid too dense of line pattern, do not use this method when the pitch is less than 1/8 inch or 3 mm.

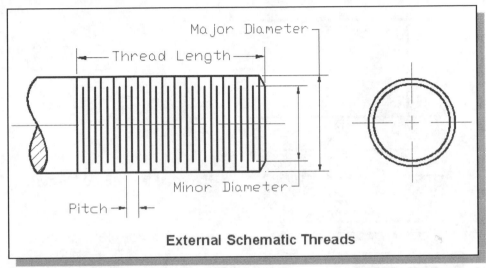

External Schematic Threads

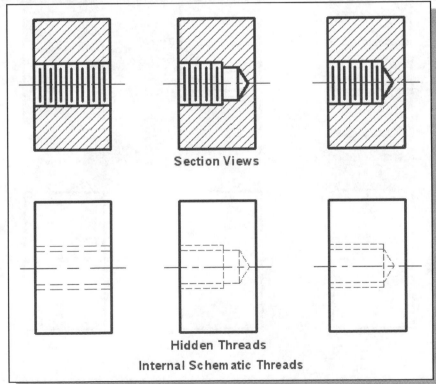

Section Views

Hidden Threads

Internal Schematic Threads

Simplified Representation

The simplified representation omits both form and crest-lines and indicates the threaded portion by dashed lines parallel to the axis at the approximate depth of thread. The simplified representation is less descriptive than the schematic representation, but they are quicker to draw and for this reason are preferred whenever feasible.

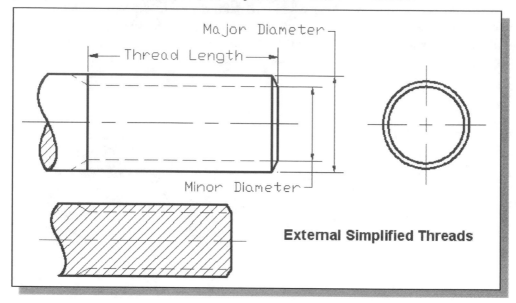

External Simplified Threads

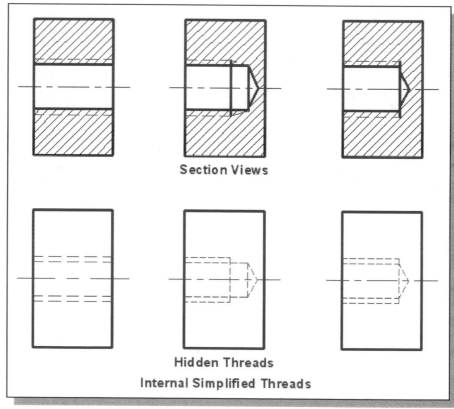

Section Views

Hidden Threads

Internal Simplified Threads

Thread Specification – English Units

The orthographic views of a thread are necessary in order to locate the position of the thread on the part. In addition, the complete thread specification is normally conveyed by means of a local note or dimensions and a local note. The essential information needed for manufacturing is **form**, **nominal (major) diameter**, **threads per inch**, and **thread class** or **toleranced dimensions**. In addition, if the thread is left-hand, the letters **LH** must be included in the specification; also, if the thread is other than single, its multiplicity must be indicated. In general, threads other than the Unified and Metric threads, Acme and buttress require tolerances.

Unified and Metric threads can be specified completely by note. The form of the specification always follows the same order: the first is the nominal size, then the number of threads per inch and the series designation (UNC, NC, etc.), and then the thread fits class. If the thread is left-hand, the letters **LH** are placed after the class. Also, when the Unified-thread classes are used, the letter **A** indicates the thread is external and letter **B** indicates the thread is internal.

For example, a thread specification of **3/4-10UNC-3A**

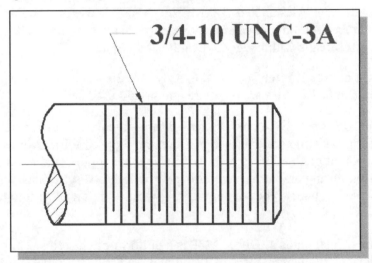

Major diameter: **3/4** in. outside diameter
Number of threads per inch: **10** threads per inch (TPI)
Thread form and Series: **UNC** stands for **Unified National Coarse** series thread
Thread Fits: Class **3** is for high accuracy
External or Internal thread: Letter **A** indicates **external thread**

The descriptions of **thread series** and **thread fits class** are illustrated in the following sections.

Unified Thread Series

Threads are classified in "series" according to the number of threads per inch used with a specific diameter. The above Unified thread example having 20 threads per inch applied to a 1/4-in. diameter results in a thread belonging to the coarse-thread series; note that one with 28 threads per inch on the same diameter describes a thread belonging to the fine-thread series. In the United States, the Unified and Metric threads, the Acme, pipe threads, buttress, and the knuckle thread all have standardized series designations. Note that only the Unified system has more than one series of standardized forms. The Unified and Metric threads standard covers eleven series of screw threads. In the descriptions of the series which follow, the letters "U" and "N" used in the series designations stand for the words "Unified" and "National," respectively. The three standard series include the **coarse-thread series**, the **fine-thread series** and the **extra fine-thread series**.

- The coarse-thread series, designated "**UNC**" or "**NC**," is recommended for general use where conditions do not require a fine thread.

- The fine-thread series, designated "**UNF**" or "**NF**," is recommended for general use in automotive and aircraft work and where special conditions require a fine thread.

- The extra fine-thread series, designated "**UNEF**" or "**NEF**," is used particularly in aircraft and aeronautical equipment, where an extremely shallow thread or a maximum number of threads within a given length is required.

There are also the eight **constant pitch series**: **4**, **6**, **8**, **12**, **16**, **20**, **28** and **32 threads** with constant pitch. The **8**, **12** and **16** series are the more commonly used series.

- The 8-thread series, designated **8UN** or **8N**, is a uniform-pitch series using eight threads per inch for any diameters. This series is commonly used for high-pressure conditions. The constant pitch allows excessive torque to be applied and maintain a proper initial tension during assembly. Accordingly, the 8-thread series has become the general series used in many types of engineering work and as a substitute for the coarse-thread series.

- The 12-thread series, designated **12UN** or **12N**, is a uniform-pitch series using 12 threads per inch for any diameters. Sizes of 12-pitch threads range from 1/2 to 13/4 in. in diameter. This series is commonly used for machine construction, where fine threads are needed for strength. It also provides continuation of the extra fine-thread series for diameters larger than 11/2 in.

- The 16-thread series, designated **16UN** or **16N**, is a uniform-pitch series using 16 threads per inch for any diameters. This series is intended for applications requiring a very fine thread, such as threaded adjusting collars and bearing retaining nuts. It also provides continuation of the extra-fine-thread series for diameters larger than 2 in.

In addition, there are three special thread series, designated **UNS**, **NS** and **UN** as covered in the standards include special combinations of diameter, pitch, and length of engagement.

Thread Fits

The classes provided by the American Standard (ANSI) are classes 1, 2 and 3. These classes are achieved through toleranced thread dimensions given in the standards.

- **Classes 1 fit**: This class of fit is intended for rapid assembly and easy production. Tolerances and allowance are largest with this class.

- **Classes 2 fit**: This class of fit is a quality standard for the bulk of screws, bolts, and nuts produced and is suitable for a wide variety of applications. A moderate allowance provides a minimum clearance between mating threads to minimize galling and seizure. A thread fit class of **2** is used as the normal production fit; this fit is assumed if none is specified.

- **Classes 3 fit**: This class of fit provides a class where accuracy and closeness of fit are important. No allowance is provided. This class of fit is only recommended where precision tools are used.

Thread Specification – Metric

Metric threads, similar to the Unified threads, can also be specified completely by note. The form of the specification always follows the same order: the first is the letter M for metric thread, and then the nominal size, then the pitch and then the tolerance class. If the thread is left-hand, the letters **LH** are placed after the class.

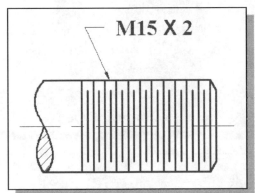

For example, the basic thread note **M15 x 2** is an adequate description for general commercial products.

Letter M: **Metric** thread
Major Diameter: **15** mm.
Pitch: **2** mm.

- **Tolerance Class**: This is used to indicate the tightness or looseness fit between the internal and external threads. In a thread note, the minor or pitch diameter tolerance is stated first followed by the major diameter tolerance if it is different. For general purpose threads, the fit 6H/6g should be used; this fit is assumed if none is specified. For a closer fit, use 6H/5g6g. Note that the number 6 in metric threads is equivalent to the *Unified threads class 2 fit*. Also, the letter g and G are used for small allowances, where h and H are used for no allowance.

Thread Notes Examples

(a) Metric

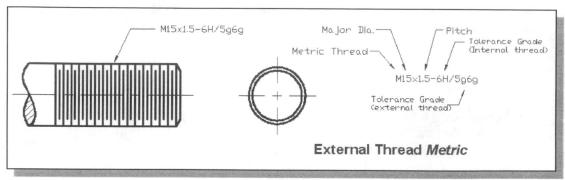

External Thread *Metric*

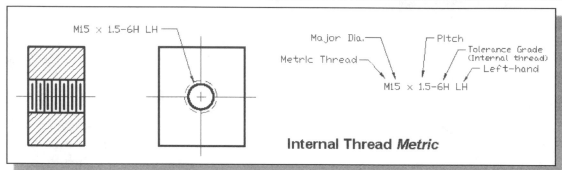

Internal Thread *Metric*

(b) Unified

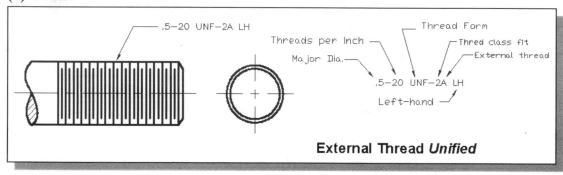

External Thread *Unified*

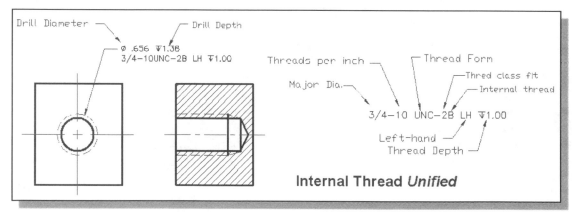

Internal Thread *Unified*

Specifying Fasteners

Fastener is a generic term that is used to describe a fairly large class of parts used to connect, fasten, or join parts together. Fasteners are generally identified by the following attributes: **Type**, **Material**, **Size**, and **Thread information**.

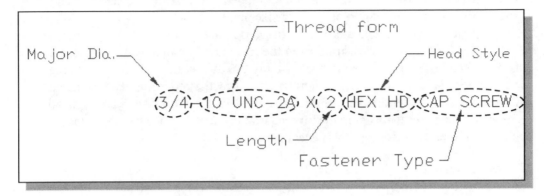

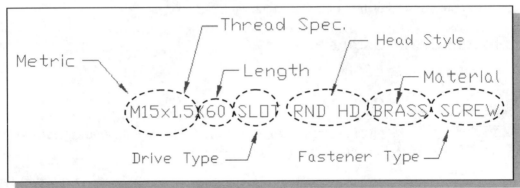

- **Type**
  Fasteners are divided into categories based on their function or design; for example, Wood Screw, Sheet Metal Screw, Hex Bolt, Washer, etc. Within the same category, some variations may exist, such as **Drive Type** and **Head Style**.

  (a) Drive Type
  Fasteners in some categories are available with different drive types such as *Phillips* or *Slotted*; for example, Phillips, Slotted, Allen/Socket, etc.

  (b) Head Style
  Many categories are also available with different head shapes or styles; for example, Flat head, Pan head, Truss head, etc.

  For some types of fasteners, there is either only one drive type or the head style is implied to be of a standard type; for example, Socket screws have the implied Allen drive.

- **Material**
  Fastener material describes the material from which the fastener was made as well as any material grade; for example, Stainless Steel, Bronze, etc.

- **Size**
  Descriptions of a fastener's size typically include its **diameter** and **length**. The fastener diameter is measured either as a size number or as a direct measurement. How fastener length is measured varies based on the type of head. As a general rule, the length of fasteners is measured from the surface of the material, to the end of the fastener. For fasteners where the head usually sits above the surface such as hex bolts and pan head screws, the measurement is from directly under the head to the end of the fastener. For fasteners that are designed to be counter sunk such as flat head screws, the fastener is measured from the point on the head where the surface of the material will be to the end of the fastener.

- **Thread Information**
  Thread pitch or thread count is used only on machine thread fasteners. The thread pitch or count describes how fine the threads are.

Commonly Used Fasteners

Hex bolts

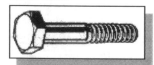

A bolt is a fastener having a head on one end and a thread on the other end. A bolt is used to hold two parts together by passing through aligned clearance holes with a nut screwed on the threaded end. (See Appendix E for specs.)

Studs

A stud is a rod with threaded ends.

Cap Screws

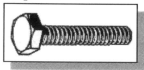

A hexagon cap screw is similar to a bolt except it is used without a nut and generally has a longer thread. Cap screws are available in a variety of head styles and materials. (See Appendix E for specs.)

Machine Screws

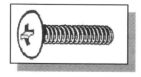

A machine screw is similar to the slot-head cap screw but smaller, available in many styles and materials. A machine screw is also commonly referred to as a stove bolt. (See Appendix E for specs.)

Wood Screws

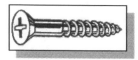

A tapered shank screw is for use exclusively in wood. Wood screws are available in a variety of head styles and materials.

Sheet Metal Screws

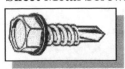

Highly versatile fasteners designed for thin materials. Sheet metal screws can be used in wood, fiberglass and metal, also called self-tapping screws, available in steel and stainless steel.

Carriage Bolts

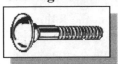

A carriage bolt is mostly used in wood with a domed shape top and a square under the head, which is pulled into the wood as the nut is tightened.

Socket Screws

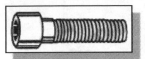

Socket screws, also known as **Allen head**, are fastened with a hexagon Allen wrench, available in several head styles and materials. (See Appendix E for specs.)

Set Screws

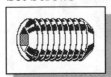

Set screws are used to prevent relative motion between two parts. A set screw is screwed into one part so that its point is pushed firmly against the other part, available in a variety of point styles and materials.

Nuts

Nuts are used to attach machine thread fasteners. (See Appendix E for specs.)

Washers

Washers provide a greater contact surface under the fastener. This helps prevent a nut, bolt or screw from breaking through the material. (See Appendix E for specs.)

Keys

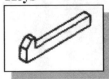

Keys are used to prevent relative motion between shafts and wheels, couplings and similar parts attached to shafts.

Rivets

Rivets are generally used to hold sheet metal parts together. Rivets are generally considered as permanent fasteners and are available in a variety of head styles and materials.

Drawing Standard Bolts

As a general rule, standard bolts and nuts are not drawn in the detail drawings unless they are non-standard sizes. The standard bolts and nuts do appear frequently in assembly drawings, and they may be drawn from the exact dimensions from the ANSI/ASME B18.2.1- 1996 standard (see Appendix E) if accuracy is important. Note that for small or unspecified chamfers, fillets, or rounds, the dimension of 1/16 in. (2mm) is generally used in creating the drawing. A simplified version of a bolt is illustrated in the following figure. Note that many companies now offer free thousands of standard fasteners and part drawings in AutoCAD DWG or DXF formats. You are encouraged to do a search on the internet and compare the downloaded drawings to the specs as listed in the appendix.

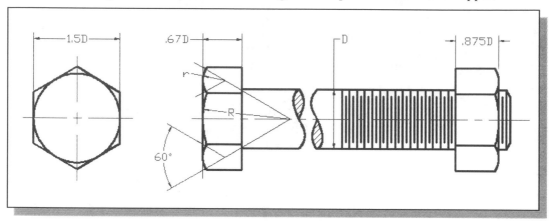

Bolt and Screw Clearances

Bolts and screws are generally used to hold parts together, and it is frequently necessary to have a clearance hole for the bolt or screw to pass through. The size of the clearance hole depends on the major diameter of the fastener and the type of fit that is desired. For bolts and screws, three standard fits are available: **normal fit**, **close fit** or **loose fit**. See Appendix for the counter-bore and countersink clearances also. (See Appendix G for more details.)

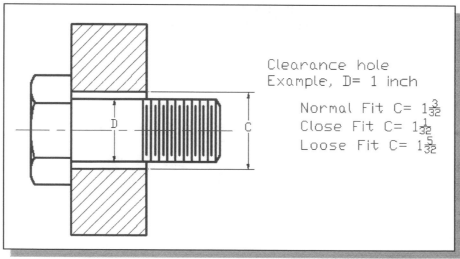

Clearance hole
Example, D= 1 inch

Normal Fit C= $1\frac{3}{32}$
Close Fit C= $1\frac{1}{32}$
Loose Fit C= $1\frac{5}{32}$

Fasteners Using Autodesk Inventor's Content Center

In Autodesk Inventor, we also have the option of using the standard parts library through what is known as the **Content Center**. The *Content Center* consists of multiple libraries of standard parts that have been created based on industry standards. Significant amounts of time can be saved by using these parts. Note that we can also create and publish our libraries so that others can reuse our parts. The *Content Center* has two modes for two distinct roles: **consumer** and **editor**. In the *consumer* mode, we access the *Content Center* libraries and use the parts as a consumer. In the *editor* mode, we can define the different categories and also define the iterations for the parts the consumer can select and use.

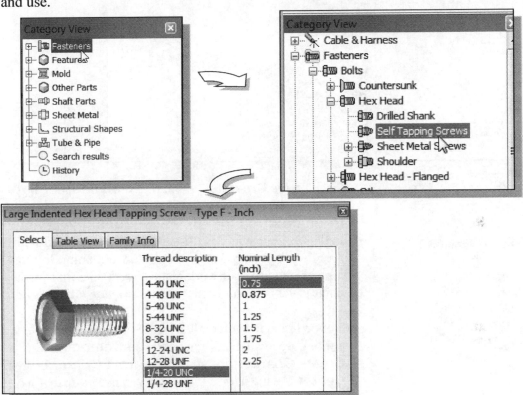

Starting Autodesk Inventor

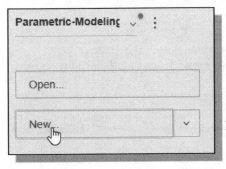

1. Select the **Autodesk Inventor** option on the *Start* menu or select the **Autodesk Inventor** icon on the desktop to start Autodesk Inventor. The Autodesk Inventor main window will appear on the screen.

2. Select the **New File** icon with a single click of the left-mouse-button in the *Launch* toolbar as shown.

3. Confirm the ***Parametric-Modeling*** project is activated; note the **Projects** button is available to view/modify the active project.

4. Select the **English** tab, and in the *Template* list select **Standard(in).iam** (*Standard Inventor Assembly Model* template file).

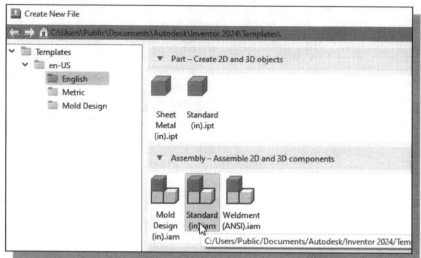

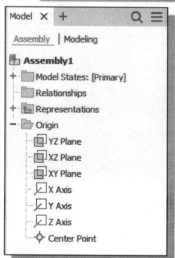

5. Click on the **Create** button in the *New File* dialog box to accept the selected settings.

- In the *browser* window, **Assembly1** is displayed with a set of work planes, work axes and a work point. In most aspects, the usage of work planes, work axes and work point is very similar to that of the *Inventor Part Modeler*.

- Notice in the *Ribbon* toolbar panels, several *component* options are available, such as **Place Component**, **Create Component** and **Place from Content Center**. As the names imply, we can use parts that have been created or create new parts within the *Inventor Assembly Modeler*.

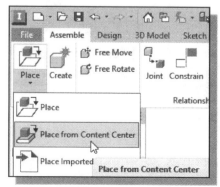

6. In the *Assemble* panel select the **Place from Content Center** command by left-clicking the icon.

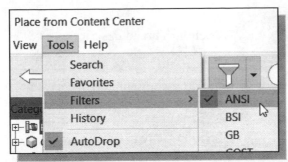

In the *Tools pull-down menu*, set the *Filters* option to **ANSI** as shown.

- Note the parts for other standards, such as GB, ISO, JIS and DIN are also included in the **Content Center**.

7. Select the **Fastener** category and then expand the **Bolts** list as shown.

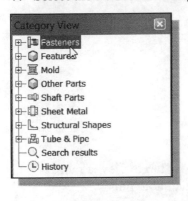

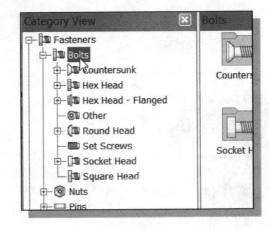

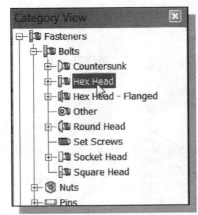

8. Choose **Hex Head** under **Bolts** as shown in the figure.

9. Select **Hex Cap Screw - Inch** as shown in the figure below.

10. Click **OK** to enter the selection dialog box.

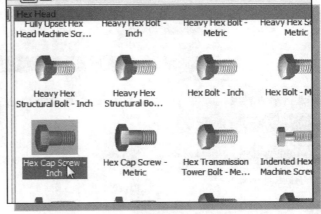

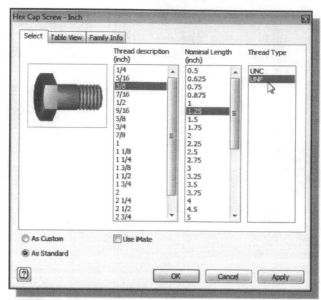

11. Set the Thread description to **3/8** inch.

12. Set the nominal length to **1.25** inch.

13. Set the Thread type to **UNF** as shown.

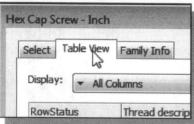

14. Click on the **Table View** tab to view a more detailed listing of the available pins.

➤ Note the selected item is highlighted in the list with additional details on the selected Cap-Screw.

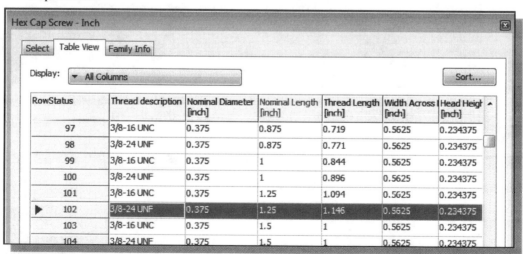

RowStatus	Thread description	Nominal Diameter [inch]	Nominal Length [inch]	Thread Length [inch]	Width Across [inch]	Head Height [inch]
97	3/8-16 UNC	0.375	0.875	0.719	0.5625	0.234375
98	3/8-24 UNF	0.375	0.875	0.771	0.5625	0.234375
99	3/8-16 UNC	0.375	1	0.844	0.5625	0.234375
100	3/8-24 UNF	0.375	1	0.896	0.5625	0.234375
101	3/8-16 UNC	0.375	1.25	1.094	0.5625	0.234375
▶ 102	3/8-24 UNF	0.375	1.25	1.146	0.5625	0.234375
103	3/8-16 UNC	0.375	1.5	1	0.5625	0.234375
104	3/8-24 UNF	0.375	1.5	1	0.5625	0.234375

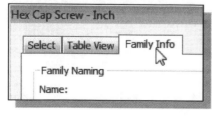

15. Click on the **Family Info** tab to view the associated information of the selected item.

> Note the selected Cap-Screw is associated to the ANSI B18.2.1 (1996) standard.

16. Click **OK** to accept the selection.

17. On your own, place a copy of the cap screw, with a single click of the left-mouse-button, inside the graphics area.

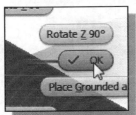

18. Click once with the right-mouse-button to bring up the option menu and select **OK** to end the placement of the cap screw.

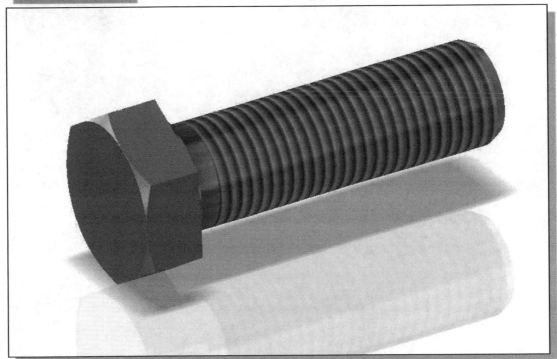

- On your own, use the content center and place a **3/8 Hex Nut – Inch** part next to the *3/8 Cap screw* as shown.

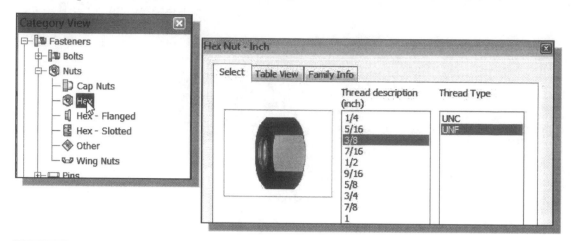

Review Questions:

1. Describe the thread specification of **3/4-16 UNF -3A**.

2. Describe the thread specification of **M25X3**.

3. List and describe four different types of commonly used fasteners.

4. Determine the sizes of a clearance hole for a **0.75** bolt using the *Close fit* option.

5. Perform an internet search and find information on the different types of **set-screws**, and create freehand sketches of four point styles of set screws you have found.

6. Perform an internet search and examine the different types of **machine screws** available, and create freehand sketches of four different styles of machine screws you have found.

7. Perform an internet search and find information on **locknuts**, and create freehand sketches of four different styles of locknuts you have found.

Notes:

Chapter 14
Assembly Modeling and Working Drawings

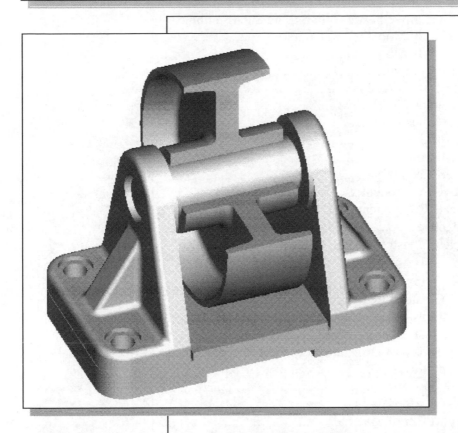

Learning Objectives

- ◆ Understand the Terminology Related to Working Drawings
- ◆ Understand the Assembly Modeling Methodology
- ◆ Understand and Control Degrees of Freedom for Assembly Components
- ◆ Understand and Utilize Autodesk Inventor Assembly Mates
- ◆ Create Exploded Assemblies
- ◆ Create Assembly Drawings
- ◆ Create and Edit a Bill of Materials

Autodesk Inventor Certified User Exam Objectives Coverage

Section 5: Assemblies

Objectives: Creating Assemblies, Viewing Assemblies, Animation Assemblies, Adaptive Features, Parts, and Subassemblies

Section 6: Drawings

Objectives: Create drawings

Autodesk Inventor Certified User Reference Guide

General Engineering Design Process

Engineering design is the ability to create and transform ideas and concepts into a product definition that meets the desired objective.

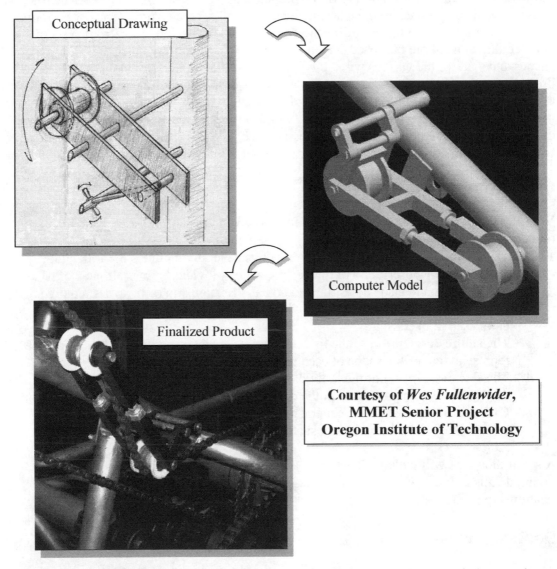

Conceptual Drawing

Computer Model

Finalized Product

**Courtesy of *Wes Fullenwider*,
MMET Senior Project
Oregon Institute of Technology**

The general procedure for the design of a new product or improving an existing product involves the following six stages:

1. **Develop and identify the desired objectives.**
2. **Conceptual design stage – concepts and ideas of possible solutions.**
3. **Engineering analysis of components.**
4. **Computer Modeling and/or prototypes.**
5. **Refine and finalize the design.**
6. **Working drawings of the finalized design.**

It is during the **conceptual design** stage when the first drawings, known as **conceptual drawings**, are usually created. The conceptual drawings are typically done in the form of freehand sketches showing the original ideas and concepts of possible solutions to the set objectives. From these conceptual drawings, engineering analyses are performed to improve and confirm the suitability of the proposed design. Working from the sketches and the results of the analyses, the design department then creates prototypes or performs computer simulations to further refine the design. Once the design is finalized, a set of detailed drawings of the proposed design is created. It is accurately made and shows the shapes and sizes of the various parts; this is known as the **detail drawing**. On a detail drawing, all the views necessary for complete shape description of a part are provided, and all the necessary dimensions and manufacturing directions are given. The set of drawings is completed with the addition of an **assembly drawing** and a parts list or bill of material. The assembly drawing is necessary as it provides the location and relationship of the parts. The completed drawing set is known as **working drawings**.

Working Drawings

Working drawings are the set of drawings used to give information for the manufacturing of a design.

The description given by a set of working drawings generally includes the following:

1. The **assembly** description of the design, which provides an overall view of what the design is and also shows the location and relationship of the parts required.
2. A **parts list** or **bill of material** provides a detailed material description of the parts used in the design.
3. The **shape description** of the individual parts, which provides the full graphical representation of the shape of each part.
4. The **size description** of the individual parts, which provides the necessary dimensions of the parts used in the design.
5. **Explanatory notes** are the general and local notes on the individual drawings, giving the specific information, such as material, heat-treatment, finish, and etc.

A set of drawings will include, in general, two classes of drawings: **detail drawings** giving details of individual parts and an **assembly drawing** giving the location and relationship of the parts.

Detail Drawings

A detail drawing is the drawing of the individual parts, giving a complete and exact description of its form, dimensions, and related construction information. A successful detail drawing will provide the manufacturing department all the necessary information to produce the parts. This is done by providing adequate orthographic views together with dimensions, notes, and a descriptive title. Information on a detail drawing usually includes the shape, size, material and finish of a part; specific information of shop operations, such as the limits of accuracy and the number of parts needed are also provided. The detail drawing should be complete and also exact in description, so that a satisfactory part can be produced. The drawings created in the previous chapters are all detail drawings.

Assembly Drawings

An *assembly drawing* is, as its name implies, a drawing of the design put together, showing the relative positions of the different parts. The *assembly drawing* of a finalized design is generally done after the detail drawings are completed. The *assembly drawing* can be made by tracing from the detail drawings. The *assembly drawing* can also be drawn from the dimensions of the detail drawings; this provides a valuable check on the correctness of the detail drawings.

The *assembly drawing* sometimes gives the overall dimensions and dimensions that can be used to aid the assembly of the design. However, many assembly drawings need no dimensions. An *assembly drawing* should not be overloaded with detail, particularly hidden detail. Unnecessary dashed lines (hidden lines) should not be used on any drawing; this is even more critical on *assembly drawings*.

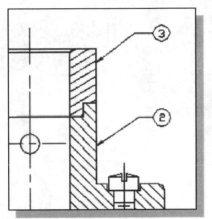

Assembly drawings usually have reference letters or numbers designating the different parts. These "numbers" are typically enclosed in circles ("balloons") with a leader pointing to the part; these numbers are used in connection with the parts list and bill of materials.

For complicated designs, besides the assembly drawing, subassembly drawings are generally used. A subassembly is a drawing showing the details of a related group of parts, as it would not be practical to include all the features on a single *assembly drawing*. Thus, a subassembly is used to aid in clarifying the relations of a subset of an assembly.

Bill of Materials (BOM) and Parts List

4	Cap screw 3/8	2	STOCK
3	Collar	1	STEEL
2	Bearing	1	STEEL
1	Base Plate	1	C.I.
No	DESCRIPTION	REQ	MATL

A bill of materials (BOM) is a table that contains information about the parts within an assembly. The BOM can include information such as part names, quantities, costs, vendors, and all of the other information related to building the part. The *parts list*, which is used in an assembly drawing, is usually a partial list of the associated BOM.

Drawing Sizes

The standard drawing paper sizes are as shown in the below tables.

American National Standard	International Standard
A – 8.5" X 11.0"	A4 – 210 mm X 297 mm
B – 11.0" X 17.0"	A3 – 297 mm X 420 mm
C – 17.0" X 22.0"	A2 – 420 mm X 594 mm
D – 22.0" X 34.0"	A1 – 594 mm X 841 mm
E – 34.0" X 44.0"	A0 – 841 mm X 1189 mm

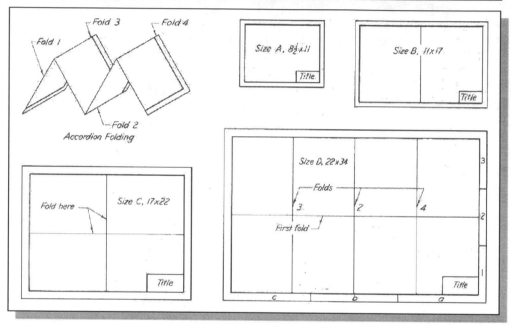

Drawing Sheet Borders and Revisions Block

The drawing sheet borders are generally drawn at a distance parallel to the edges of the sheet, typically with distance varying from 0.25" to 0.5" or 5 mm to 10 mm.

Once a drawing has been released to the shop, any alterations or changes should be recorded on the drawing and new prints should be issued to the production facilities. In general, the upper right corner of a working drawing sheet is the designated area for such records; this is known as the **revisions** block.

		REVISIONS			
ZONE	REV	DESCRIPTION		DATE	APPROVED

Title Blocks

The title of a working drawing is usually placed in the lower right corner of the sheet. The spacing and arrangement of the space depend on the information to be given. In general, the title of a working drawing should contain the following information:

1. **Name of the company and its location.**
2. **Name of the part represented.**
3. **Signature of the person who made the drawing and the date of completion.**
4. **Signature of the checker and the date of completion.**
5. **Signature of the approving personnel and the date of approval.**
6. **Scale of the drawing.**
7. **Drawing number.**

Other information may also be given in the title blocks area, such as material, heat treatment, finish, hardness, and general tolerances; it depends on the company and the peculiarities of the design.

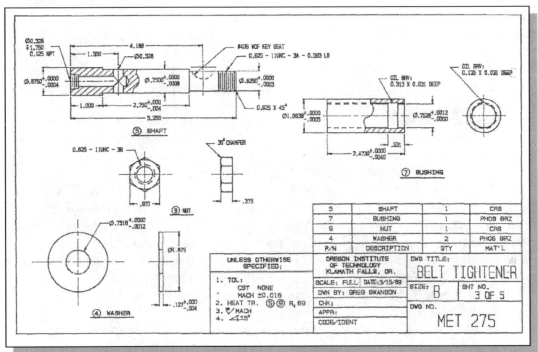

Working Drawings with Autodesk Inventor

In the previous lessons, we have gone over the fundamentals of creating basic parts and drawings. In this lesson, we will examine the assembly modeling functionality of Autodesk Inventor. We will start with a demonstration on how to create and modify assembly models. The main task in creating an assembly is establishing the assembly relationships between parts. To assemble parts into an assembly, we will need to consider the assembly relationships between parts. It is a good practice to assemble parts based on the way they would be assembled in the actual manufacturing process. We should also consider breaking down the assembly into smaller subassemblies, which helps the management of parts. In Autodesk Inventor, a subassembly is treated the same way as a single part during assembling. Many parallels exist between assembly modeling and part modeling in parametric modeling software such as Autodesk Inventor.

Autodesk Inventor provides full associative functionality in all design modules, including assemblies. When we change a part model, Autodesk Inventor will automatically reflect the changes in all assemblies that use the part. We can also modify a part in an assembly. **Bi-directional full associative functionality** is the main feature of parametric solid modeling software that allows us to increase productivity by reducing design cycle time.

One of the key features of Autodesk Inventor is the use of an assembly-centric paradigm, which enables users to concentrate on the design without depending on the associated parameters or constraints. Users can specify how parts fit together and the Autodesk Inventor *assembly-based fit function* automatically determines the parts' sizes and positions. This unique approach is known as the **Direct Adaptive Assembly** approach, which defines part relationships directly with no order dependency.

In this lesson, we will also illustrate the basic concept of Autodesk Inventor's **Adaptive Design** approach. The key element in doing **Adaptive Design** is to *under-constrain* features or parts. The applied *assembly constraints* in the assembly modeler are used to control the sizes, shapes, and positions of *under-constrained* sketches, features, and parts. No equations are required, and this approach is extremely flexible when performing modifications and changes to the design. We can modify adaptive assemblies at any point, in any order, regardless of how the parts were originally placed or constrained.

In Autodesk Inventor, features and parts can be made adaptive at any time during the creation or assembly. The features of a part can be defined as adaptive when they are created in the part file. When we place such a part in an assembly, the features will then resize and change shape based on the applied assembly constraints. We can make features and parts adaptive from either the part modeling or assembly modeling environments.

The *Adaptive Design approach* is a unique design methodology that can only be found in Autodesk Inventor. The goal of this methodology is to improve the design process and allows you, the designer, to *design the way you think*.

Assembly Modeling Methodology

The Autodesk Inventor assembly modeler provides tools and functions that allow us to create 3D parametric assembly models. An assembly model is a 3D model with any combination of multiple part models. *Parametric assembly constraints* can be used to control relationships between parts in an assembly model.

Autodesk Inventor can work with any of the assembly modeling methodologies:

The Bottom Up Approach

The first step in the *bottom up* assembly modeling approach is to create the individual parts. The parts are then pulled together into an assembly. This approach is typically used for smaller projects with very few team members.

The Top Down Approach

The first step in the *top down* assembly modeling approach is to create the assembly model of the project. Initially, individual parts are represented by names or symbols. The details of the individual parts are added as the project gets further along. This approach is typically used for larger projects or during the conceptual design stage. Members of the project team can then concentrate on the particular section of the project to which he/she is assigned.

The Middle Out Approach

The *middle out* assembly modeling approach is a mixture of the bottom-up and top-down methods. This type of assembly model is usually constructed with most of the parts already created, and additional parts are designed and created using the assembly for construction information. Some requirements are known, and some standard components are used, but new designs must also be produced to meet specific objectives. This combined strategy is a very flexible approach for creating assembly models.

The different assembly modeling approaches described above can be used as guidelines to manage design projects. Keep in mind that we can start modeling our assembly using one approach and then switch to a different approach without any problems.

In this lesson, the *bottom up* assembly modeling approach is illustrated. All of the parts (components) required to form the assembly are created first. Autodesk Inventor's assembly modeling tools allow us to create complex assemblies by using components that are created in part files or are placed in assembly files. A component can be a subassembly or a single part, where features and parts can be modified at any time. The sketches and profiles used to build part features can be fully or partially constrained. Partially constrained features may be adaptive, which means the size or shape of the associated parts are adjusted in an assembly when the parts are constrained to other parts. The basic concept and procedure of using the adaptive assembly approach is demonstrated in the tutorial.

The Shaft Support Assembly

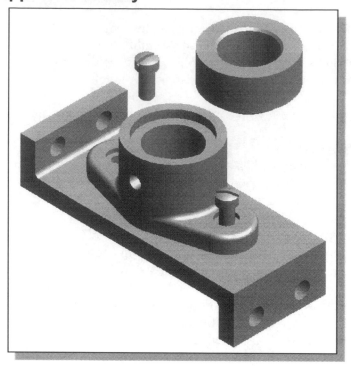

Additional Parts

Four parts are required for the assembly: (1) **Collar**, (2) **Bearing**, (3) **Base-Plate** and
(4) **Cap-Screw**. On your own, create the four parts as shown below; save the models as
separate part files: *Collar, Bearing, Base-Plate*, and *Cap-Screw*. (Place all parts in the
Chapter14 folder and close all part files or exit *Autodesk Inventor* after you have
created the parts.)

(1) *Collar*

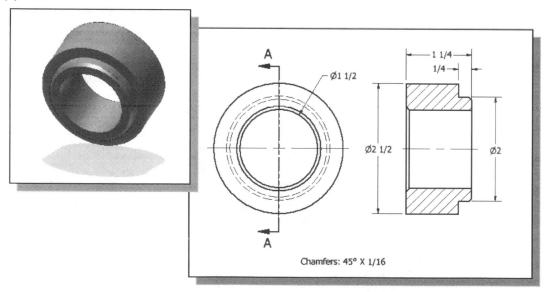

(2) **_Bearing_** (Construct the part with the datum origin aligned to the bottom center.)

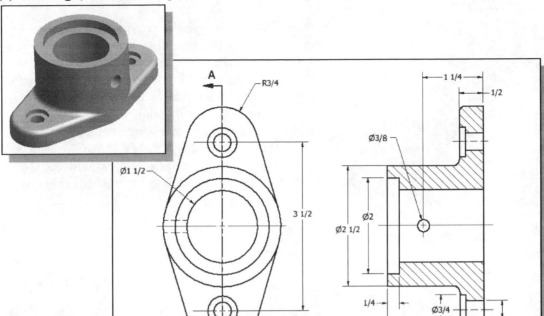

(3) **_Base-Plate_** (Construct the part with the datum origin aligned to the bottom center of the large hole.)

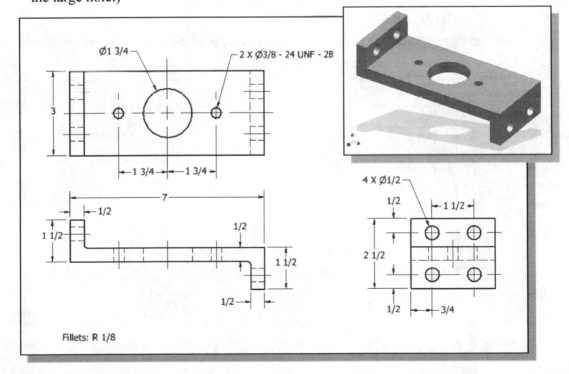

(4) *Cap-Screw*

- Autodesk Inventor provides two options for creating threads: **Thread** and **Coil**. The **Thread** command does not create true 3D threads; a pre-defined thread image is applied on the selected surface, as shown in the figure. The **Coil** command can be used to create true threads, which contain complex three-dimensional curves and surfaces. You are encouraged to experiment with the **Coil** command and/or the **Thread** command to create threads.

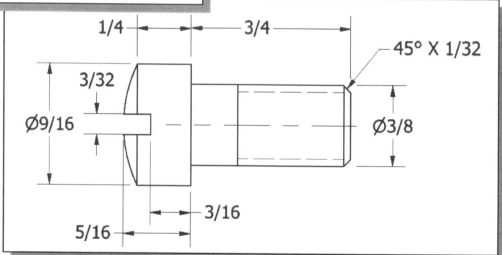

- Hint: First create a revolved feature using the profile shown below.

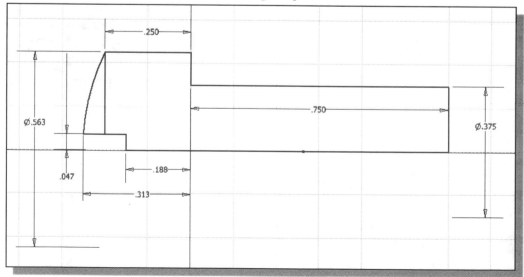

Starting Autodesk Inventor

1. Select the **Autodesk Inventor** option on the *Start* menu or select the **Autodesk Inventor** icon on the desktop to start Autodesk Inventor. The Autodesk Inventor main window will appear on the screen.

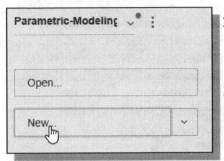

2. Select the **New File** icon with a single click of the left-mouse-button in the *Launch* toolbar as shown.

3. Confirm the *Parametric-Modeling-Exercises* project is activated; note the **Projects** button is available to view/modify the active project.

4. Select the **English** tab, and in the *Template* list select **Standard(in).iam** (*Standard Inventor Assembly Model* template file).

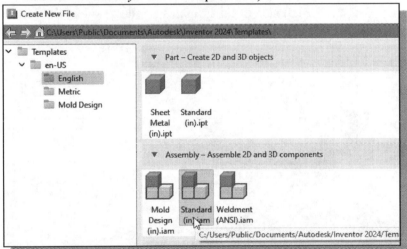

5. Click on the **Create** button in the *New File* dialog box to accept the selected settings.

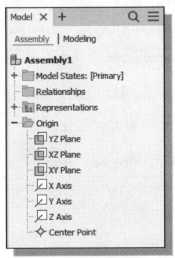

- In the *browser* window, **Assembly1** is displayed with a set of work planes, work axes and a work point. In most aspects, the usage of work planes, work axes and work point is very similar to that of the *Inventor Part Modeler*.

- Notice, in the *Ribbon* toolbar panels, several *component* options are available, such as **Place Component**, **Create Component** and **Place from Content Center**. As the names imply, we can use parts that have been created or create new parts within the *Inventor Assembly Modeler*.

Placing the First Component

The first component placed in an assembly should be a fundamental part or subassembly. The first component in an assembly file sets the orientation of all subsequent parts and subassemblies. The origin of the first component is aligned to the origin of the assembly coordinates and the part is grounded (all degrees of freedom are removed). The rest of the assembly is built on the first component, the *base component*. In most cases, this *base component* should be one that is **not likely to be removed** and **preferably a non-moving part** in the design. Note that there is no distinction in an assembly between components; the first component we place is usually considered as the *base component* because it is usually a fundamental component to which others are constrained. We can change the base component to a different base component by placing a new base component, specifying it as grounded, and then re-constraining any components placed earlier, including the first component. For our project, we will use the ***Base-Plate*** as the base component in the assembly.

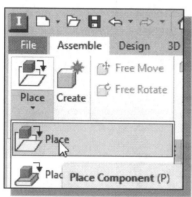

1. In the *Assemble* panel (the toolbar that is located to the left side of the graphics window) select the **Place Component** command by left-clicking the icon.

2. Select the ***Base-Plate*** (part file: ***Base-Plate.ipt***) in the list window.

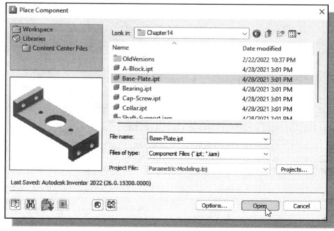

3. Click on the **Open** button to retrieve the model.

4. Right-click once to bring up the option menu and select **Place Grounded at Origin.**

5. Right-click again to bring up the option menu and select **OK** to end the placement of the *Base-Plate* part.

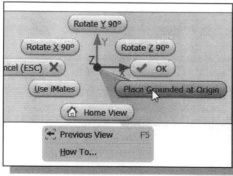

Placing the Second Component

We will retrieve the *Bearing* part as the second component of the assembly model.

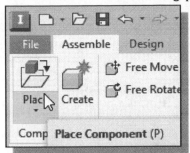

1. In the *Assemble* panel (the toolbar that is located to the left side of the graphics window) select the **Place Component** command by left-clicking the icon.

2. Select the **Bearing** design (part file: ***Bearing.ipt***) in the list window. And click on the **Open** button to retrieve the model.

3. Place the *Bearing* toward the upper right corner of the graphics window, as shown in the figure.

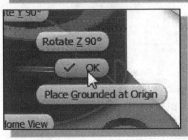

4. Inside the graphics window, right-click once to bring up the option menu and select **OK** to end the placement of the *Bearing* part.

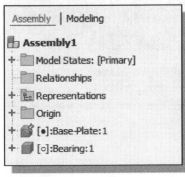

• Inside the *browser* window, the retrieved parts are listed in their corresponding order. The **Pin** icon in front of the *Base-Plate* filename signifies the part is grounded and all *six degrees of freedom* are restricted. The number behind the filename is used to identify the number of copies of the same component in the assembly model.

Degrees of Freedom and Constraints

Each component in an assembly has six **degrees of freedom (DOF)**, or ways in which rigid 3D bodies can move: movement along the X, Y, and Z axes (translational freedom), plus rotation around the X, Y, and Z axes (rotational freedom). *Translational DOFs* allow the part to move in the direction of the specified vector. *Rotational DOFs* allow the part to turn about the specified axis.

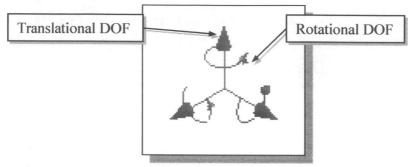

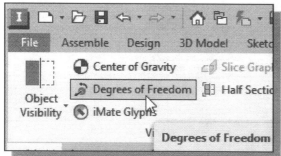

➤ Select the **Degrees of Freedom** option in the **View** tab to display the DOF of the unconstrained component.

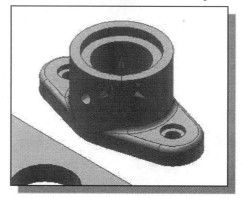

In Autodesk Inventor, the degrees-of-freedom symbol shows the remaining degrees of freedom (both translational and rotational) for one or more components of the active assembly. When a component is fully constrained in an assembly, the component cannot move in any direction. The position of the component is fixed relative to other assembly components. All of its degrees of freedom are removed. When we place an assembly constraint between two selected components, they are positioned relative to one another. Movement is still possible in the unconstrained directions.

It is usually a good idea to fully constrain components so that their behavior is predictable as changes are made to the assembly. Leaving some degrees of freedom open can sometimes help retain design flexibility. As a general rule, we should use only enough constraints to ensure predictable assembly behavior and avoid unnecessary complexity.

Assembly Constraints

We are now ready to assemble the components together. We will start by placing assembly constraints on the **Bearing** and the **Base-Plate**.

To assemble components into an assembly, we need to establish the assembly relationships between components. It is a good practice to assemble components the way they would be assembled in the actual manufacturing process. **Assembly constraints** create a parent/child relationship that allows us to capture the design intent of the assembly. Because the component that we are placing actually becomes a child to the already assembled components, we must use caution when choosing constraint types and references to make sure they reflect the intent.

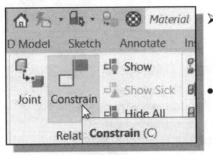

➢ Switch back to the *Assemble* panel; select the **Constrain** command by left-clicking once on the icon.

• The *Place Constraints* dialog box appears on the screen. Five types of assembly constraints are available.

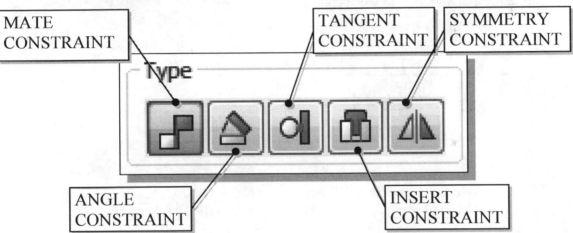

MATE CONSTRAINT

TANGENT CONSTRAINT

SYMMETRY CONSTRAINT

ANGLE CONSTRAINT

INSERT CONSTRAINT

• Assembly models are created by applying proper *assembly constraints* to the individual components. The constraints are used to restrict the movement between parts. Constraints eliminate rigid body degrees of freedom (**DOF**). A 3D part has *six degrees of freedom* since the part can rotate and translate relative to the three coordinate axes. Each time we add a constraint between two parts, one or more DOF is eliminated. The movement of a fully constrained part is restricted in all directions. Five basic types of assembly constraints are available in Autodesk Inventor: Mate, Angle, Tangent, Insert and Symmetry. Each type of constraint removes different combinations of rigid body degrees of freedom. Note that it is possible to apply different constraints and achieve the same results.

➢ **Mate** – Constraint positions components face-to-face, or adjacent to one another, with faces flush. Removes one degree of linear translation and two degrees of angular rotation between planar surfaces. Selected surfaces point in opposite directions and can be **offset** by a specified distance. **Mate** constraint positions selected faces normal to one another, with faces coincident.

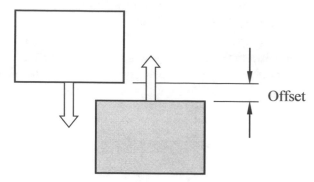

➢ **Flush** – Makes two planes coplanar with their faces aligned in the same direction. Selected surfaces point in the same direction and are offset by a specified distance. Flush constraint aligns components adjacent to one another with faces flush and positions selected faces, curves, or points so that they are aligned with surface normals pointing in the same direction. (Note that the Flush constraint is listed as a selectable option in the Mate constraint.)

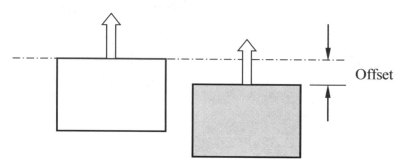

➢ **Angle** – Creates an angular assembly constraint between parts, subassemblies, or assemblies. Selected surfaces point in the direction specified by the angle.

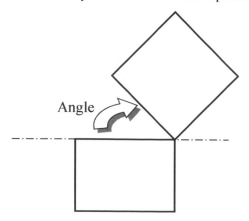

➢ **Tangent** – Aligns selected faces, planes, cylinders, spheres, and cones to contact at the point of tangency. Tangency may be on the inside or outside of a curve, depending on the selection of the direction of the surface normal. A Tangent constraint removes one degree of translational freedom.

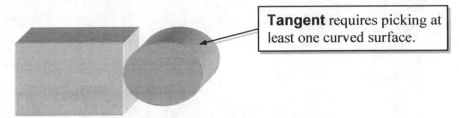

Tangent requires picking at least one curved surface.

➢ **Insert** – Aligns two circles, including their center axes and planes. Selected circular surfaces become co-axial. Insert constraint is a combination of a face-to-face Mate constraint between planar faces and a Mate constraint between the axes of the two components. A rotational degree of freedom remains open. The surfaces do not need to be full 360-degree circles. Selected surfaces can point in opposite directions or in the same direction and can be offset by a specified distance.

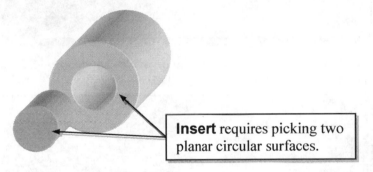

Insert requires picking two planar circular surfaces.

• **Symmetry** – The Symmetry constraint positions two objects symmetrically according to a plane or planar face. The Symmetry constraint is available in the Place Constraint dialog box.

Symmetry requires picking two objects and a plane.

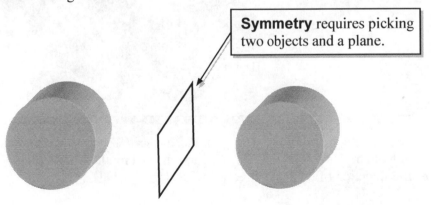

Apply the First Assembly Constraint

1. In the *Place Constraint* dialog box, confirm the constraint type is set to **Mate** constraint and select the **top horizontal surface** of the base part as the first part for the **Mate** alignment command.

2. On your own, dynamically rotate the displayed model to view the bottom of the *Bearing* part, as shown in the figure below.

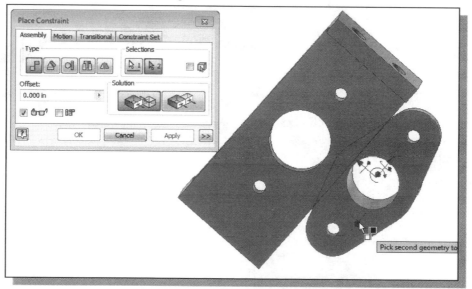

3. Click on the bottom face of the *Bearing* part as the second part selection to apply the constraint. Note the direction normals shown in the figure; the **Mate** constraint requires the selection of opposite direction of surface normals.

4. Click on the **Apply** button to accept the selection and apply the **Mate** constraint.

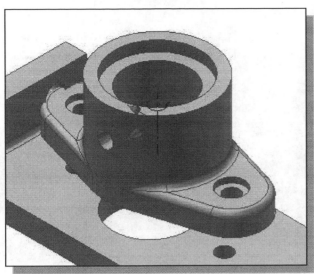

5. On your own, reset the display to **Isometric view** by clicking on the home view in the View Cube.

❖ Notice the DOF symbol is adjusted automatically in the graphics window. The **Mate** constraint removes one degree of linear translation and two degrees of angular rotation between the selected planar surfaces. The *Bearing* part can still move along two axes and rotate about the third axis.

Apply a Second Mate Constraint

The **Mate** constraint can also be used to align axes of cylindrical features.

1. In the *Place Constraint* dialog box, confirm the constraint type is set to **Mate** constraint and move the cursor near the cylindrical surface of the right counter bore hole of the *Bearing* part. Select the axis when it is displayed as shown. (Hint: Use the *Dynamic Rotation* option to assist the selection.)

2. Move the cursor near the cylindrical surface of the small hole on the *Base-Plate* part. Select the axis when it is displayed as shown.

3. In the *Place Constraint* dialog box, set the *Solution* option to **Aligned** (if not aligned) and click on the **Apply** button to accept the selection and apply the **Mate** constraint.

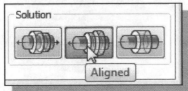

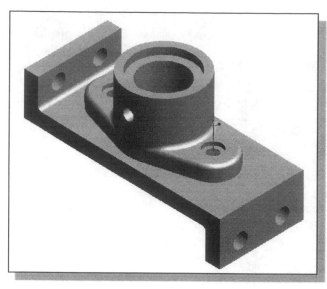

4. In the *Place Constraint* dialog box, click on the **Cancel** button to exit the Place Constraint command.

❖ The *Bearing* part appears to be placed in the correct position. But the DOF symbol indicates that this is not the case; the bearing part can still rotate about the displayed vertical axis.

Constrained Move

To see how well a component is constrained, we can perform a constrained move. A constrained move is done by dragging the component in the graphics window with the left-mouse-button. A constrained move will honor previously applied assembly constraints. That is, the selected component and parts constrained to the component move together in their constrained positions. A grounded component remains grounded during the move.

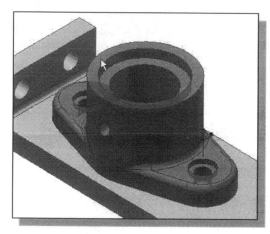

1. Inside the *graphics window*, move the cursor on top of the top surface of the *Bearing* part as shown in the figure.

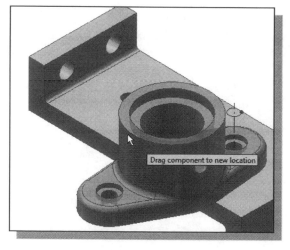

2. Press and hold down the left-mouse-button and drag the *Bearing* part downward.

❖ The *Bearing* part can freely rotate about the displayed axis.

3. On your own, use the dynamic rotation command to view the alignment of the *Bearing* part.

4. Rotate the *Bearing* part and adjust the display roughly as shown in the figure.

Apply a Flush Constraint

Besides selecting the surfaces of solid models to apply constraints, we can also select the established work planes to apply the assembly constraints. This is an additional advantage of using the *BORN technique* in creating part models. For the *Bearing* part, we will apply a **Mate** constraint to two of the work planes and eliminate the last rotational DOF.

1. On your own, inside the *browser* window toggle *ON* the **Visibility** for the two corresponding work planes, shown on the following page, which will be used for alignment of the *Base-Plate* and the *Bearing* parts.

2. In the *Relationship* panel, select the **Constrain** command by left-clicking once on the icon.

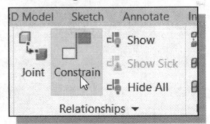

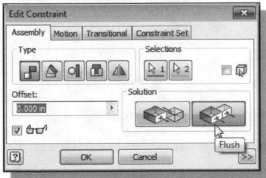

3. In the *Place Constraint* dialog box, switch the *Solution* option to **Flush** as shown.

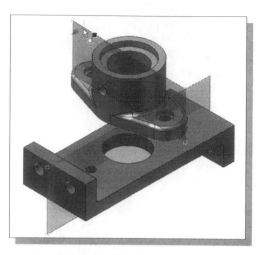

4. Select the *work plane* of the *Base-Plate* part as the first part for the Flush alignment command.

5. Select the corresponding *work plane* of the *Bearing* part as the second part for the Flush alignment command.

❖ Note that the Flush constraint makes two planes coplanar with their faces aligned in the same direction.

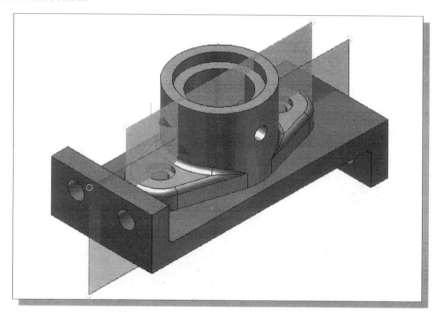

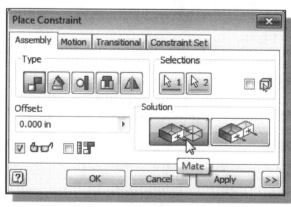

6. In the *Place Constraint* dialog box, switch the *Solution* option to **Mate** and notice the *Bearing* part is rotated 180 degrees to satisfy the Mate constraint.

❖ Note that the *Show Preview* option allows us to preview the result before accepting the selection.

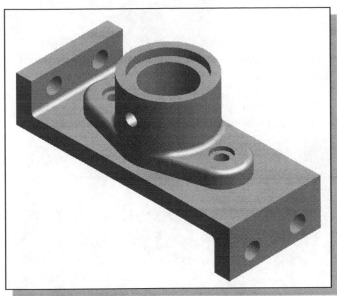

7. On your own, align the two parts as shown and click on the **Apply** button to accept the settings.

8. In the *Place Constraint* dialog box, click on the **Cancel** button to exit the Place Constraint command.

❖ Note the DOF symbol disappears, which indicates the assembly is fully constrained.

Placing the Third Component

We will retrieve the *Collar* part as the third component of the assembly model.

1. In the *Assemble* panel (the toolbar that is located to the left side of the graphics window) select the **Place Component** command by left-clicking the icon.

2. Select the **Collar** design (part file: **Collar.ipt**) in the list window. And click on the **Open** button to retrieve the model.

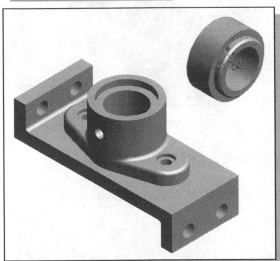

3. Place the *Collar* part toward the upper right corner of the graphics window, as shown in the figure.

4. Inside the *graphics window*, **right-click** once to bring up the option menu and select **OK** to end the placement of the *Collar* part.

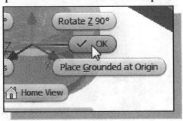

❖ Notice the DOF symbol displayed on the screen. The *Collar* part can move linearly and rotate about the three axes (six degrees of freedom).

Applying an Insert Constraint

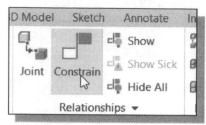

1. In the *Relationships* panel, select the **Constrain** command by left-clicking once on the icon.

2. In the *Place Constraint* dialog box, switch to the **Insert** constraint.

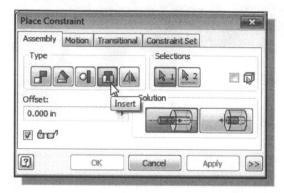

3. Select the inside corner of the *Collar* part as the first surface to apply the **Insert** constraint, as shown in the figure.

4. Select the inside circle on the top surface of the *Bearing* part as the second surface to apply the **Insert** constraint, as shown in the figure.

5. Click on the **Apply** button to accept the settings.

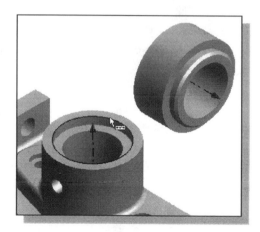

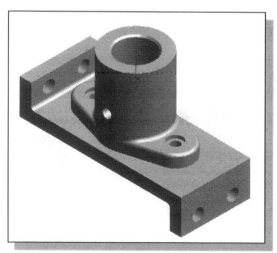

➤ Note that one rotational degree of freedom remains open; the *Collar* part can still freely rotate about the displayed DOF axis.

Assemble the Cap-Screws

We will place two of the *Cap-Screw* parts to complete the assembly model.

1. In the *Assemble* panel (the toolbar that is located to the left side of the graphics window) select the **Place Component** command by left-clicking once on the icon.

2. Select the ***Cap-Screw*** design (part file: ***Cap-Screw.ipt***) in the list window. And click on the **Open** button to retrieve the model.

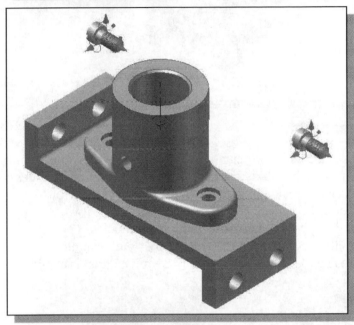

3. Place two copies of the *Cap-Screw* part on both sides of the *Collar* by clicking twice on the screen as shown in the figure.

4. Inside the *graphics window*, **right-click** once to bring up the option menu and select **OK** to end the Place Component command.

❖ Notice the DOF symbols displayed on the screen. Each *Cap-Screw* has six degrees of freedom. Both parts are referencing the same external part file, but each can be constrained independently.

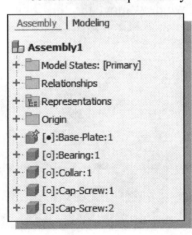

• Inside the *browser* window, the retrieved parts are listed in the order they are placed. The number behind the part name is used to identify the number of copies of the same part in the assembly model. Move the cursor to the last part name and notice the corresponding part is highlighted in the graphics window.

➢ On your own, use the **Place Constraints** command and assemble the *Cap-Screws* in place as shown in the figure below.

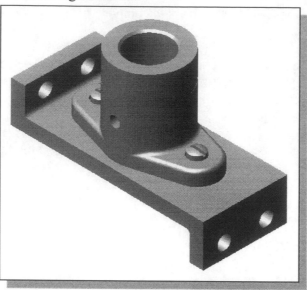

Exploded View of the Assembly

Exploded assemblies are often used in design presentations, catalogs, sales literature, and in the shop to show all of the parts of an assembly and how they fit together. In Autodesk Inventor, an exploded assembly can be created by two methods: (1) using the **Move Component** and **Rotate Component** commands in the *Assembly Modeler*, which contains only limited options for the operation but can be done very quickly; (2) transferring the assembly model into the *Presentation Modeler*. For our example, we will create an exploded assembly by using the **Move Component** command that is available in the *Assembly Modeler*.

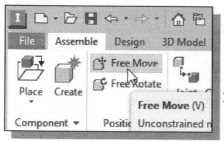

1. In the *Position* toolbar panel, select the **Free Move** command by left-clicking once on the icon.

2. Inside the graphics window, move the cursor on top of the top surface of the *Collar* part as shown in the figure.

3. Press and hold down the left-mouse-button and drag the *Collar* part toward the right side of the assembly as shown.

4. On your own, repeat the above steps and create an exploded assembly by repositioning the components as shown in the figure below.

5. Inside the graphics window, right-click once to bring up the option menu and select **OK** to end the **Move Component** command.

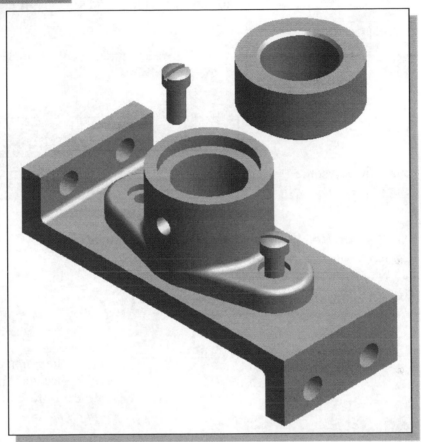

❖ The **Move Component** and **Rotate Component** commands are used to temporarily reposition the components in the graphics window. The displayed image is temporary, but it can be printed with the **Print** command through the pull-down menu.

6. Click on the **Update** button in the *Standard* toolbar area.

❖ Note that the components are reset back to their assembled positions, based on the applied assembly constraints.

Editing the Components

The *associative functionality* of Autodesk Inventor allows us to change the design at any level, and the system reflects the changes at all levels automatically.

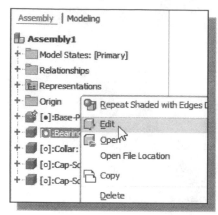

1. Inside the *Desktop Browser*, move the cursor on top of the **Bearing** part. Right-click once to bring up the option menu and select **Edit** in the option list.

❖ Note that we are automatically switched back to *Part Editing Mode*.

2. On your own, adjust the diameter of the small *Drill Hole* to **0.25** as shown.

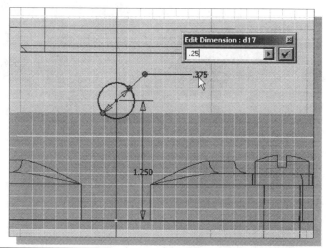

3. Click on the **Update** button in the *Standard* toolbar area to proceed with updating the model.

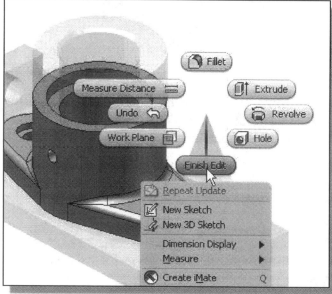

4. Inside the *graphics window*, click once with the **right-mouse-button** to display the *option menu*.

5. Select **Finish Edit** in the pop-up menu to exit *Part Editing Mode* and return to *Assembly Mode*.

➢ Autodesk Inventor has updated the part in all levels and in the current *Assembly Mode*. On your own, open the *Bearing* part file to confirm the modification is performed.

Adaptive Design Approach

Autodesk Inventor's ***Adaptive Design*** approach allows us to use the applied *assembly constraints* to control the sizes, shapes, and positions of ***underconstrained*** sketches, features, and parts. In this section, we will examine the procedure to apply the constraints directly on 3D parts; note that the adaptive design approach is also applicable to 2D sketches.

1. In the *Assemble* panel, select the **Create Component** command by left-clicking once on the icon.

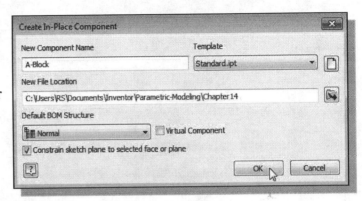

2. In the *Create In-Place Component* dialog box, enter **A-Block** as the new file name.

3. Set the *File Location* to **Chapter14** as shown.

4. Click on the **OK** button to accept the settings.

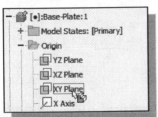

5. Click on the **XY Plane** of the *Base-Plate* part, inside the *browser* window, to apply a Flush constraint.

6. Activate the *Model* toolbar and select the **2D Sketch** command by left-clicking once on the icon.

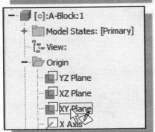

7. Select the **XY Plane**, in the *browser* window of the *A-Block* part, to align the sketch plane.

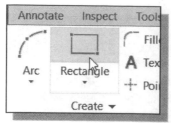

8. Select the **Two point rectangle** command by clicking once with the left-mouse-button on the icon in the *2D Sketch* panel.

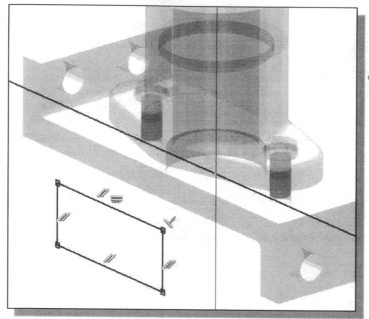

9. On your own, create a rectangle of arbitrary size below the assembly model as shown in the figure.

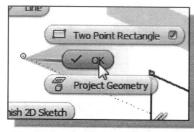

10. Inside the *graphics window*, click once with the **right-mouse-button** to display the option menu and select **OK** to end the Rectangle command.

11. Exit the sketch mode by clicking the **Finish Sketch** button.

12. In the *Ribbon* toolbar panel, select the **Extrude** command in the *Model tab* as shown in the figure below.

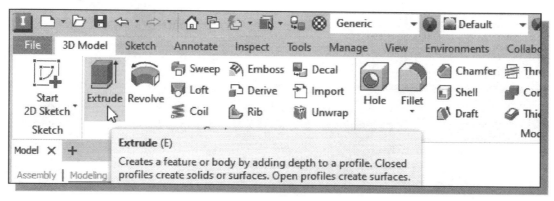

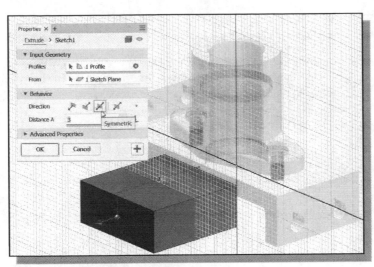

13. Inside the *Extrude* dialog box, select the **Symmetric** option and set the *Extents Distance* to **3 in** as shown.

14. Click on the **OK** button to proceed with creating the feature.

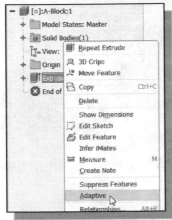

15. Right-click on the ***Extrusion1*** feature of the *A-Block* part, inside the *browser* window, and select **Adaptive** to allow the use of the adaptive design approach.

16. Click inside the graphics window to deselect any part or assembly.

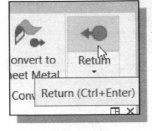

17. Select **Return** in the Ribbon toolbar to exit *Part Editing Mode* and return to *Assembly Modeling Mode*.

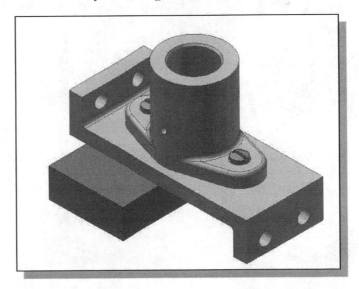

• Note that the rectangular block is created with only one dimension, the extrude distance. The 2D sketch of the part is intentionally under-constrained.

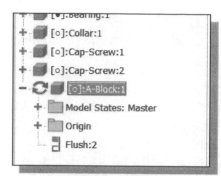

- Note the **Adaptive** icon, the two-rotate-arrows symbol, appears in front of the part name in the *browser* window.

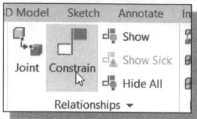

18. In the *Assemble* panel, select the **Constrain** command by left-clicking once on the icon.

19. On your own, use the **Flush** constraint to align the *A-block* to the left-edge of the *Base-Plate* as shown.

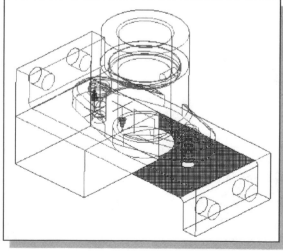

20. Create a **Mate** constraint to align the top of the *A-block* to the bottom of the *Base-Plate* part as shown.

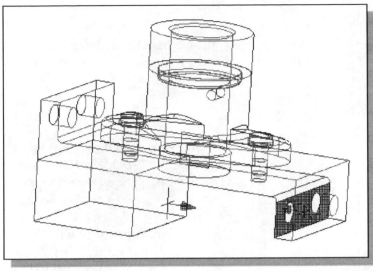

21. Use the **Mate** constraint and align the right surface of the *A-block* to the bottom left surface of the *Base-Plate* part as shown.

- Note that the length of the *A-block* part is adjusted to fit the defined constraint.

22. In the *Place Constraint* dialog box, click on the **Cancel** button to end the Place Constraint command.

Delete and Re-apply Assembly Constraints

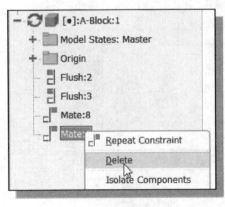

1. Inside the *browser* window, right-click on the last **Mate** constraint of the *A-Block* part to bring up the option menu.

2. Select **Delete** to remove the applied constraint.

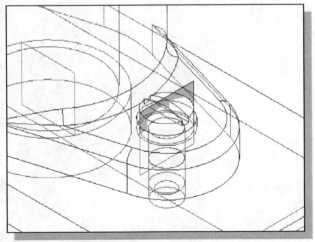

3. On your own, switch *ON* the **Visibility** of the vertical work plane of the *Cap-Screw* part, the plane that is perpendicular to the length of the *Base-Plate* part, as shown.

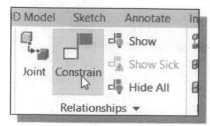

4. In the *Assembly* position panel, select the **Constrain** command by left-clicking once on the icon.

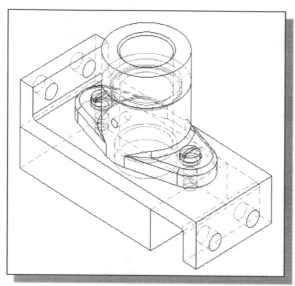

5. On your own, align the right vertical surface of the *A-block* part to the vertical work plane as shown in the figure.

6. On your own, experiment with aligning the right vertical surface of the *A-block* part to other vertical surfaces of the assembly model.

• As can be seen, the length of the *A-block* is adjusted to the newly applied constraint. The ***Adaptive Design*** approach allows us to have greater flexibility and simplifies the design process.

7. On your own, save the assembly model as **Shaft-Support.iam** under the Chapter14 folder.

Set up a Drawing of the Assembly Model

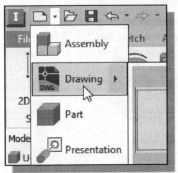

1. Click on the **drop-down arrow** next to the **New File** icon in the *Quick Access* toolbar area to display the available New File options.

2. Select **Drawing** from the option list.

3. Click on the **Base View** in the *Drawing Views* panel to create a base view.

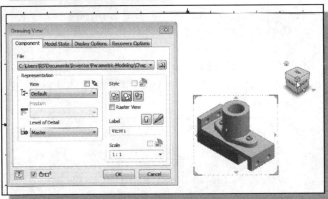

4. In the *Drawing View* dialog box, set *Orientation* to **Iso Top Right View, Scale 1:1** and **Hidden Line Removed** as shown in the figure.

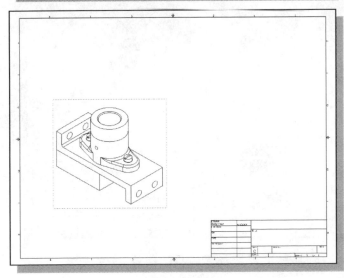

5. Move the cursor inside the graphics window and place the **base** view near the lower left side of the *Border* as shown.

6. Click the **OK** button to end the command.

❖ Note that the default sheet size is much bigger than the created view. *Inventor* allows us to adjust the sheet size even when views have been created.

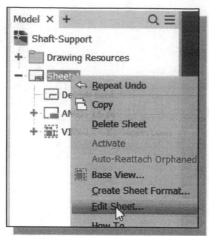

7. Inside the *Drawing Browser* window, right-click on **Sheet:1** to display the option menu.

8. Select **Edit Sheet** in the option menu to display the settings for the drawing.

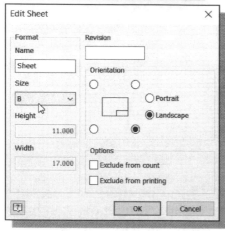

9. Set the sheet size to **B-size** as shown.

10. Click on the **OK** button to accept the settings and exit the Edit Sheet command.

11. On your own, reposition the Isometric view as shown in the figure.

❖ Note that the *Border* and *Title Block* are automatically replaced as the sheet size is adjusted.

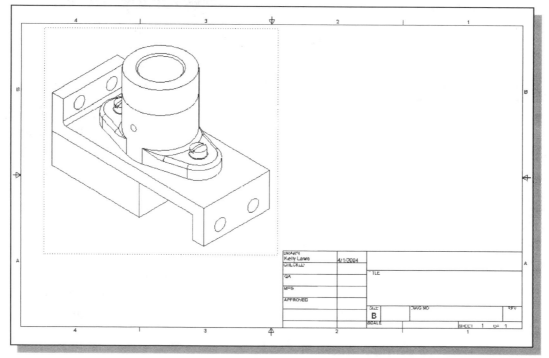

Creating a Parts List

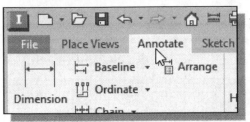

1. In the *Ribbon Toolbar*, click on the **Annotate** tab as shown.

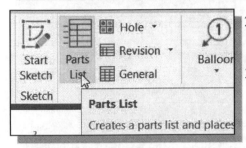

2. In the *Drawing Annotation* window, click on the **Parts List** button.

3. In the *prompt area*, the message "*Select a view*" is displayed. Items in the selected view will be listed in the *Parts List*. Select the **base view**.

❖ The *Parts List* dialog box appears on the screen; options are available to make adjustments to the numbering system and table wrap settings.

➢ The *Parts List* – BOM options are as follows:

Structured: Creates a parts list in which subassemblies are assigned using a nested numbering system (for example, 1, 1.1, 1.1.1). The nested number extends as many levels as needed for the assembly levels in the model.

Only Parts: Creates a parts list that sequentially numbers all parts in the assembly, including parts that are contained in subassemblies.

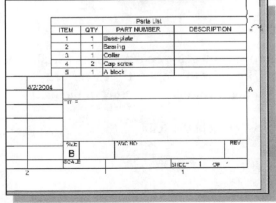

❖ Note that no subassemblies are currently used in the assembly; we will accept the default settings.

4. Click **OK** to accept the default settings and also enable the **BOM View** option.

5. Place the *Parts List* above the *Title Block* as shown.

Edit the Parts List

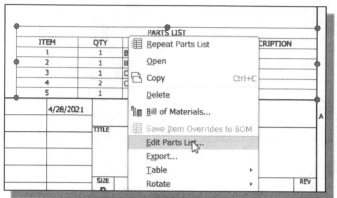

1. Move the cursor on top of the *Parts List* and **right-click** once to display the option menu.

2. Choose **Edit Parts List** in the option menu as shown.

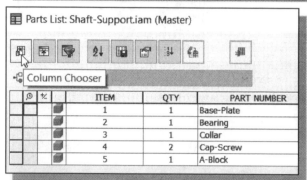

3. Click on the **Column Chooser** button as shown.

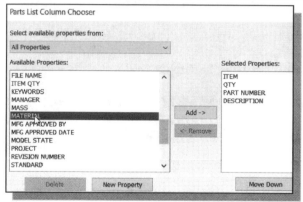

4. Select **MATERIAL** in the *Available Properties* list as shown.

5. Click **Add** to add the selected item to the *Selected Properties* list.

6. On your own, adjust the *Selected Properties* list as shown. (Hint: Use the **Move Up** and **Move Down** buttons to arrange the order of the list.)

7. Click **OK** to accept the settings.

❖ The *Parts List* is adjusted using the new settings. Note that, currently, all of the parts are using the **Generic** material type.

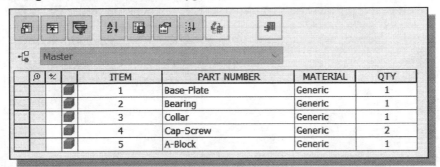

			ITEM	PART NUMBER	MATERIAL	QTY
			1	Base-Plate	Generic	1
			2	Bearing	Generic	1
			3	Collar	Generic	1
			4	Cap-Screw	Generic	2
			5	A-Block	Generic	1

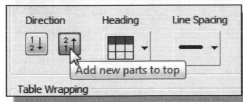

8. Click on the **Table Layout** button as shown.

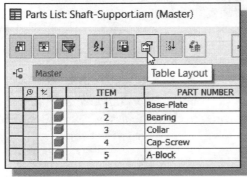

9. Set the *Table Direction* to **Add new parts to top** as shown.

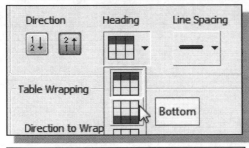

10. Set the *Heading Placement* to **Bottom** as shown.

11. Turn **OFF** the *Parts List Title*.

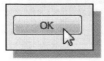

12. Click **OK** twice to accept the *Parts List* settings.

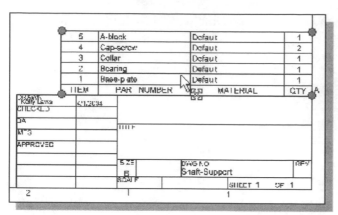

13. On your own, adjust the position of the *Parts List* so that it is aligned to the top edge of the *Title Block* as shown. (Hint: Look for the MOVE symbol next to the cursor.)

Change the Material Type

We will switch back to the assembly model to change the assignments of the material type.

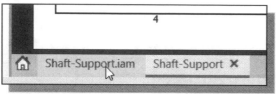

1. Click on the *Shaft-Support.iam* tab to switch back to the assembly model.

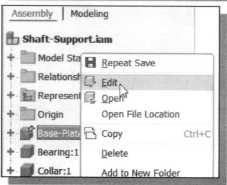

2. Inside the *Model Browser* window, right-click on **Base-Plate:1** to display the option menu.

3. Select **Edit** in the option menu to enter the *Edit Mode* for the selected part.

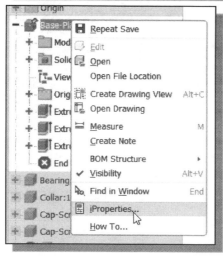

4. Inside the *Model Browser* window, select the **Base-plate** part by clicking once with the left-mouse-button.

5. Right-click on **Base-plate:1** to display the option menu and choose **iProperties** in the options list.

6. Click on the **Physical** tab in the *Base-plate.ipt Properties* window.

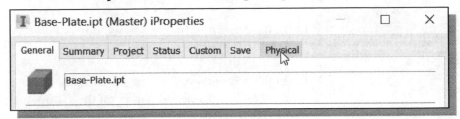

7. Choose the **Steel, Mild** in the *Material* list. Note the properties of the selected material are also displayed in the *Properties* list as shown in the figure.

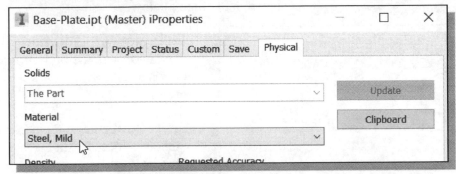

❖ Autodesk Inventor comes with many materials that have pre-entered information; additional material types/properties can also be added/changed as well.

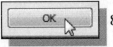

8. Click **OK** to accept the setting and exit the *Materials* dialog box.

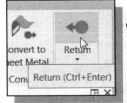

9. Select **Return** in the Standard toolbar to exit the *Part Editing Mode*.

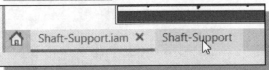

10. On your own, switch back to the *Shaft-Support.idw* window and notice the *Material* information for the **Base-plate** part is now updated.

11. On your own, repeat the above steps to change the material information for the other parts as shown in the figure below.

5	A-Block	Aluminum 6061	1
4	Cap-Screw	Steel, Mild	2
3	Collar	Bronze, Soft Tin	1
2	Bearing	Iron, Cast	1
1	Base-Plate	Steel, Mild	1
ITEM	PART NUMBER	MATERIAL	QTY

Add the Balloon Callouts

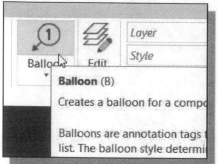

1. In the *Drawing Annotation* window, click on the **Balloon** button.

2. In the prompt area, the message "*Select a location*" is displayed. Click on the ***Collar*** part to attach an arrowhead to the part.

3. Pick another location to **place the balloon** as shown in the figure below.

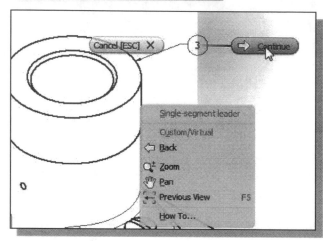

4. Inside the graphics window, click once with the right-mouse-button to display the **option menu**.

5. Select **Continue** in the pop-up menu to proceed with the creation of the balloon.

Completing the Title Block Using the iProperties option

1. Inside the *Model Browser* window, select the **Shaft-Support** drawing by clicking once with the left-mouse-button.

2. Right-click on **Shaft-Support** to display the option menu and choose **iProperties** in the options list.

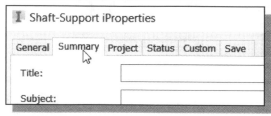

3. Click on the **Summary** tab to view the list of general information regarding the design.

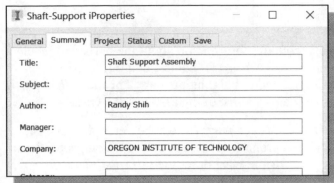

4. Enter the **Title**, **Author** and **Company** information as shown.

5. Click on the Project tab and enter the Part Number, Project name as shown.

6. Click **OK** to accept the setting and exit the *iProperties* dialog box.

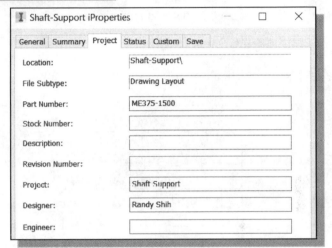

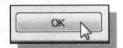

7. On your own, add the additional balloons and complete the drawing as shown.

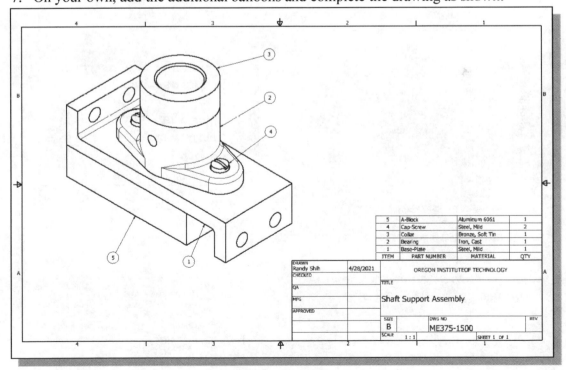

Bill of Materials

A bill of materials (BOM) is a table that contains information about the parts within an assembly. The BOM can include information such as part names, quantities, costs, vendors, and all of the other information related to building the part. The *parts list*, which is used in an assembly drawing, is usually a partial list of the associated BOM.

In Autodesk Inventor, both the *bill of materials* and *parts list* can be derived directly from data generated by the assembly and the part properties. We can select which properties to be included in the *bill of materials* or *parts list*, in what order the information is presented, and in what format to export the information. The exported file can be used in an application such as a spreadsheet or text editor.

(a) BOM from Parts List

1. Move the cursor on top of the *Parts List* and right-click once to display the option menu.

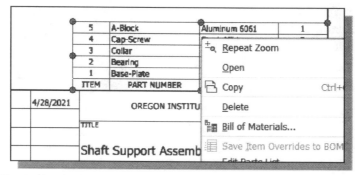

2. Choose **Export...** in the option menu as shown.

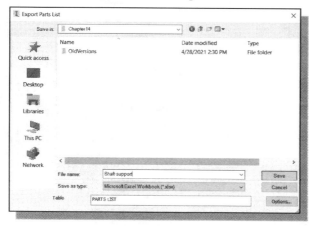

3. Confirm the *Save as type* list is set to **Microsoft Excel**.

4. Enter *Shaft-support* as the filename and click **Save** to export the BOM.

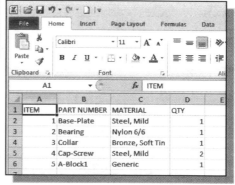

❖ On your own, examine the exported BOM by opening up the file in *Excel*.

(b) BOM from Assembly Model

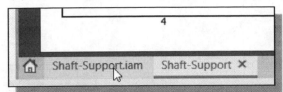

1. Click on the ***Shaft-Support.iam*** tab to switch back to the assembly model.

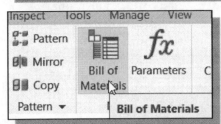

2. Select **Bill of Materials** in the *Manage* tab of the *Ribbon* toolbar panel.

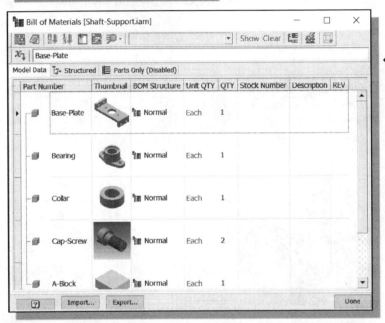

❖ Note that many of the controls and options are similar to those of the **Parts List** command in the *Drawing Mode*.

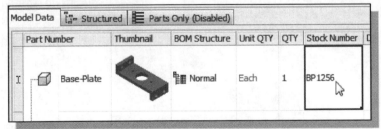

3. Click inside the ***Stock Number*** box to enter the *Edit Mode*.

4. Enter **BP1256** as the new *Stock Number*.

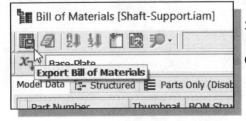

5. Click the **Export Bill of Materials** button.

6. On your own, export using the Microsoft Excel format and examine the *BOM* in Microsoft Excel.

Review Questions: (Time: 35 minutes)

1. What is the purpose of using *assembly constraints*?

2. List three of the commonly used *assembly constraints*.

3. Describe the difference between the **Mate** constraint and the **Flush** constraint.

4. In an assembly, can we place more than one copy of a part? How is it done?

5. How should we determine the assembly order of different parts in an assembly model?

6. How do we adjust the information listed in the **parts list** of an assembly drawing?

7. In Autodesk Inventor, describe the procedure to create a **bill of materials** (BOM)?

8. Create sketches showing the steps you plan to use to create the four parts required for the assembly shown on the next page:

Ex.1)

Ex.2)

Ex.3)

Ex.4)

Exercises: (Time: 180 minutes.)

1. **Wheel Assembly** (Create a set of detail and assembly drawings. All dimensions are in mm.)

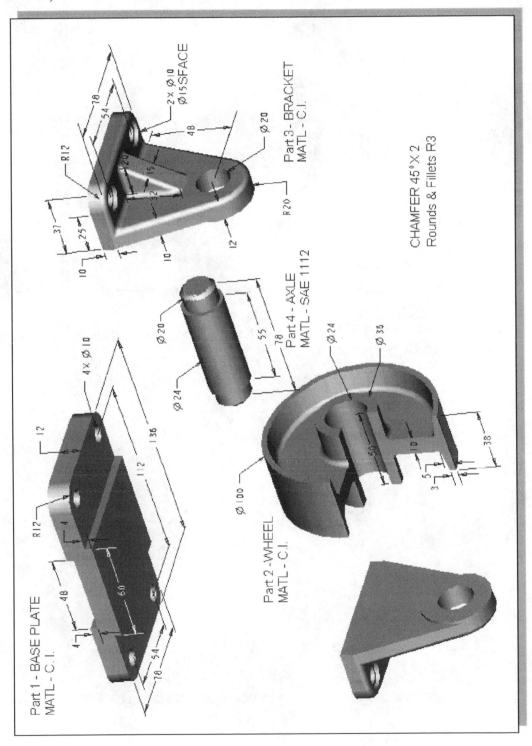

2. **Vise Assembly** (Create a set of detail and assembly drawings. All dimensions are in inches.)

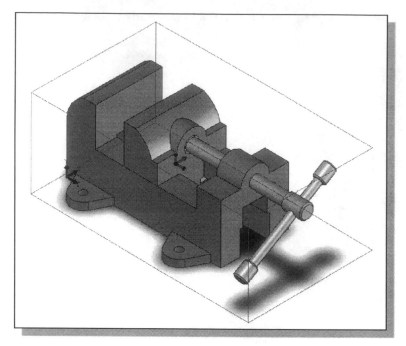

(a) Base: The 1.5 inch wide and 1.25 inch wide slots are cut through the entire base. Material: **Gray Cast Iron**.

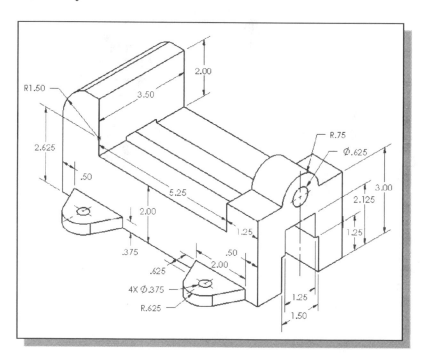

(b) Jaw: The shoulder of the jaw rests on the flat surface of the base and the jaw opening is set to 1.5 inches. Material: **Gray Cast Iron**.

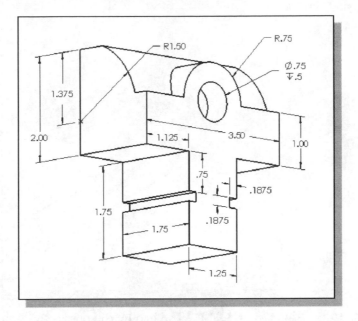

(c) Key: 0.1875 inch H x 0.3125 inch W x 1.75 inch L. The keys fit into the slots on the jaw with the edge faces flush as shown in the sub-assembly to the right. Material: **Alloy Steel**.

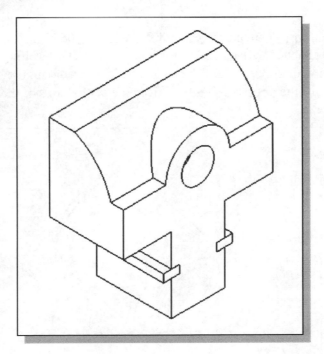

(d) Screw: There is one chamfered edge (0.0625 inch x 45°). The flat ∅ 0.75″ edge of the screw is flush with the corresponding recessed ∅ 0.75 face on the jaw. Material: **Alloy Steel**.

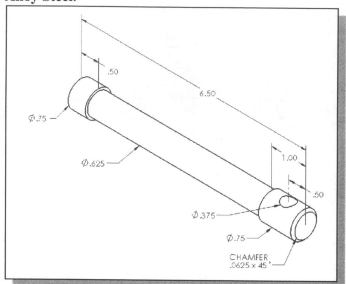

(e) Handle Rod: ∅ **0.375″ x 5.0″ L**. The handle rod passes through the hole in the screw and is rotated to an angle of 30° with the horizontal as shown in the assembly view. The flat ∅ 0.375″ edges of the handle rod are flush with the corresponding recessed ∅ 0.735 faces on the handle knobs. Material: **Alloy Steel**.

(f) Handle Knob: There are two chamfered edges (0.0625 inch x 45°). The handle knobs are attached to each end of the handle rod. The resulting overall length of the handle with knobs is 5.50″. The handle is aligned with the screw so that the outer edge of the upper knob is 2.0″ from the central axis of the screw. Material: **Alloy Steel**.

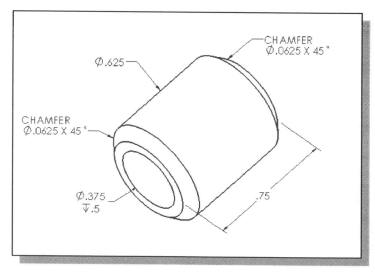

3. **Leveling Assembly** (Create a set of detail and assembly drawings. All dimensions are in mm.)

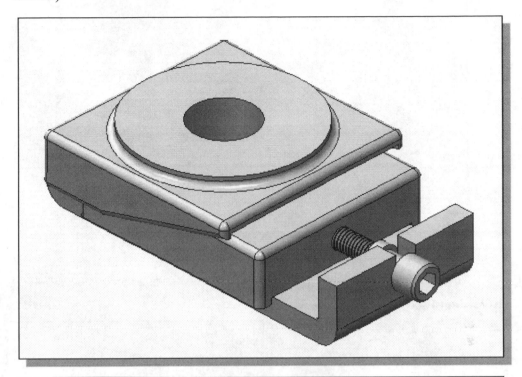

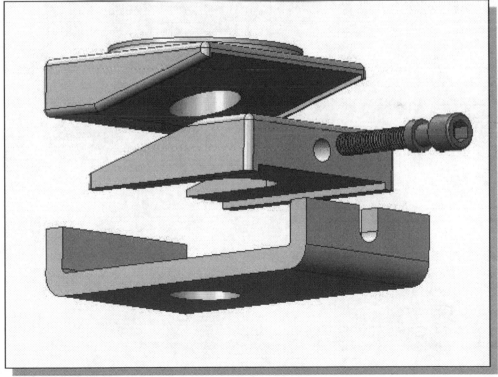

(a) Base Plate

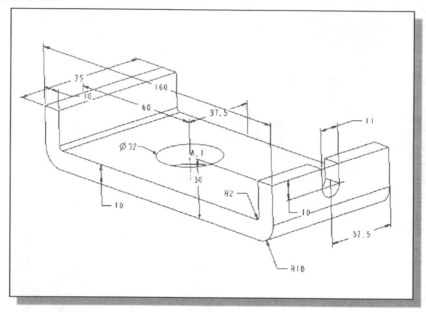

(b) Sliding Block (Rounds & Fillets: R3)

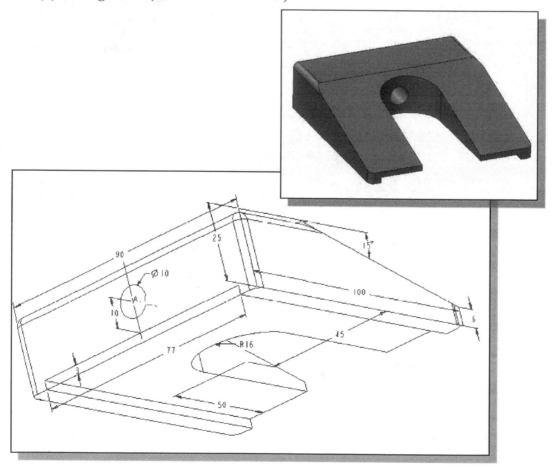

(c) Lifting Block (Rounds & Fillets: R3)

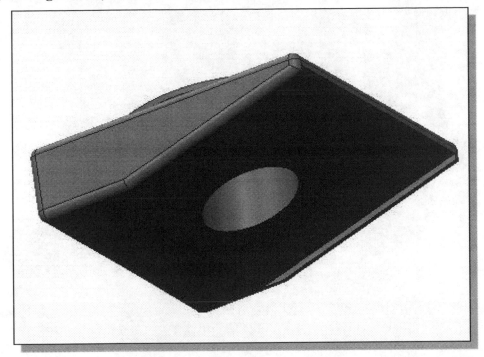

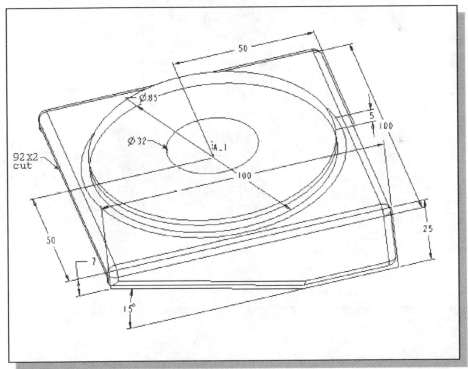

(d) Adjusting Screw (M10 × 1.5) (Use the **Threads** or **Coil** command to create the threads.)

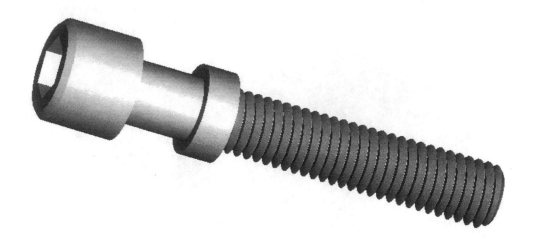

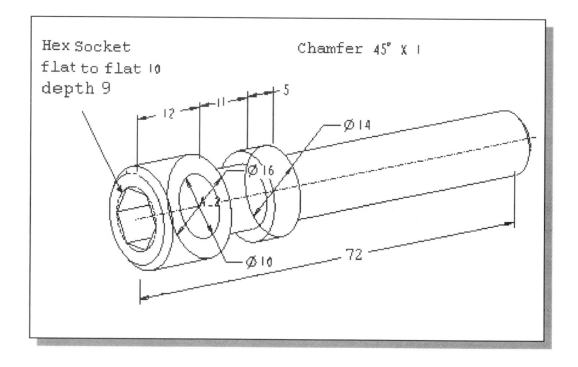

Chapter 15
Introduction to Stress Analysis

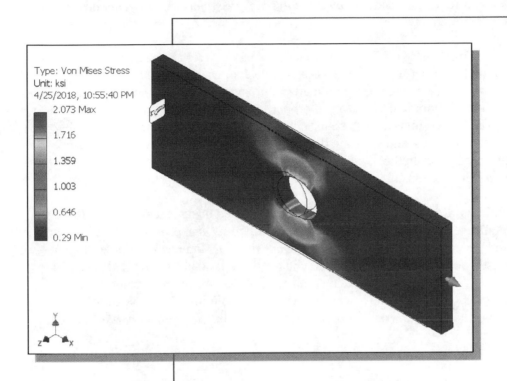

Type: Von Mises Stress
Unit: ksi
4/25/2018, 10:55:40 PM

2.073 Max

1.716

1.359

1.003

0.646

0.29 Min

Learning Objectives

- ◆ **Create FEA Study**
- ◆ **Apply Fixtures and Loads**
- ◆ **Perform Basic Stress Analysis**
- ◆ **View Results**
- ◆ **Assess Accuracy of Results**
- ◆ **Output the Associated FEA Report File**

Introduction

In this chapter we will explore basic design analysis using the *Inventor Stress Analysis Module*. The *Stress Analysis Module* is a special module available for part, sheet metal, and assembly documents. The *Stress Analysis Module* has commands unique to its purpose. With Autodesk Inventor, *contact analysis, frame analysis* and *dynamic analysis* can also be performed.

Inventor Stress Analysis Module provides a tool for basic stress analysis, allowing the user to examine the effects of applied forces on a design. Displacements, strains and stresses in a part are calculated based on material properties, fixtures, and applied loads. Stress results can be compared to material properties, such as yield strength, to perform failure analysis. The results can also be used to identify critical areas, calculate safety factors at various regions, and simulate deformation. *Inventor Stress Analysis Module* provides an easy-to-use method within the Autodesk Inventor's *Stress Analysis Module* to perform an initial stress analysis. The results can be used to improve the design.

In *Inventor Stress Analysis Module*, stresses are calculated using **linear static analysis** based on the **finite element method**. *Linear static analysis* is appropriate if deflections are small and vary only slowly. *Linear static analysis* omits time as a variable. It also excludes plastic action and deflections that change the way loads are applied. The *finite element method (FEM)* is a numerical method for finding approximate solutions to complex systems. The technique is widely used for the solution of complex problems in engineering mechanics. Analysis using the method is called *finite element analysis (FEA)*.

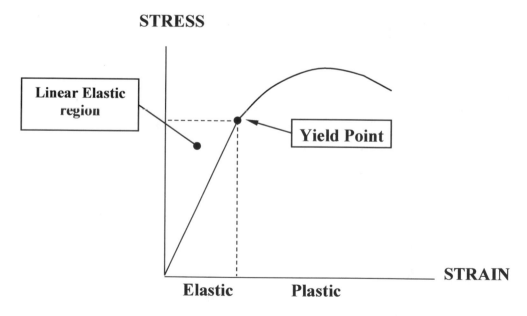

Stress-Strain diagram of typical ductile material

In the finite element method, a complex system is modeled as an equivalent system of smaller bodies of simple shapes, or *elements*, which are interconnected at common points called *nodes*. This process is called *discretization*; an example is shown in the figures below. The mathematical equations for the system are formulated first for each finite element, and the resulting system of equations is solved simultaneously to obtain an approximate solution for the entire system. In general, a better approximation is obtained by increasing the number of elements, which will require more computing time and resources.

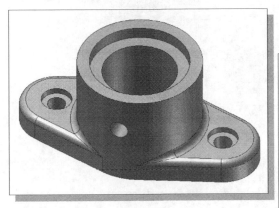
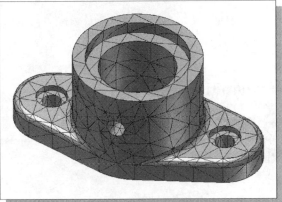

Inventor Stress Analysis Module utilizes tetrahedral elements for which the edges and faces can be curvilinear and which allow the modeling of curved surfaces, as seen in the figure above. The behavior of these elements is analyzed using linear static analysis and the appropriate material properties to relate local coordinate nodal displacements to local forces. The motion of each node is described by displacements in the X, Y, and Z directions, called *degrees of freedom* (DOFs). The equations describing the behavior of each element are assembled into a global system of equations, incorporating compatibility requirements based on connectivity among elements. Using the known material properties, supports, and loads, *Inventor Stress Analysis Module* solves the system of equations for the unknown displacements at each node. These displacements are used in the results stage to calculate strains and stresses.

While *Inventor Stress Analysis Module* is a powerful and easy-to-use tool, it is important to appreciate that it is the designer's responsibility to properly assess the accuracy of the results. A better FEA approximation is generally obtained by increasing the number of elements. An assessment must be made regarding the mesh used to discretize the model to ensure it is adequate. There are other important factors affecting the accuracy of the results. The material properties used in the analysis must accurately characterize the behavior of the material. The supports and loads must be applied in a manner which accurately reflects the actual conditions. Proper meshing and application of boundary conditions often require significant experience in FEA and may require tools and capabilities not available in *Inventor Stress Analysis Module*. *Inventor Stress Analysis Module* is an easy-to-use tool for a quick stress analysis.

Problem Statement

Determine the maximum normal stress that loading produces in the aluminum-6061 plate.

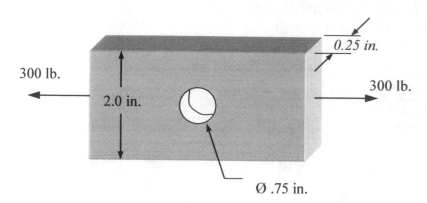

Preliminary Analysis

- **Maximum Normal Stress**

The nominal normal stress developed at the smallest cross section (through the center of the hole) in the plate is

$$\sigma_{nominal} = \frac{P}{A} = \frac{300}{(2 - 0.75) \times .25} = 960 \text{ psi.}$$

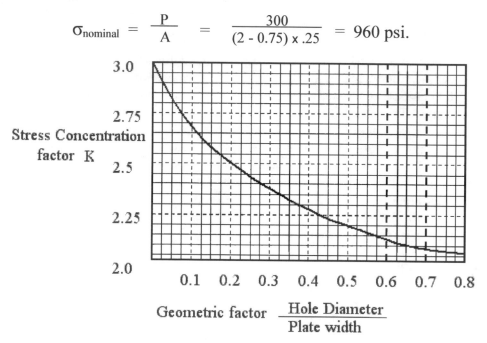

Geometric factor = .75/2 = 0.375

Stress concentration factor K is obtained from the graph, **K = 2.27**

$$\sigma_{MAX} = K \, \sigma_{nominal} = 2.27 \times 960 = 2180 \text{ psi.}$$

- **Maximum Displacement**

We will also estimate the displacement under the loading condition. For a statically determinant system, the stress results depend mainly on the geometry. The material properties can be in error and still the FEA analysis comes up with the same stresses. However, the displacements always depend on the material properties. Thus, it is necessary to always estimate both the stress and displacement prior to a computer FEA analysis.

The classic one-dimensional displacement can be used to estimate the displacement of the problem:

$$\delta = \frac{PL}{EA}$$

Where P=force, L=length, A=area, E= elastic modulus, and δ = deflection.

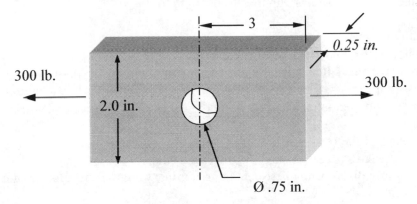

A lower bound of the displacement of the right edge, measured from the center of the plate, is obtained by using the full area:

$$\delta_{lower} = \frac{PL}{EA} = \frac{300 \times 3}{10E6 \times (2 \times 0.25)} = 1.8E\text{-}4 \text{ in.}$$

and an upper bound of the displacement would come from the reduced section:

$$\delta_{upper} = \frac{PL}{EA} = \frac{300 \times 3}{10E6 \times (1.25 \times 0.25)} = 2.88E\text{-}4 \text{ in.}$$

but the best estimate is a sum from the two regions:

$$\delta_{average} = \frac{PL}{EA} = \frac{300 \times 0.375}{10E6 \times (1.25 \times 0.25)} + \frac{300 \times 2.625}{10E6 \times (2.0 \times 0.25)}$$

$$= 3.6E\text{-}5 + 1.58E\text{-}4 = 1.94E\text{-}4 \text{ in.}$$

Finite Element Analysis Procedure

In the previous section, an approximate preliminary analysis was performed prior to carrying out the finite element analysis; this will help us to gain some insights into the problem and also serves as a means of checking the finite element analysis results.

For a typical linear static analysis problem, the finite element analysis requires the following steps:

1. Preliminary Analysis.

2. Preparation of the finite element model:
    a. Model the problem into finite elements.
    b. Prescribe the geometric and material information of the system.
    c. Prescribe how the system is supported.
    d. Prescribe how the loads are applied to the system.

3. Perform calculations:
    a. Generate a stiffness matrix of each element.
    b. Assemble the individual stiffness matrices to obtain the overall, or global, stiffness matrix.
    c. Solve the global equations and compute displacements, strains, and stresses.

4. Post-processing of the results:
    a. Viewing the stress contours and the displaced shape.
    b. Checking any discrepancy between the preliminary analysis results and the FEA results.

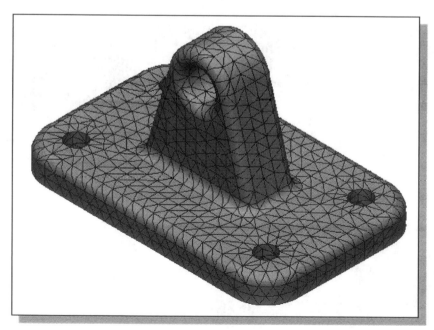

Create the Autodesk Inventor Part

1. Select the **Autodesk Inventor** option on the *Start* menu or select the **Autodesk Inventor** icon on the desktop to start Autodesk Inventor. The Autodesk Inventor main window will appear on the screen.

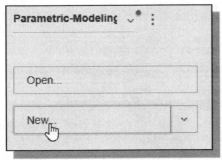

2. Select the **New File** icon with a single click of the left-mouse-button as shown.

3. Select the **English** tab, and in the *New File* area select **Standard(in).ipt**.

4. Pick **Create** in the *New File* dialog box to accept the selected settings.

Create the 2D Sketch for the Plate

1. In the *3D Model* tab select the **Start 2D Sketch** command by left-clicking once on the icon.

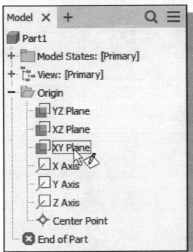

2. In the *Status Bar* area, the message "*Select plane to create sketch or an existing sketch to edit.*" is displayed. Select the **XY Plane** by clicking the associated item in the graphics area or inside the *Model* history tree window or in the graphics window.

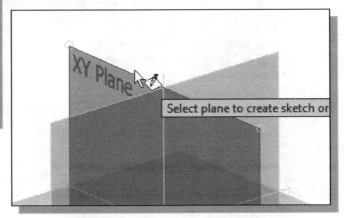

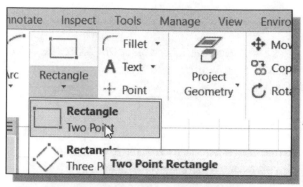

3. Select the **Two point rectangle** command by clicking once with the left-mouse-button on the icon in the *Sketch* toolbar.

4. Create a rectangle of arbitrary size positioned near the center of the screen.

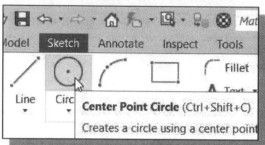

5. Select the **Center Point Circle** command by clicking once with the left-mouse-button on the icon in the *Sketch* toolbar.

6. Create a **circle** of arbitrary size and aligned to the *center point* as shown.

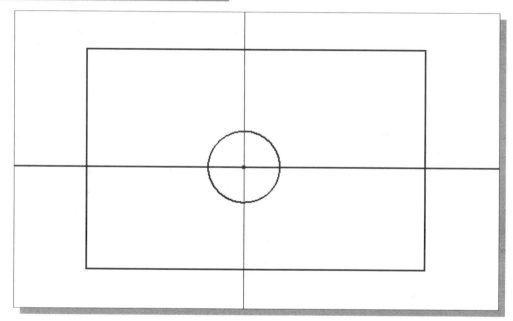

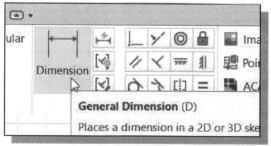

7. Select the **General Dimension** command in the *Constrain* panel.

8. On your own, create and adjust the dimensions of the rectangle and circle as shown. (Hint: Use parametric equations to position the geometry.)

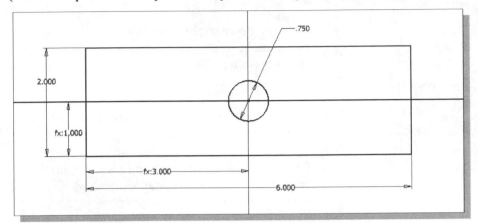

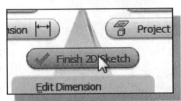

9. Inside the *graphics window*, click once with the right-mouse-button to display the option menu. Select **Finish 2D Sketch** in the pop-up menu to end the Sketch option.

10. In the *3D Model* toolbar, select the **Extrude** command by left-clicking on the icon.

11. On your own, create an **Extruded** feature with a thickness of **0.25 in** as shown.

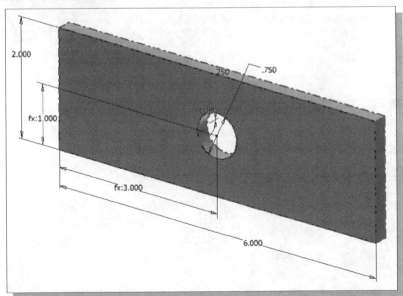

Assigning the Material Properties

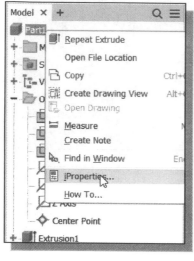

1. In the *browser*, **right-click** once on the *part name* to bring up the option menu, and then pick **iProperties** in the *pop-up* menu.

2. On your own, look at the different information listed in the *iProperties* dialog box.

3. Click on the **Physical** tab; this is the page that contains the physical properties of the selected model.

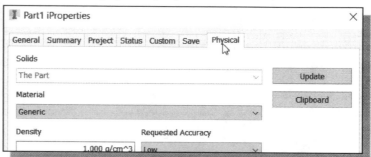

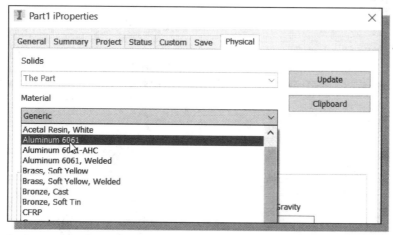

4. Click the down-arrow in the *Material* option to display the material list, and select **Aluminum-6061** as shown.

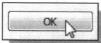

5. Click **OK** to accept the settings.

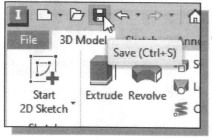

6. On your own, save the part with the filename **FEA_Al_Plate.ipt**.

Switch to the Stress Analysis Module

1. In the *Ribbon* toolbar, select the **Environments** tab as shown.

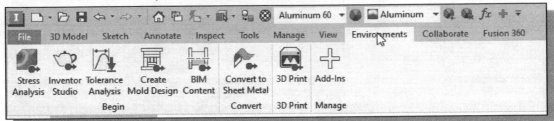

2. Click **Stress Analysis** to enter the *Inventor Stress Analysis Module*.

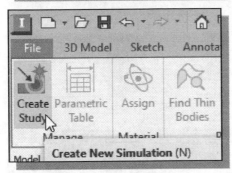

3. Click **Create New Simulation** to start a new simulation.

4. Note the default settings include (1) *Simulation Type* set to perform **Static Analysis** and (2) *Design Objective* set to **Single Point**. These settings are used for basic *linear static analysis*.

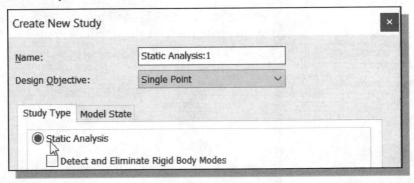

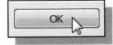
5. Click **OK** to accept the default settings.

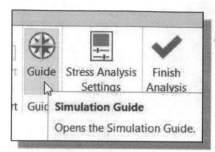

6. In the *Ribbon* toolbar, select the **Simulation Guide** icon as shown.

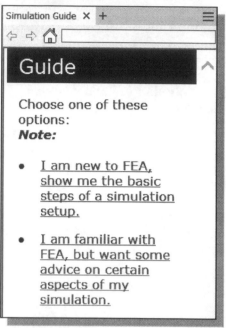

➢ Note the *Simulation Guide* provides an overall view of the Inventor FEA procedure. You are encouraged to read through the *Simulation Guide* to get familiar with the general procedure in performing the Inventor FEA simulation.

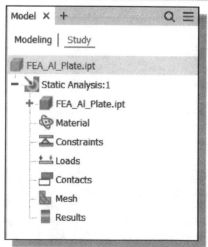

➢ Notice the *Ribbon* toolbar and the *browser* window, to the left side of the graphics area, now display items associated with the *Stress Analysis Simulation Module*. The list in the *browser* window shows the elements necessary to perform the *Finite Element Stress Analysis*.

7. In the *Ribbon* toolbar area, select **Assign Materials** to examine the material assignment for the part.

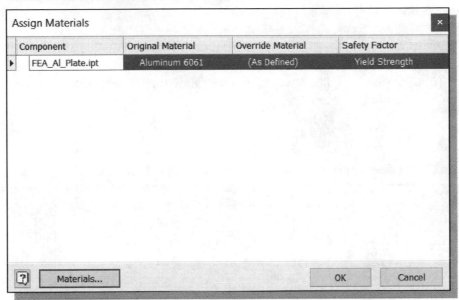

- Note that the assigned material, **Aluminum-6061**, is shown under the *Original Material*; the *Override Material* option is available for us to examine the effects of using different materials.

8. Click **OK** to accept the settings.

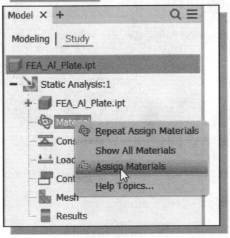

- Note that the same command can also be accessed through the *browser* window.

Apply Constraints and Loads

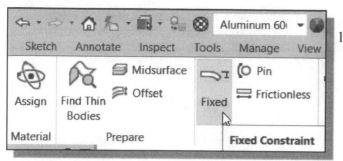

1. In the *Ribbon* toolbar area, select **Fixed Constraint** to assign the support condition of the plate.

2. Click on the **top left corner** of the *ViewCube* to rotate the display; the plate will be rotated 90 degrees showing the left vertical face of the plate.

3. Select the small **vertical surface** of the left-end of the plane as shown.

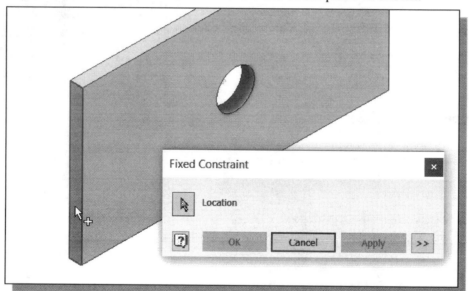

4. Note the selection of the surface is recognized as the selection label is changed to **Faces**. Click **OK** to accept the selection.

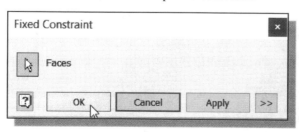

5. In the *Ribbon* toolbar area, select the **Force** command as shown.

6. Click on the **Home** icon above the *ViewCube* to reset the display back to the isometric view.

7. Click on the small vertical surface to the right end of the plate. In the *Force* dialog box, enter **300 lbf** as the force value and check the **Reverse Direction** option box. Notice the direction of the force load is now outward.

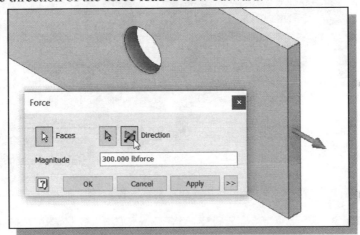

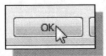

8. Click **OK** to accept the selection and exit the Force command.

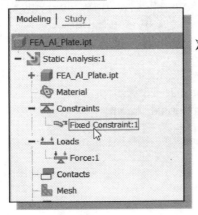

> Notice the Fixed constraint and Force load appear in the *browser* window to the left of the graphics area with the default names Constraint:1 and Force:1.

Create a Mesh and Run the Solver

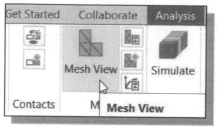

1. In the *Ribbon* toolbar area, select the **Mesh View** command as shown.

• Note that with the default settings, *Inventor Simulation* generated 642 nodes and 276 elements, which is a relatively coarse mesh.

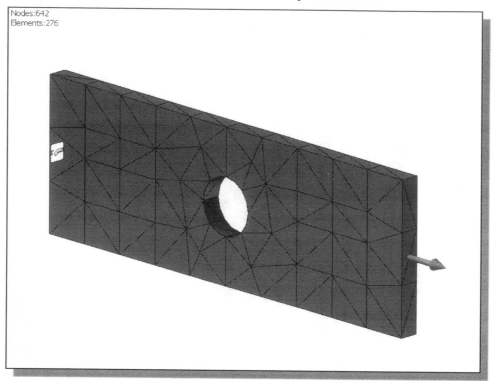

• As a rule in creating the first FEA mesh, start with a relatively small number of elements and progressively move to more refined models. The main objective of the first coarse mesh analysis is to obtain a rough idea of the overall stress distribution. In most cases, use of a complex and/or a very refined FEA model is not justifiable since it most likely provides computational accuracy at the expense of unnecessarily increased processing time.

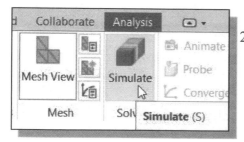

2. In the *Ribbon* toolbar area, select the **Simulate** command as shown.

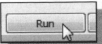

3. Click the **Run simulation** button to proceed with the FEA simulation.

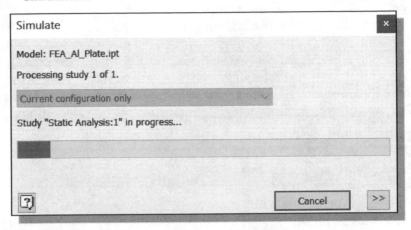

- Note that if the stress display does not look good, you might need to choose a different option of the *Visual Style* to find a better viewing of the stress result. Depending on the computer hardware, some settings might provide better viewing than others.

➢ A plot of equivalent stress (Von Mises stress) is generated and displayed in the graphics area. A color-coded scale is used with the associated scale bar displayed at the right. Red represents the regions of highest stress. The maximum stress is 2051 psi, which is well below the yield strength of the material (3999.3 psi) and is located at the stress concentration points at the upper and lower quadrants of the hole.

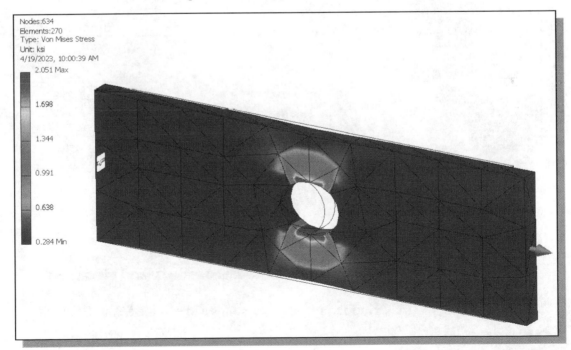

Refinement of the FEA Mesh – Global Element Size

In order to gain accuracy for complex geometries or to represent a highly varying stress distribution, finer elements must be used. The increase in the number of elements will also increase the solution time and computer disk space. This is why the refinement of mesh around the high stress areas is required. The process of mesh refinement is called convergence analysis. As our first analysis confirmed the stress concentration points at the upper and lower quadrants of the hole, we will next refine the mesh to obtain a more accurate FEA result. One way of refinement is simply to adjust the element size to a smaller value. This can be done through adjusting the **Mesh Settings**.

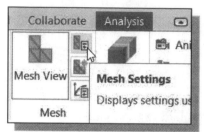

1. In the *Ribbon* toolbar area, select the **Mesh Settings** command as shown.

2. In the *Mesh Settings* dialog box, adjust the *Average Element Size* to **0.05**, which will create twice as many elements. Note the number is a fraction of the length dimension of the plate.

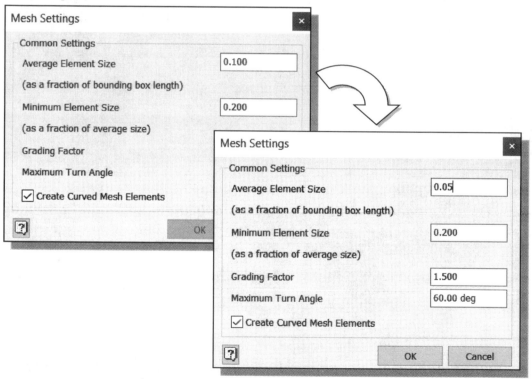

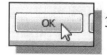

3. Click **OK** to accept the new settings and exit the **Mesh Settings** command.

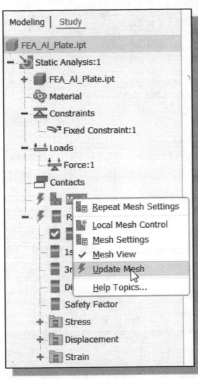

4. It is necessary to update the mesh with the new settings through the *browser* window. Right-click on the **Mesh** item to bring up the **option list**.

5. Select **Update Mesh** in the option list to activate the mesh update.

- Note that with the new settings, *Inventor Simulation* generated 2711 nodes and 1479 elements, which is almost three times more than the original coarse mesh.

6. On your own, perform the FEA analysis.

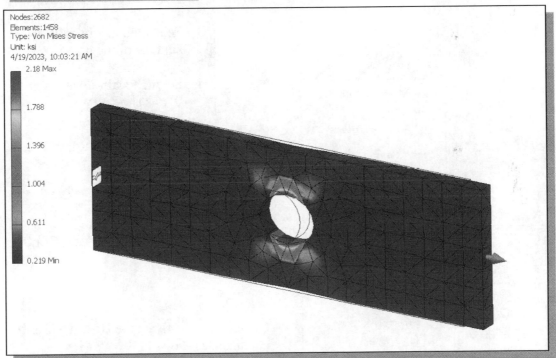

> The maximum stress with the refinement is now 2180psi, which is higher than the previous FEA analysis, and very close to the preliminary analysis. Next, we will further refine the mesh in the high stress area.

Refinement of the FEA Mesh – Local Element Size

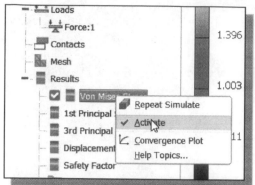

1. In the *browser* window, right-click on the **Von Mises Stress** item to bring up the *option list* and **de-activate** the result as shown.

2. In the *Ribbon* toolbar area, select the **Local Mesh Control** command as shown.

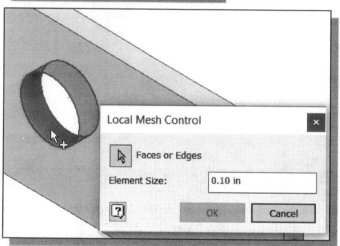

3. Select the **inside cylindrical surface** of the hole as shown.

4. Set the local *Element Size* to **0.1** as shown.

5. Click **OK** to accept the new setting and exit the Local Mesh Control command.

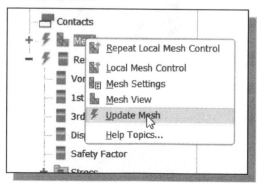

6. On your own, activate the **Update Mesh** command as shown.

- Note that with the new settings, *Inventor Simulation* generated 4378 nodes and 2404 elements, which is about seven times the original coarse mesh.

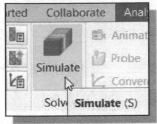

7. On your own, perform the FEA analysis.

- The maximum stress with the refinement is now 2203 psi, which is just a bit higher than the previous FEA analysis. The refinement caused only a small difference compared to the previous mesh.

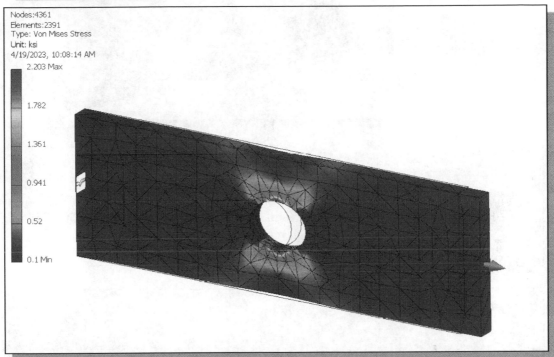

Nodes:4361
Elements:2391
Type: Von Mises Stress
Unit: ksi
4/19/2023, 10:08:14 AM

2.203 Max

1.782

1.361

0.941

0.52

0.1 Min

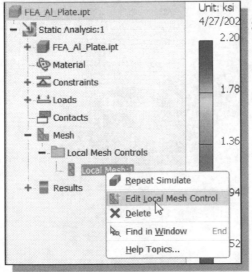

8. On your own, adjust the element size to 0.05 of the Local Mesh Control value as shown.

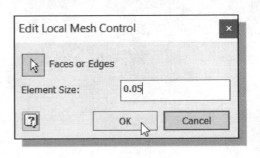

Edit Local Mesh Control

Faces or Edges

Element Size: 0.05

OK   Cancel

9. On your own, perform another FEA with the current settings.

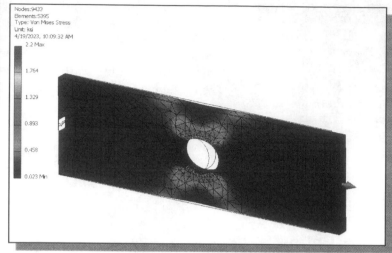

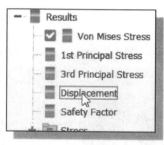

10. Double-click on the **Displacement** item in the *browser* window to display the displacement of the plate.

*   The FEA result showed the total displacement of 4.058e-4 inch, which matches quite well with the 3.88e-4 inch preliminary analysis estimate.

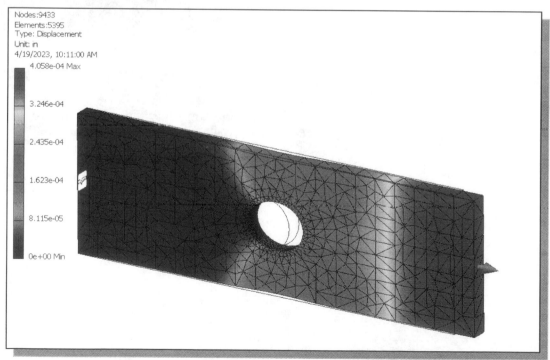

Comparison of Results

The accuracy of the *Inventor Simulation* results for this problem can be checked by comparing them to the analytical results presented earlier. In the Preliminary Analysis section, the maximum stress was calculated using a stress concentration factor and the value obtained was **2180 psi**. One should realize the analytical result is obtained through the use of charts from empirical data and therefore involves some degrees of error. The maximum stress obtained by finite element analysis using *Inventor Simulation* ranges from **2051** to **2180 psi**. In the Preliminary Analysis section, the maximum displacement was also estimated to be around **1.94E-4 inches**, measured from the center of the hole to one end of the plate. The maximum displacement obtained by finite element analysis using *Inventor Simulation* was around **2.0285E-4 inches**. The agreement between the analytical results and those from *Inventor Simulation* demonstrate the potential of *Inventor Simulation* as a very powerful design tool.

In FEA, the process of mesh refinement is called convergence analysis. For our analysis, the refinement of the mesh does show the FEA results converging near the analytical results. The refinement to the third mesh is quite adequate for our analysis. Any further refinement does not provide any additional insight and is therefore not necessary.

Number of Elements	σ_{max} (psi)	D_{max} (in)
270	2051	2.024e-4
1458	2180	2.027e-4
2391	2203	2.028e-4
5395	2200	2.029e-4

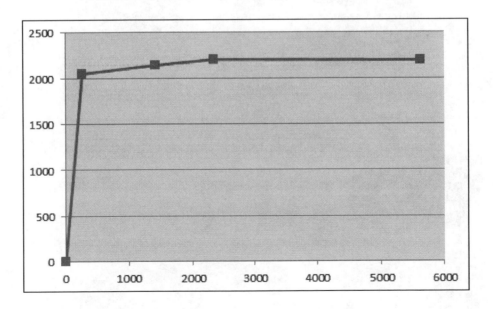

Create an HTML Report

Inventor Stress Analysis Simulation module also includes options to create a report in HTML format which contains data related to the simulation.

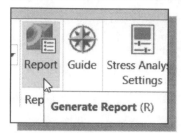

1. Select the **Report** option as shown.

2. On your own, examine the different options available for the report. Click **OK** to generate a report.

➢ *Inventor Stress Analysis Environment* will create an HTML file and automatically open it in your default Web browser. The HTML file is saved in the location shown in the *File Information* section at the beginning of the report.

3. On your own, read through the report. Notice that all the relevant data are recorded.

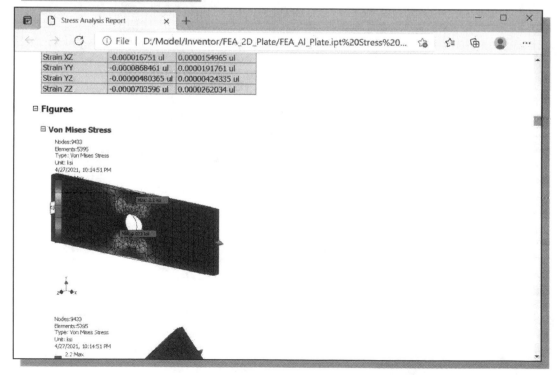

Geometric Considerations of Finite Elements

In the previous sections, the entire plate was created and analyzed, but a closer examination of the associated geometry suggests a more effective approach can be used to analyze the plate.

For *linear statics analysis*, designs with symmetrical features can often be reduced to expedite the analysis.

For our plate problem, there are two planes of symmetry. Thus, we only need to create an FE model that is one-fourth of the actual system. By taking advantage of symmetry, we can use a finer subdivision of elements that will provide more accurate and faster results.

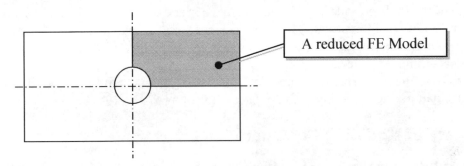

A reduced FE Model

In performing a stress analysis, it is necessary to consider the constraints in all directions. For our plate model, deformations will occur along the axes of symmetry; we will therefore place roller constraints along the two center lines as shown in the figure below. You are encouraged to perform the FEA on this more simplified model and compare the results obtained in the previous sections.

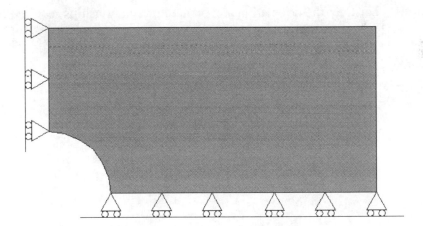

❖ One should also be cautious of using symmetrical characteristics in FEA. The symmetry characteristics of boundary conditions and loads should be considered. Also, note the symmetry characteristic that is used in *linear statics analysis* does not imply similar symmetrical results in vibration or buckling modes.

Conclusion

Design includes all activities involved from the original conception to the finished product. Design is the process by which products are created and modified. For many years designers sought ways to describe and analyze three-dimensional designs without building physical models. With advancements in computer technology, the creation of parametric models on computers offers a wide range of benefits. Parametric models are easier to interpret and can be easily altered. Parametric models can be analyzed using finite element analysis software, and simulation of real-life loads can be applied to the models and the results graphically displayed.

Throughout this text, various modeling techniques have been presented. Mastering these techniques will enable you to create intelligent and flexible solid models. The goal is to make use of the tools provided by Autodesk Inventor and to successfully capture the *DESIGN INTENT* of the product. In many instances, only one approach to the modeling tasks was presented; you are encouraged to repeat all of the lessons and develop different ways of accomplishing the same tasks. We have only scratched the surface of Autodesk Inventor's functionality. The more time you spend using the system, the easier it will be to perform parametric modeling with Autodesk Inventor.

Summary of Modeling Considerations

- **Design Intent** – determine the functionality of the design; select features that are central to the design.

- **Order of Features** – consider the parent/child relationships necessary for all features.

- **Dimensional and Geometric Constraints** – the way in which the constraints are applied determines how the components are updated.

- **Relations** – consider the orientation and parametric relationships required between features and in an assembly.

Review Questions:

1. Describe the required steps in performing a stress analysis using *Inventor Stress Analysis Environment*.

2. Describe two ways the material properties can be defined or edited.

3. What is meant by the term *Constraints* in *Inventor Stress Analysis Module*?

4. How do we control whether a load applied to a face is in the *outward* or *inward* direction?

5. Define *degrees of freedom* (DOF).

6. How do we end the *Inventor Stress Analysis Module* and return to the model in Autodesk Inventor?

Exercises:

1. The shaft shown below is fixed at the large end and a 50 kN force is applied to the small end. Find the maximum stress and maximum deflection in the shaft. The material is AISI 1020. (Dimensions in mm.)

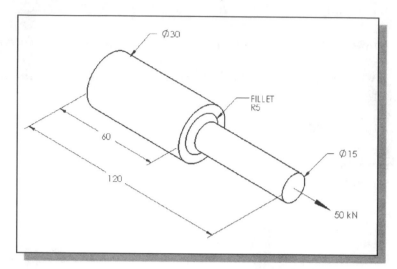

2. For the hanging bracket, the top face is fixed and a 100 psi pressure load is applied to the horizontal surface as shown. Find the maximum stress and maximum deflection in the bracket. The material is Alloy Steel. (Dimensions in inches.)

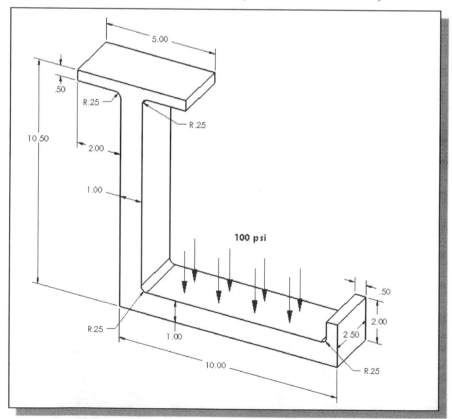

APPENDIX A

Running and Sliding Fits – American National Standard

Basic Hole System, Limits are in thousandths of an inch.
Limits for hole and shaft are applied to the basic size to obtain
the limits of sizes for the parts.

Nominal Size Range, Inches Over To	Class RC1			Class RC2			Class RC3			Class RC4		
	Limits of Clearance	Standard Limits		Limits of Clearance	Standard Limits		Limits of Clearance	Standard Limits		Limits of Clearance	Standard Limits	
		Hole H5	Shaft g4		Hole H6	Shaft g5		Hole H7	Shaft f6		Hole H8	Shaft f7
0 – 0.12	0.1	+0.2	-0.1	0.1	+0.25	-0.1	0.3	+0.4	-0.3	0.3	+0.6	-0.3
	0.45	+0	-0.25	0.55	0	-0.3	0.95	0	-0.55	1.3	0	-0.7
0.12 – 0.24	0.15	+0.2	-0.15	0.15	+0.3	-0.15	0.4	+0.5	-0.4	0.4	+0.7	-0.4
	0.5	0	-0.3	0.65	0	-0.35	1.12	0	-0.7	1.6	0	-0.9
0.24 – 0.40	0.2	+0.25	-0.2	0.2	+0.4	-0.2	0.5	+0.6	-0.5	0.5	+0.9	-0.5
	0.6	0	-0.35	0.85	0	-0.45	1.5	0	-0.9	2.0	0	-1.1
0.40 – 0.71	0.25	+0.3	-0.25	0.25	+0.4	-0.25	0.6	+0.7	-0.6	0.6	+1.0	-0.6
	0.75	+0	-0.45	0.95	0	-0.55	1.7	0	-1.0	2.3	0	-1.3
0.71 – 1.19	0.3	+0.4	-0.3	0.3	+0.5	-0.3	0.8	+0.8	-0.8	0.8	+1.2	-0.8
	0.95	0	-0.55	1.2	0	-0.7	2.1	0	-1.3	2.8	0	-1.6
1.19 – 1.97	0.4	+0.4	-0.4	0.4	+0.6	-0.4	1.0	+1.0	-1.0	1.0	+1.6	-1.0
	1.1	0	-0.7	1.4	0	-0.8	2.6	0	-1.6	3.6	0	-2.0
1.97 – 3.15	0.4	+0.5	-0.4	0.4	+0.7	-0.4	1.2	+1.2	-1.2	1.2	+1.8	-1.2
	1.2	0	-0.7	1.6	0	-0.9	3.1	0	-1.9	4.2	0	-2.4
3.15 – 4.73	0.5	+0.6	-0.5	0.5	+0.9	-0.5	1.4	+1.4	-1.4	1.4	+2.2	-1.4
	1.5	0	-0.9	2.0	0	-1.1	3.7	0	-2.3	5.0	0	-2.8
4.73 – 7.09	0.6	+0.7	-0.6	0.6	+1.0	-0.6	1.6	+1.6	-1.6	1.6	+2.5	-1.6
	1.8	0	-1.1	2.3	0	-1.3	4.2	0	-2.6	5.7	0	-3.2
7.09 – 9.85	0.6	+0.8	-0.6	0.6	+1.2	-0.6	2.0	+1.8	-2.0	2.0	+2.8	-2.0
	2.0	0	-1.2	2.6	0	-1.4	5.0	0	-3.2	6.6	0	-3.8
9.85 – 12.41	0.8	+0.9	-0.8	0.8	+1.2	-0.7	2.5	+2.0	-2.5	2.5	+3.0	-2.2
	2.3	0	-1.4	2.9	0	-1.6	5.7	0	-3.7	7.5	0	-4.2
12.41– 15.75	1.0	+1.0	-1.0	1.0	+1.4	-0.7	3.0	+2.2	-3.0	3.0	+3.5	-2.5
	2.7	0	-1.7	3.4	0	-1.7	6.6	0	-4.4	8.7	0	-4.7
15.75– 19.69	1.2	+1.0	-1.2	0.8	+1.6	-0.8	4.0	+2.5	-4.0	2.8	+4.0	-2.8
	3.0	0	-2.0	3.4	0	-1.8	8.1	0	-5.6	9.3	0	-5.3

USAS/ASME B4.1 – 1967 (R2004) Standard. For larger diameters, see the standard. ASME/ANSI
B18.3.5M – 1986 (R2002) Standard. Reprinted from the standard listed by permission of the American
Society of Mechanical Engineers. All rights reserved.

APPENDIX A (Continued)

Running and Sliding Fits – American National Standard

Basic Hole System, Limits are in thousandths of an inch.
Limits for hole and shaft are applied to the basic size to obtain
the limits of sizes for the parts.

Nominal size Range, Inches Over To	Class RC5			Class RC6			Class RC7			Class RC8			Class RC9		
	Limits of Clearance	Hole H8	Shaft e7	Limits of Clearance	Hole H9	Shaft e8	Limits of Clearance	Hole H9	Shaft d8	Limits of Clearance	Hole H10	Shaft C9	Limits of Clearance	Hole H11	Shaft
0 – 0.12	0.6 / 1.6	+0.6 / 0	-0.6 / -1.0	0.6 / 2.2	+1.0 / 0	-0.6 / -1.2	1.0 / 2.6	+1.0 / 0	-1.0 / -1.6	2.5 / 5.1	+1.6 / 0	-2.5 / -3.5	4.0 / 8.1	+2.5 / 0	-4.0 / -5.6
0.12 – 0.24	0.8 / 2.0	+0.7 / 0	-0.8 / -1.3	0.8 / 2.7	+1.2 / 0	-0.8 / -1.5	1.2 / 3.1	+1.2 / 0	-1.2 / -1.9	2.8 / 5.8	+1.8 / 0	-2.8 / -4.0	4.5 / 9.0	+3.0 / 0	-4.5 / -6.0
0.24 – 0.40	1.0 / 2.5	+0.9 / 0	-1.0 / -1.6	1.0 / 3.3	+1.4 / 0	-1.0 / -1.9	1.6 / 3.9	+1.4 / 0	-1.6 / -2.5	3.0 / 6.6	+2.2 / 0	-3.0 / -4.4	5.0 / 10.7	+3.5 / 0	-5.0 / -7.2
0.40 – 0.71	1.2 / 2.9	+1.0 / 0	-1.2 / -1.9	1.2 / 3.8	+1.6 / 0	-1.2 / -2.2	2.0 / 4.6	+1.6 / 0	-2.0 / -3.0	3.5 / 7.9	+2.8 / 0	-3.5 / -5.1	6.0 / 12.8	+4.0 / 0	-6.0 / -8.8
0.71 – 1.19	1.6 / 3.6	+1.2 / 0	-1.6 / -2.4	1.6 / 4.8	+2.0 / 0	-1.6 / -2.8	2.5 / 5.7	+2.0 / 0	-2.5 / -3.7	4.5 / 10.0	+3.5 / 0	-4.5 / -6.5	7.0 / 15.5	+5.0 / 0	-7.0 / -10.5
1.19 – 1.97	2.0 / 4.6	+1.6 / 0	-2.0 / -3.0	2.0 / 6.1	+2.5 / 0	-2.0 / -3.6	3.0 / 7.1	+2.5 / 0	-3.0 / -4.6	5.0 / 11.5	+4.0 / 0	-5.0 / -7.5	8.0 / 18.0	+6.0 / 0	-8.0 / 12.0
1.97 – 3.15	2.5 / 5.5	+1.8 / 0	-2.5 / -3.7	2.5 / 7.3	+3.0 / 0	-2.5 / -4.3	4.0 / 8.8	+3.0 / 0	-4.0 / -5.8	6.0 / 13.5	+4.5 / 0	-6.0 / -9.0	9.0 / 20.5	+7.0 / 0	-9.0 / -13.5
3.15 – 4.73	3.0 / 6.6	+2.2 / 0	-3.0 / -4.4	3.0 / 8.7	+3.5 / 0	-3.0 / -5.2	5.0 / 10.7	+3.5 / 0	-5.0 / -7.2	7.0 / 15.5	+5.0 / 0	-7.0 / -10.5	10.0 / 24.0	+9.0 / 0	-10.0 / -15.0
4.73 – 7.09	3.5 / 7.6	+2.5 / 0	-3.5 / -5.1	3.5 / 10.0	+4.0 / 0	-3.5 / -6.0	6.0 / 12.5	+4.0 / 0	-6.0 / -8.5	8.0 / 18.0	+6.0 / 0	-8.0 / -12.0	12.0 / 28.0	+10.0 / 0	-12.0 / -18.0
7.09 – 9.85	4.0 / 8.6	+2.8 / 0	-4.0 / -5.8	4.0 / 11.3	+4.5 / 0	-4.0 / -6.8	7.0 / 14.3	+4.5 / 0	-7.0 / -9.8	10.0 / 21.5	+7.0 / 0	-10.0 / -14.5	15.0 / 34.0	+12.0 / 0	-15.0 / -22.0
9.85 – 12.41	5.0 / 10.0	+3.0 / 0	-5.0 / -7.0	5.0 / 13.0	+5.0 / 0	-5.0 / -8.0	8.0 / 16.0	+5.0 / 0	-8.0 / -11	12.0 / 25.0	+8.0 / 0	-12.0 / -17.0	18.0 / 38.0	+12.0 / 0	-18.0 / -26.0
12.41 – 15.75	6.0 / 11.7	+3.5 / 0	-6.0 / -8.2	6.0 / 15.5	+6.0 / 0	-6.0 / -9.5	10.0 / 19.5	+6.0 / 0	-10 / -13.5	14.0 / 29.0	+9.0 / 0	-14.0 / -20.0	22.0 / 45.0	+14.0 / 0	-22.0 / -31.0
15.75 – 19.69	8.0 / 14.5	+4.0 / 0	-8.0 / -10.5	8.0 / 18.0	+6.0 / 0	-8.0 / -12.0	12.0 / 22.0	+6.0 / 0	-12.0 / -16.0	16.0 / 32.0	+10.0 / 0	-16.0 / -22.0	25.0 / 51.0	+16.0 / 0	-25.0 / -35.0

USAS/ASME B4.1 – 1967 (R2004) Standard. For larger diameters, see the standard. ASME/ANSI B18.3.5M – 1986 (R2002) Standard. Reprinted from the standard listed by permission of the American Society of Mechanical Engineers. All rights reserved.

Locational clearance fits – American National Standard (Inches)

Basic Hole System, Limits are in thousandths of an inch.
Limits for hole and shaft are applied to the basic size to obtain the limits of sizes for the parts.

Nominal Size Range Inches		Class LC1		Class LC2		Class LC3		Class LC4	
		Standard Limits		Standard Limits		Standard Limits		Standard Limits	
Over	To	Hole	Shaft	Hole	Shaft	Hole	Shaft	Hole	Shaft
0 - 0.12		+0.25	0	+0.4	0	+0.6	0	+1.6	0
		0	-0.2	0	-0.25	0	-0.4	0	-1.0
0.12- 0.24		+0.3	0	+0.5	0	+0.7	0	+1.8	0
		0	-0.2	0	-0.3	0	-0.5	0	-1.2
0.24- 0.40		+0.4	0	+0.6	0	+0.9	0	+2.2	0
		0	-0.25	0	-0.4	0	-0.6	0	-1,4
0,40- 0.71		+0.4	0	+0.7	0	+1,0	0	+2.6	0
		0	-0.3	0	-0.4	0	-0.7	0	-1.6
0.71 - 1.10		+0.5	0	+0 .8	0	+1.2	0	+3.5	0
		0	-0.4	0	-0.5	0	-0.8	0	-2.0
1.19- 1.97		+0.6	0	+1 0	0	+1.6	0	+4.0	0
		0	-0.4	0	-0.6	0	-1.0	0	-2.5
1.97- 3.15		+0.7	0	+1.2	0	+1.8	0	+4.5	0
		0	-0.5	o	-0.7	0	-1.2	0	-3.0
3.15- 4.73		+0.9	0	+1.4	0	+2.2	0	+5.0	0
		0	-0.6	0	-0.9	0	-1.4	0	-3.5
4.73- 7.09		+1.0	0	+1.6	0	+2.5	0	+6.0	0
		0	-0.7	0	-1.0	0	-1.6	0	-4.0
7.09- 9.85		+1.2	0	+1.8	0	+2.8	0	+7.0	0
		0	-0.8	0	-1.2	0	-1.8	0	-4.5
9.85 - 12.41		+1.2	0	+2.0	0	+3.0	0	+8,0	0
		0	-0.9	0	-1.2	0	-2.0	0	-5.0
12.41 - 15.75		+1.4	0	+2.2	0	+3.5	0	+9.0	0
		0	-1.0	0	-1.4	0	-2.2	0	-6.0
15.75 - 19.69		+1.6	0	+2.5	0	+4.0	0	+10.0	0
		0	-1.0	0	-1.6	0	-2.5	0	-6.0

USAS/ASME B4.1 – 1967 (R2004) Standard. For larger diameters, see the standard. ASME/ANSI B18.3.5M – 1986 (R2002) Standard. Reprinted from the standard listed by permission of the American Society of Mechanical Engineers. All rights reserved.

APPENDIX A (Locational clearance fits Continued)

Locational clearance fits – American National Standard (Inches)

Basic Hole System, Limits are in thousandths of an inch.
Limits for hole and shaft are applied to the basic size to obtain
the limits of sizes for the parts.

Nominal Size Range Inches	Class LC5		Class LC6		Class LC7		Class LC8	
	Standard Limits		Standard Limits		Standard Limits		Standard Limits	
Over To	Hole	Shaft	Hole	Shaft	Hole	Shaft	Hole	Shaft
0 - 0.12	+0.4	-0.1	+1.0	-0.3	+1.6	-0.8	+1.6	-1.0
	0	-0.35	0	-0.9	0	-1.6	0	-2.0
0.12 - 0.24	+0.5	-0.15	+1.2	-0.4	+1.8	-0.8	+1.8	-1.2
	0	-0.45	0	-1.1	0	-2.0	0	-2.4
0.24 - 0.40	+0.6	-0.2	+1.4	-0.5	+2.2	-1.0	+2.2	-1.6
	0	-0.6	0	-1.4	0	-2.4	0	-3.0
0.40- 0.71	+0.7	-0.25	+1.6	-0.6	+2.8	-1.2	+2.8	-2.0
	0	-0.65	0	-1.6	0	-2.8	0	-3.6
0.71 - 1.19	+0.8	-0.3	+2.0	-0.8	+3.5	-1.6	-3.5	-2.5
	0	-0.8	0	-2.0	0	-3.6	0	-4.5
1.19- 1.97	+1.0	-0.4	+2.5	-1.0	+4.0	-2.0	+4.0	-3.0
	0	-1.0	0	-2.6	0	-4.5	0	-5.5
1.97- 3.15	+1.2	-0.4	+3.0	-1.2	+4.5	-2.5	+4.5	-4.0
	0	-1 1	0	-3.0	0	-5.5	0	-7.0
3.15- 4.73	+1.4	-0.5	+3.5	-1.4	+5.0	-3.0	+5.0	-5.0
	0	-1.4	0	-3.6	0	-6.5	0	-8.5
4.73- 7.09	+1.6	-0.6	+4.0	-1.6	+6.0	-3.5	+6.0	-6.0
	0	-1.6	0	-4.1	0	-7.5	0	-10.0
7.09- 9.85	+1.6	-0.6	+4.5	-2.0	+7.0	-4.0	+7.0	-7.0
	0	-1.8	0	-4.8	0	-8.5	0	-11.5
9.85 - 12.41	+2.0	-0.7	+5.0	-2.2	+8.0	-4.5	+8.0	-7.0
	0	-1.9	0	-5.2	0	-9.5	0	-12.0
12 41 - 15.75	+2.2	-0.7	+6.0	-2.5	+9.0	-5.0	+9.0	-8.0
	0	-2.1	0	-6.0	0	-11.0	0	-14.0
15.75 - 19.69	+2.5	-0.8	+6.0	-2.8	+10.0	-5.0	+10.0	-9.0
	0	-2.4	0	-6.8	0	-11.0	0	-15.0

APPENDIX A (Locational clearance fits Continued)

Locational clearance fits – American National Standard (Inches)

Basic Hole System, Limits are in thousandths of an inch.
Limits for hole and shaft are applied to the basic size to obtain
the limits of sizes for the parts.

Nominal Size Range Inches		Class LC9		Class LC10		Class LC11	
		Standard Limits		Standard Limits		Standard Limits	
Over	To	Hole	Shaft	Hole	Shaft	Hole	Shaft
0	- 0.12	+2.5	-2.5	+4.0	-4.0	+6.0	-5.0
		0	-4.1	0	-8.0	0	-11.0
0.12-	0.24	+3.0	-3.0	+5.0	-4.5	+7.0	-6.0
		0	-5.2	0	-9.5	0	-13.0
0.25-	0.40	+3.5	-3.5	+6.0	-5.0	+9.0	-7.0
		0	-6.3	0	-11.0	0	-16.0
0.40-	0.71	+4.0	-4.5	+7.0	-6.0	+10.0	-8.0
		0	-8.0	0	-13.0	0	-18.0
0.71 -	1.19	+5.0	-5.0	+8.0	-7.0	+12.0	-10.0
		0	-9.0	0	-15.0	0	-22.0
1.19-	1.97	+6.0	-6.0	+10.0	-8.0	+16.0	-12.0
		0	-10.5	0	-18.0	0	-28.0
1.97-	3.15	+7.0	-7.0	+12.0	-10.0	+18.0	-14.0
		0	-12.0	0	-22.0	0	-32.0
3.15-	4.73	+9.0	-8.0	+14.0	-11.0	+22.0	-16.0
		0	-14.0	0	-25.0	0	-38.0
4.73-	7.09	+10.0	-10.0	+16.0	-12.0	+25.0	-18.0
		0	-17.0	0	-28.0	0	-43.0
7.09-	9.85	+12.0	-12.0	+18.0	-16.0	+25.0	-22.0
		0	-20.0	0	-34.0	0	-50.0
9.85 -	12.41	+12.0	-0.7	+20.0	-20.0	+30.0	-28.0
		0	-1.9	0	-40.0	0	-58.0
12 41 -	15.75	+14.0	-14.0	+22.0	-22.0	+35.0	-30.0
		0	-23.0	0	-44.0	0	-65.0
15.75 -	19.69	+16.0	-16.0	+25.0	-25.0	+40.0	-35.0
		0	-26.0	0	-50.0	0	-75.0

USAS/ASME B4.1 – 1967 (R2004) Standard. For larger diameters, see the standard. ASME/ANSI
B18.3.5M – 1986 (R2002) Standard. Reprinted from the standard listed by permission of the American
Society of Mechanical Engineers. All rights reserved.

APPENDIX A

Force and Shrink fits – American National Standard (Inches) - Partial List

Basic Hole System, Limits are in thousandths of an inch.
Limits for hole and shaft are applied to the basic size to obtain the limits of sizes for the parts.

❖ Values shown below are in thousandths of an inch																
		Class FN 1			Class FN 2			Class FN 3			Class FN 4			Class FN 5		
Nominal Size Range, Inches		Interference Limits	Standard Limits		Interference Limits	Standard Limits		Interference Limits	Standard Limits		Interference Limits	Standard Limits		Interference Limits	Standard Limits	
Over	To		Hole	Shaft		Hole	Shaft		Hole	Shaft		Hole	Shaft		Hole	Shaft
			H6			H7	s6		H7	t6		H7	u6		H8	x7
0	0.12	+0.05	+0.25	+0.5	+0.2	+0.4	+0.85				+0.3	+0.4	+0.95	+0.3	+0.6	+1.3
		+0.5	0	+0.3	+0.85	0	+0.6				+0.95	0	+0.7	+1.3	0	+0.9
0.12	0.24	+0.1	+0.3	+0.6	+0.2	+0.5	+1.0				+0.4	+0.5	+1.2	+0.5	+0.7	+1.7
		+0.6	0	+0.4	+1.0	0	+0.7				+1.2	0	+0.9	+1.7	0	+1.2
0.24	0.4	+0.1	+0.4	+0.75	+0.4	+0.6	+1.4				+0.6	+0.6	+1.6	+0.5	+0.9	+2.0
		+0.75	0	+0.5	+1.4	0	+1.0				+1.6	0	+1.2	+2.0	0	+1.4
0.40	0.56	+0.1	+0.4	+0.8	+0.5	+0.7	+1.6				+0.7	+0.7	+1.8	+0.6	+1.0	+2.3
		+0.8	0	+0.5	+1.6	0	+1.2				+1.8	0	+1.4	+2.3	0	+1.6
0.56	0.71	+0.2	+0.4	+0.9	+0.5	+0.7	+1.6				+0.7	+0.7	+1.8	+0.8	+1.0	+2.5
		+0.9	0	+0.6	+1.6	0	+1.2				+1.8	0	+1.4	+2.5	0	+1.8
0.71	0.95	+0.2	+0.5	+1.1	+0.6	+0.8	+1.9				+0.8	+0.8	+2.1	+1.0	+1.2	+3.0
		+1.1	0	+0.7	+1.9	0	+1.4				+2.1	0	+1.6	+3.0	0	+2.2
0.95	1.19	+0.3	+0.5	+1.2	+0.6	+0.8	+1.9	+0.8	+0.8	+2.1	+1.0	+0.8	+2.3	+1.3	+1.2	+3.3
		+1.2	0	+0.8	+1.9	0	+1.4	+2.1	0	+1.6	+2.3	0	+1.8	+3.3	0	+2.5
1.19	1.58	+0.3	+0.6	+1.3	+0.8	+1.0	+2.4	+1.0	+1.0	+2.6	+1.5	+1.0	+3.1	+1.4	+1.6	+4.0
		+1.3	0	+0.9	+2.4	0	+1.8	+2.6	0	+2.0	+3.1	0	+2.5	+4.0	0	+3.0
1.58	1.97	+0.4	+0.6	+1.4	+0.8	+1.0	+2.4	+1.2	+1.0	+2.8	+1.8	+1.0	+3.4	+2.4	+1.6	+5.0
		+1.4	0	+1.0	+2.4	0	+1.8	+2.8	0	+2.2	+3.4	0	+2.8	+5.0	0	+4.0
1.97	2.56	+0.6	+0.7	+1.8	+0.8	+1.2	+2.7	+1.3	+1.2	+3.2	+2.3	+1.2	+4.2	+3.2	+1.8	+6.2
		+1.8	0	+1.3	+2.7	0	+2.0	+3.2	0	+2.5	+4.2	0	+3.5	+6.2	0	+5.0
2.56	3.15	+0.7	+0.7	+1.9	+1.0	+1.2	+2.9	+1.8	+1.2	+3.7	+2.8	+1.2	+4.7	+4.2	+1.8	+7.2
		+1.9	0	+1.4	+2.9	0	+2.2	+3.7	0	+3.0	+4.7	0	+4.0	+7.2	0	+6.0
3.15	3.94	+0.9	+0.9	+2.4	+1.4	+1.4	+3.7	+2.1	+1.4	+4.4	+3.6	+1.4	+5.9	+4.8	+2.2	+8.4
		+2.4	0	+1.8	+3.7	0	+2.8	+4.4	0	+3.5	+5.9	0	+5.0	+8.4	0	+7.0
3.94	4.73	+1.1	+0.9	+2.6	+1.6	+1.4	+3.9	+2.6	+1.4	+4.9	+4.6	+1.4	+6.9	+5.8	+2.2	+9.4
		+2.6	0	+2.0	+3.9	0	+3.0	+4.9	0	+4.0	+6.9	0	+6.0	+9.4	0	+8.0

APPENDIX B – METRIC LIMITS AND FITS

Hole Basis Clearance Fits

Preferred Hole Basis Clearance Fits. Dimensions in mm.

Basic Size	Loose Running		Free Running		Close Running		Sliding		Locational Clearance	
	Hole H11	Shaft c11	Hole H9	Shaft d9	Hole H8	Shaft f7	Hole H7	Shaft g6	Hole H7	Shaft h6
1 max	1.060	0.940	1.025	0.980	1.014	0.994	1.010	0.998	1.010	1.000
min	1.000	0.880	1.000	0.955	1.000	0.984	1.000	0.992	1.000	0.994
1.2 max	1.260	0.940	1.225	1.180	1.214	1.194	1.210	1.198	1.210	1.200
min	1.200	0.880	1.200	1.155	1.200	1.184	1.200	1.192	1.200	1.194
1.6 max	1.660	1.540	1.625	1.580	1.614	1.594	1.610	1.598	1.610	1.600
min	1.600	1.480	1.600	1.555	1.600	1.584	1.600	1.592	1.600	1.594
2 max	2.060	1.940	2.025	1.980	2.014	1.994	2.010	1.998	2.010	2.000
min	2.000	1.880	2.000	1.955	2.000	1.984	2.000	1.992	2.000	1.994
2.5 max	2.560	2.440	2.525	2.480	2.514	2.494	2.510	2.498	2.510	2.500
min	2.500	2.380	2.500	2.455	2.500	2.484	2.500	2.492	2.500	2.494
3 max	3.060	2.940	3.025	2.980	3.014	2.994	3.010	2.998	3.010	3.000
min	3.000	2.880	3.000	2.955	3.000	2.984	3.000	2.992	3.000	2.994
4 max	4.075	3.930	4.030	3.970	4.018	3.990	4.012	3.996	4.012	4.000
min	4.000	3.855	4.000	3.940	4.000	3.987	4.000	3.988	4.000	3.992
5 max	5.075	4.930	5.030	4.970	5.018	4.990	5.012	4.998	5.012	5.000
min	5.000	4.855	5.000	4.940	5.000	4.978	5.000	4.988	5.000	4.992
6 max	6.075	5.930	6.030	5.970	6.018	5.990	6.012	5.996	6.012	6.000
min	6.000	5.855	6.000	5.940	6.000	5.978	6.000	5.988	6.000	5.992
8 max	8.090	7.920	8.036	7.960	8.022	7.987	8.015	7.995	8.015	8.000
min	8.000	7.830	8.000	7.924	8.000	7.972	8.000	7.986	8.000	7.991
10 max	10.090	9.920	10.036	9.960	10.022	9.987	10.015	9.995	10.015	10.000
min	10.000	9.830	10.000	9.924	10.000	9.972	10.000	9.986	10.000	9.991
12 max	12.110	11.905	12.043	11.950	12.027	11.984	12.018	11.994	12.018	12.000
min	12.000	11.795	12.000	11.907	12.000	11.966	12.000	11.983	12.000	11.989
16 max	16.110	15.905	16.043	15.950	16.027	15.984	16.018	15.994	16.018	16.000
min	16.000	15.795	16.000	15.907	16.000	15.966	16.000	15.983	16.000	15.989
20 max	20.130	19.890	20.052	19.935	20.033	19.980	20.021	19.993	20.021	20.000
min	20.000	19.760	20.000	19.883	20.000	19.959	20.000	19.980	20.000	19.987
25 max	25.130	24.890	25.052	24.935	25.033	24.980	24.993	25.021	25.021	25.000
min	25.000	24.760	25.000	24.883	25.000	24.959	24.980	25.000	25.000	24.987
30 max	30.130	29.890	30.052	29.935	30.033	29.980	30.021	29.993	30.021	30.000
min	30.000	29.760	30.000	29.883	30.000	29.959	30.000	29.980	30.000	29.987
40 max	40.160	39.880	40.062	39.920	40.039	39.975	40.025	39.991	40.025	40.000
min	40.000	39.720	40.000	39.858	40.000	39.950	40.000	39.975	40.000	39.984

APPENDIX B – METRIC LIMITS AND FITS (Continued)

Hole Basis Transition and Interference Fits

Preferred Hole Basis Clearance Fits. Dimensions in mm.

Basic Size		Locational Transition		Locational Transition		Locational Transition		Medium Drive		Force	
		Hole H7	Shaft k6	Hole H7	Shaft n6	Hole H7	Shaft p6	Hole H7	Shaft s6	Hole H7	Shaft u6
1	max	1.010	1.006	1.010	1.010	1.010	1.012	1.010	1.020	1.010	1.024
	min	1.000	1.000	1.000	1.004	1.000	1.006	1.000	1.014	1.000	1.018
1.2	max	1.210	1.206	1.210	1.210	1.210	1.212	1.210	1.220	1.210	1.224
	min	1.200	1.200	1.200	1.204	1.200	1.206	1.200	1.214	1.200	1.218
1.6	max	1.610	1.606	1.610	1.610	1.610	1.612	1.610	1.620	1.610	1.624
	min	1.600	1.600	1.600	1.604	1.600	1.606	1.600	1.614	1.600	1.618
2	max	2.010	2.006	2.010	2.020	2.010	2.012	2.010	2.020	2.010	2.024
	min	2.000	2.000	2.000	2.004	2.000	2.006	2.000	1.014	2.000	2.018
2.5	max	2.510	2.510	2.510	2.510	2.510	2.512	2.510	2.520	2.510	2.524
	min	2.500	2.500	2.500	2.504	2.500	2.506	2.500	2.514	2.500	2.518
3	max	3.010	3.010	3.010	3.010	3.010	3.012	3.010	3.020	3.010	3.024
	min	3.000	3.000	3.000	3.004	3.000	3.006	3.000	3.014	3.000	3.018
4	max	4.012	4.012	4.012	4.016	4.012	4.020	4.012	4.027	4.012	4.031
	min	4.000	4.000	4.000	4.008	4.000	4.012	4.000	4.019	4.000	4.023
5	max	5.012	5.009	5.012	5.016	5.012	5.020	5.012	5.027	5.012	5.031
	min	5.000	5.001	5.000	5.008	5.000	5.012	5.000	5.019	5.000	5.023
6	max	6.012	6.009	6.012	6.016	6.012	6.020	6.012	6.027	6.012	6.031
	min	6.000	6.001	6.000	6.008	6.000	6.012	6.000	6.019	6.000	6.023
8	max	8.015	8.010	8.015	8.019	8.015	8.024	8.015	8.032	8.015	8.037
	min	8.000	8.001	8.000	8.010	8.000	8.015	8.000	8.023	8.000	8.028
10	max	10.015	10.010	10.015	10.019	10.015	10.024	10.015	10.032	10.015	10.037
	min	10.000	10.001	10.000	10.010	10.000	10.015	10.000	10.023	10.000	10.028
12	max	12.018	12.012	12.018	12.023	12.018	12.029	12.018	12.039	12.018	12.044
	min	12.000	12.001	12.000	12.012	12.000	12.018	12.000	12.028	12.000	12.033
16	max	16.018	16.012	16.018	16.023	16.018	16.029	16.018	16.039	16.018	16.044
	min	16.000	16.001	16.000	16.012	16.000	16.018	16.000	16.028	16.000	16.033
20	max	20.021	20.015	20.021	20.028	20.021	20.035	20.021	20.048	20.021	20.054
	min	20.000	20.002	20.000	20.015	20.000	20.022	20.000	20.035	20.000	20.041
25	max	25.021	25.015	25.021	25.028	25.021	25.035	25.021	25.048	25.021	25.061
	min	25.000	25.002	25.000	25.015	25.000	25.022	25.000	25.035	25.000	25.048
30	max	30.021	30.015	30.021	30.028	30.021	30.035	30.021	30.048	30.021	30.061
	min	30.000	30.002	30.000	30.015	30.000	30.022	30.000	30.035	30.000	30.048
40	max	40.025	40.018	40.025	40.033	40.025	40.042	40.025	40.059	40.025	40.076
	min	40.000	40.002	40.000	40.017	40.000	40.026	40.000	40.043	40.000	40.060

ANSI B4.2 – 1978 (R2004) Standard. ASME/ANSI B18.3.5M – 1986 (R2002) Standard. Reprinted from the standard listed by permission of the American Society of Mechanical Engineers. All rights reserved.

APPENDIX B – METRIC LIMITS AND FITS (Continued)

Shaft Basis Clearance Fits

Preferred Shaft Basis Clearance Fits. Dimensions in mm.

Basic Size	Loose Running		Free Running		Close Running		Sliding		Locational Clearance	
	Hole C11	Shaft h11	Hole D9	Shaft h9	Hole F8	Shaft h7	Hole G7	Shaft h6	Hole H7	Shaft h6
1 max	1.120	1.000	1.045	1.000	1.020	1.000	1.012	1.000	1.010	1.000
min	1.060	0.940	1.020	0.975	1.006	0.990	1.002	0.994	1.000	0.994
1.2 max	1.320	1.200	1.245	1.200	1.220	1.200	1.212	1.200	1.210	1.200
min	1.260	1.140	1.220	1.175	1.206	1.190	1.202	1.194	1.200	1.194
1.6 max	1.720	1.600	1.645	1.600	1.620	1.600	1.612	1.600	1.610	1.600
min	1.660	1.540	1.620	1.575	1.606	1.590	1.602	1.594	1.600	1.594
2 max	2.120	2.000	2.045	2.000	2.020	2.000	2.012	2.000	2.010	2.000
min	2.060	1.940	2.020	1.975	2.006	1.990	2.002	1.994	2.000	1.994
2.5 max	2.620	2.500	2.545	2.500	2.520	2.500	2.512	2.500	2.510	2.500
min	2.560	2.440	2.520	2.475	2.506	2.490	2.502	2.494	2.500	2.494
3 max	3.120	3.000	3.045	3.000	3.020	3.000	3.012	3.000	3.010	3.000
min	3.060	2.940	3.020	2.975	3.006	2.990	3.002	2.994	3.000	2.994
4 max	4.145	4.000	4.060	4.000	4.028	4.000	4.016	4.000	4.012	4.000
min	4.070	3.925	4.030	3.970	4.010	3.988	4.004	3.992	4.000	3.992
5 max	5.145	5.000	5.060	5.000	5.028	5.000	5.016	5.000	5.012	5.000
min	5.070	4.925	5.030	4.970	5.010	4.988	5.004	4.992	5.000	4.992
6 max	6.145	6.000	6.060	6.000	6.028	6.000	6.016	6.000	6.012	6.000
min	6.070	5.925	6.030	5.970	6.010	5.988	6.004	5.992	6.000	5.992
8 max	8.170	8.000	8.076	8.000	8.035	8.000	8.020	8.000	8.015	8.000
min	8.080	7.910	8.040	7.964	8.013	7.985	8.005	7.991	8.000	7.991
10 max	10.170	10.000	10.076	10.000	10.035	10.000	10.020	10.000	10.015	10.000
min	10.080	9.910	10.040	9.964	10.013	9.985	10.005	9.991	10.000	9.991
12 max	12.205	12.000	12.093	12.000	12.043	12.000	12.024	12.000	12.018	12.000
min	12.095	11.890	12.050	11.957	12.016	11.982	12.006	11.989	12.000	11.989
16 max	16.205	16.000	16.093	16.000	16.043	16.000	16.024	16.000	16.018	16.000
min	16.095	15.890	16.050	15.957	16.016	15.982	16.006	15.989	16.000	15.989
20 max	20.240	20.000	20.117	20.000	20.053	20.000	20.028	20.000	20.021	20.000
min	20.110	19.870	20.065	19.948	20.020	19.979	20.007	19.987	20.000	19.987
25 max	25.240	25.000	25.117	25.000	25.053	25.000	25.028	25.000	25.021	25.000
min	25.110	24.870	25.065	24.948	25.020	24.979	25.007	24.987	25.000	24.987
30 max	30.240	30.000	30.117	30.000	30.053	30.000	30.028	30.000	30.021	30.000
min	30.110	29.870	30.065	29.948	30.020	29.979	30.007	29.987	30.000	29.987
40 max	40.280	40.000	40.142	40.000	40.064	40.000	40.034	40.000	40.025	40.000
min	40.120	39.840	40.080	39.938	40.025	39.975	40.009	39.984	40.000	39.984

ANSI B4.2 – 1978 (R2004) Standard. ASME/ANSI B18.3.5M – 1986 (R2002) Standard. Reprinted from the standard listed by permission of the American Society of Mechanical Engineers. All rights reserved.

APPENDIX B - METRIC LIMITS AND FITS (Continued)

Shaft Basis Transition and Interference Fits

Preferred Shaft Basis Transition and Interference Fits. Dimensions in mm.

Basic Size	Locational Transition		Locational Transition		Locational Interference		Medium Drive		Force	
	Hole K7	Shaft h6	Hole N7	Shaft h6	Hole P7	Shaft h6	Hole S7	Shaft h6	Hole U7	Shaft h6
1 max	1.000	1.000	0.996	1.000	0.994	1.000	0.986	1.000	0.982	1.000
min	0.990	0.994	0.986	0.994	0.984	0.994	0.976	0.994	0.972	0.994
1.2 max	1.200	1.200	1.196	1.200	1.194	1.200	1.186	1.200	1.182	1.200
min	1.190	1.194	1.186	1.194	1.184	1.194	1.176	1.194	1.172	1.194
1.6 max	1.600	1.600	1.596	1.600	1.594	1.600	1.586	1.600	1.582	1.600
min	1.590	1.594	1.586	1.594	1.584	1.594	1.576	1.594	1.572	1.594
2 max	2.000	2.000	1.996	2.000	1.994	2.000	1.986	2.000	1.982	2.000
min	1.990	1.994	1.986	1.994	1.984	1.994	1.976	1.994	1.972	1.994
2.5 max	2.500	2.500	2.496	2.500	2.494	2.500	2.486	2.500	2.482	2.500
min	2.490	2.494	2.486	2.494	2.484	2.494	2.476	2.494	2.472	2.494
3 max	3.000	3.000	2.996	3.000	2.994	3.000	2.986	3.000	2.982	3.000
min	2.990	2.994	2.986	2.994	2.984	2.994	2.976	2.994	2.972	2.994
4 max	4.003	4.000	3.996	4.000	3.992	4.000	3.985	4.000	3.981	4.000
min	3.991	5.992	3.984	5.992	3.980	5.992	3.973	5.992	3.969	5.992
5 max	5.003	5.000	4.996	5.000	4.992	5.000	4.985	5.000	4.981	5.000
min	4.991	4.992	4.984	4.992	4.980	4.992	4.973	4.992	4.969	4.992
6 max	6.003	6.000	5.996	6.000	5.992	6.000	5.985	6.000	5.981	6.000
min	5.991	5.992	5.984	5.992	5.980	5.992	5.973	5.992	5.969	5.992
8 max	8.005	8.000	7.996	8.000	7.991	8.000	7.983	8.000	7.978	8.000
min	7.990	7.991	7.981	7.991	7.976	7.991	7.968	7.991	7.963	7.991
10 max	10.005	10.000	9.996	10.000	9.991	10.000	9.983	10.000	9.978	10.000
min	9.990	9.991	9.981	9.991	9.976	9.991	9.968	9.991	9.963	9.991
12 max	12.006	12.000	11.995	12.000	11.989	12.000	11.979	12.000	11.974	12.000
min	11.988	11.989	11.977	11.989	11.971	11.989	11.961	11.989	11.956	11.989
16 max	16.006	16.000	15.995	16.000	15.989	16.000	15.979	16.000	15.974	16.000
min	15.988	15.989	15.977	15.989	15.971	15.989	15.961	15.989	15.956	15.989
20 max	20.006	20.000	19.993	20.000	19.986	20.000	19.973	20.000	19.967	20.000
min	19.985	19.987	19.972	19.987	19.965	19.987	19.952	19.987	19.946	19.987
25 max	25.006	25.000	24.993	25.000	24.986	25.000	24.973	25.000	24.960	25.000
min	24.985	24.987	24.972	24.987	24.965	24.987	24.952	24.987	24.939	24.987
30 max	30.006	30.000	29.993	30.000	29.986	30.000	29.973	30.000	29.960	30.000
min	29.985	29.987	29.972	29.987	29.965	29.987	29.952	29.987	29.939	29.987
40 max	40.007	40.000	39.992	40.000	39.983	40.000	39.966	40.000	39.949	40.000
min	39.982	39.984	39.967	39.984	39.958	39.984	39.941	39.984	39.924	39.984

APPENDIX C – UNIFIED NATIONAL THREAD FORM

(External Threads) Approximate Minor diameter = D – 1.0825P   P = Pitch

Nominal Size, in.	Basic Major Diameter (D)	Coarse UNC		Fine UNF		Extra Fine UNEF	
		Thds. Per in.	Tap Drill Dia.	Thds Per in.	Tap Drill. Dia.	Thds. Per in.	Tap Drill Dia.
#0	0.060	...	...	80	3/64	...	...
#1	0.0730	64	0.0595	72	0.0595	...	...
#2	0.0860	56	0.0700	64	0.0700	...	...
#3	0.0990	48	0.0785	56	0.0820	...	...
#4	0.1120	40	0.0890	48	0.0935	...	...
#5	0.1250	40	0.1015	44	0.1040	...	...
#6	0.1380	32	0.1065	40	0.1130	...	...
#8	0.1640	32	0.1360	36	0.1360	...	...
#10	0.1900	24	0.1495	32	0.1590	...	...
#12	0.2160	24	0.1770	28	0.1820	32	0.1850
1/4	0.2500	20	0.2010	28	0.2130	32	7/32
5/16	0.3125	18	0.257	24	0.272	32	9/32
3/8	0.3750	16	5/16	24	0.332	32	11/32
7/16	0.4375	14	0.368	20	25/64	28	13/32
1/2	0.5000	13	27/64	20	29/64	28	15/32
9/16	0.5625	12	31/64	18	33/64	24	33/64
5/8	0.6250	11	17/32	18	37/64	24	37/64
11/16	0.675	...	...	...	...	24	41/64
3/4	0.7500	10	21/32	16	11/16	20	45/64
13/16	0.8125	...	...	...	...	20	49/64
7/8	0.8750	9	49/64	14	13/16	20	53/64
15/16	0.9375	...	...	...	...	20	57/64
1	1.0000	8	7/8	12	59/64	20	61/64
1 1/8	1.1250	7	63/64	12	1 3/64	18	1 5/64
1 1/4	1.2500	7	1 7/64	12	1 11/64	18	1 3/16
1 3/8	1.3750	6	1 7/32	12	1 19/64	18	1 5/16
1 1/2	1.5000	6	1 11/32	12	1 27/64	18	1 7/16
1 5/8	1.6250	...	...	...	...	18	1 9/16
1 3/4	1.7500	5	1 9/16	...	...	...	...
1 7/8	1.8750	...	...	...	...	...	...
2	2.0000	4 1/2	1 25/32	...	...	...	...
2 1/4	2.2500	4 1/2	2 1/32	...	...	...	...
2 1/2	2.5000	4	2 1/4	...	...	...	...
2 3/4	2.7500	4	2 1/2	...	...	...	...

APPENDIX D – METRIC THREAD FORM

(External Threads) Approximate Minor diameter $= D - 1.2075P$ $P = $ Pitch
Preferred sizes for commercial threads and fasteners are shown in boldface type.

Coarse (general purpose)		Fine	
Nominal Size & Thread Pitch	**Tap Drill Diameter, mm**	**Nominal Size & Thread Pitch**	**Tap Drill Diameter, mm**
M1.6 x 0.35	1.25	---	---
M1.8 x 0.35	1.45	---	---
M2 x 0.4	1.6	---	---
M2.2 x 0.45	1.75	---	---
M2.5 x 0.45	2.05	---	---
M3 x 0.5	2.5	---	---
M3.5 x 0.6	2.9	---	---
M4 x 0.7	3.3	---	---
M4.5 x 0.75	3.75	---	---
M5 x 0.8	4.2	---	---
M6 x 1	5.0	---	---
M7 x 1	6.0	---	---
M8 x 1.25	6.8	**M8 x 1**	7.0
M9 x 1.25	7.75	---	---
M10 x 1.5	8.5	**M10 x 1.25**	8.75
M11 x 1.5	9.50	---	---
M12 x 1.75	10.30	**M12 x 1.25**	10.5
M14 x 2	12.00	**M14 x 1.5**	12.5
M16 x 2	14.00	**M16 x 1.5**	14.5
M18 x 2.5	15.50	**M18 x 1.5**	16.5
M20 x 2.5	17.5	**M20 x 1.5**	18.5
M22 x 2.5*	19.5	**M22 x 1.5**	20.5
M24 x 3	21.0	**M24 x 2**	22.0
M27 x 3*	24.0	**M27 x 2**	25.0
M30 x 3.5	26.5	**M30 x 2**	28.0
M33 x 3.5	29.5	M33 x 2	31.0
M36 x 4	32.0	**M36 x 2**	33.0
M39 x 4	35.0	M39 x 2	36.0
M42 x 4.5	37.5	**M42 x 2**	39.0
M45 x 4.5	40.5	M45 x 1.5	42.0
M48 x 5	43.0	**M48 x 2**	45.0
M52 x 5	47.0	M52 x 2	49.0
M56 x 5.5	50.5	**M56 x 2**	52.0
M60 x 5.5	54.5	M60 x 1.5	56.0
M64 x 6	58.0	**M64 x 2**	60.0
M68 x 6	62.0	M68 x 2	64.0
M72 x 6	66.0	**M72 x 2**	68.0
M80 x 6	74.0	**M80 x 2**	76.0
M90 x 6	84.0	**M90 x 2**	86.0
M100 x 6	94.0	**M100 x 2**	96.0

*Only for high strength structural steel fasteners
ASME B1.13M – 2001 Standard. Reprinted from the standard listed by permission of the American Society of Mechanical Engineers. All rights reserved.

APPENDIX E – FASTENERS (INCH SERIES)

Dimensions of Hex Cap Screws (Finished Hex Bolts)

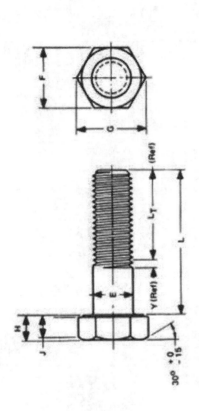

Nominal Size or Basic Product Dia	E Body Dia		F Width Across Flats			G Width Across Corners		H Height			J Wrenching Height	L_T Thread Length For Screw Lengths		Y Transition Thread Length	Runout of Bearing Surface FIM
	Max	Min	Basic	Max	Min	Max	Min	Basic	Max	Min	Min	6 in. and Shorter Basic	Over 6 in. Basic	Max	Max
1/4 0.2500	0.2500	0.2450	7/16	0.438	0.428	0.505	0.488	5/32	0.163	0.150	0.106	0.750	1.000	0.250	0.010
5/16 0.3125	0.3125	0.3065	1/2	0.500	0.489	0.577	0.557	13/64	0.211	0.195	0.140	0.875	1.125	0.278	0.011
3/8 0.3750	0.3750	0.3690	9/16	0.562	0.551	0.650	0.628	15/64	0.243	0.226	0.160	1.000	1.250	0.312	0.012
7/16 0.4375	0.4375	0.4305	5/8	0.625	0.612	0.722	0.698	9/32	0.291	0.272	0.195	1.125	1.375	0.357	0.013
1/2 0.5000	0.5000	0.4930	3/4	0.750	0.736	0.866	0.840	5/16	0.323	0.302	0.215	1.250	1.500	0.385	0.014
9/16 0.5625	0.5625	0.5545	13/16	0.812	0.798	0.938	0.910	23/64	0.371	0.348	0.250	1.375	1.625	0.417	0.015
5/8 0.6250	0.6250	0.6170	15/16	0.938	0.922	1.083	1.051	25/64	0.403	0.378	0.289	1.500	1.750	0.455	0.017
3/4 0.7500	0.7500	0.7410	1 1/8	1.125	1.100	1.299	1.254	15/32	0.483	0.455	0.324	1.750	2.000	0.500	0.020
7/8 0.8750	0.8750	0.8660	1 5/16	1.312	1.285	1.516	1.465	35/64	0.563	0.531	0.378	2.000	2.250	0.556	0.023
1 1.0000	1.0000	0.9900	1 1/2	1.500	1.469	1.732	1.675	39/64	0.627	0.591	0.416	2.250	2.500	0.625	0.026
1 1/8 1.1250	1.1250	1.1140	1 11/16	1.688	1.631	1.949	1.859	11/16	0.718	0.658	0.461	2.500	2.750	0.714	0.029
1 1/4 1.2500	1.2500	1.2390	1 7/8	1.875	1.812	2.165	2.066	25/32	0.813	0.749	0.530	2.750	3.000	0.714	0.033
1 3/8 1.3750	1.3750	1.3630	2 1/16	2.062	1.994	2.382	2.273	27/32	0.878	0.810	0.569	3.000	3.250	0.833	0.036
1 1/2 1.5000	1.5000	1.4880	2 1/4	2.230	2.175	2.598	2.480	15/16	0.974	0.902	0.640	3.250	3.500	0.833	0.039
1 3/4 1.7500	1.7500	1.7380	2 5/8	2.625	2.538	3.031	2.893	1 3/32	1.134	1.054	0.748	3.750	4.000	1.000	0.046
2 2.0000	2.0000	1.9880	3	3.000	2.900	3.464	3.306	1 7/32	1.263	1.175	0.825	4.250	4.500	1.111	0.052
2 1/4 2.2500	2.2500	2.2380	3 3/8	3.375	3.262	3.897	3.719	1 3/8	1.423	1.327	0.933	4.750	5.000	1.111	0.059
2 1/2 2.5000	2.5000	2.4880	3 3/4	3.750	3.625	4.330	4.133	1 17/32	1.583	1.479	1.042	5.250	5.500	1.250	0.065
2 3/4 2.7500	2.7500	2.7380	4 1/8	4.125	3.988	4.763	4.546	1 11/16	1.744	1.632	1.151	5.750	6.000	1.250	0.072
3 3.0000	3.0000	2.9880	4 1/2	4.500	4.350	5.196	4.959	1 7/8	1.935	1.815	1.290	6.250	6.500	1.250	0.079

APPENDIX E – FASTENERS (INCH SERIES) Continued

DIMENSIONS OF HEXAGON AND SPLINE SOCKET HEAD CAP SCREWS (1960 SERIES)

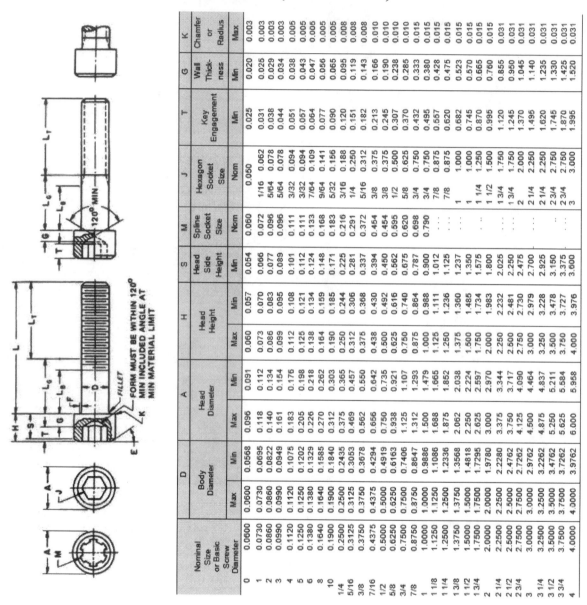

Nominal Size or Basic Screw Diameter	D Body Diameter Max	D Body Diameter Min	A Head Diameter Max	A Head Diameter Min	H Head Height Max	H Head Height Min	S Head Side Height Min	M Spline Socket Size Nom	J Hexagon Socket Size Nom	T Key Engagement Min	G Wall Thickness Min	K Chamfer or Radius Max
0 — 0.0600	0.0600	0.0568	0.096	0.091	0.060	0.057	0.054	0.060	0.050	0.025	0.020	0.003
1 — 0.0730	0.0730	0.0695	0.118	0.112	0.073	0.070	0.066	0.072	1/16 0.062	0.031	0.025	0.003
2 — 0.0860	0.0860	0.0822	0.140	0.134	0.086	0.083	0.077	0.096	5/64 0.078	0.038	0.029	0.003
3 — 0.0990	0.0990	0.0949	0.161	0.154	0.099	0.095	0.089	0.096	5/64 0.078	0.044	0.034	0.003
4 — 0.1120	0.1120	0.1075	0.183	0.176	0.112	0.108	0.101	0.111	3/32 0.094	0.051	0.038	0.005
5 — 0.1250	0.1250	0.1202	0.205	0.198	0.125	0.121	0.112	0.111	3/32 0.094	0.057	0.043	0.005
6 — 0.1380	0.1380	0.1329	0.226	0.218	0.138	0.134	0.124	0.133	7/64 0.109	0.064	0.047	0.005
8 — 0.1640	0.1640	0.1585	0.270	0.262	0.164	0.159	0.148	0.168	9/64 0.141	0.077	0.056	0.005
10 — 0.1900	0.1900	0.1840	0.312	0.303	0.190	0.185	0.171	0.183	5/32 0.156	0.090	0.065	0.005
1/4 — 0.2500	0.2500	0.2435	0.375	0.365	0.250	0.244	0.225	0.216	3/16 0.188	0.120	0.095	0.008
5/16 — 0.3125	0.3125	0.3053	0.469	0.457	0.312	0.306	0.281	0.291	1/4 0.250	0.151	0.119	0.008
3/8 — 0.3750	0.3750	0.3678	0.562	0.550	0.375	0.368	0.337	0.372	5/16 0.312	0.182	0.143	0.008
7/16 — 0.4375	0.4375	0.4294	0.656	0.642	0.438	0.430	0.394	0.454	3/8 0.375	0.213	0.166	0.010
1/2 — 0.5000	0.5000	0.4919	0.750	0.735	0.500	0.492	0.450	0.454	3/8 0.375	0.245	0.190	0.010
5/8 — 0.6250	0.6250	0.6163	0.938	0.921	0.625	0.616	0.562	0.595	1/2 0.500	0.307	0.238	0.010
3/4 — 0.7500	0.7500	0.7406	1.125	1.107	0.750	0.740	0.675	0.620	5/8 0.625	0.370	0.285	0.010
7/8 — 0.8750	0.8750	0.8647	1.312	1.293	0.875	0.864	0.787	0.698	3/4 0.750	0.432	0.333	0.015
1 — 1.0000	1.0000	0.9886	1.500	1.479	1.000	0.988	0.900	0.790	3/4 0.750	0.495	0.380	0.015
1 1/8 — 1.1250	1.1250	1.1086	1.688	1.665	1.125	1.111	1.012	…	7/8 0.875	0.557	0.428	0.015
1 1/4 — 1.2500	1.2500	1.2336	1.875	1.852	1.250	1.236	1.125	…	7/8 0.875	0.620	0.475	0.015
1 3/8 — 1.3750	1.3750	1.3568	2.062	2.038	1.375	1.360	1.237	…	1 1.000	0.682	0.523	0.015
1 1/2 — 1.5000	1.5000	1.4818	2.250	2.224	1.500	1.485	1.350	…	1 1.000	0.745	0.570	0.015
1 3/4 — 1.7500	1.7500	1.7295	2.625	2.597	1.750	1.734	1.575	…	1 1/4 1.250	0.870	0.665	0.015
2 — 2.0000	2.0000	1.9780	3.000	2.970	2.000	1.983	1.800	…	1 1/2 1.500	0.995	0.760	0.015
2 1/4 — 2.2500	2.2500	2.2280	3.375	3.344	2.250	2.232	2.025	…	1 3/4 1.750	1.120	0.855	0.031
2 1/2 — 2.5000	2.5000	2.4762	3.750	3.717	2.500	2.481	2.250	…	1 3/4 1.750	1.245	0.950	0.031
2 3/4 — 2.7500	2.7500	2.7262	4.125	4.090	2.750	2.730	2.475	…	2 2.000	1.370	1.045	0.031
3 — 3.0000	3.0000	2.9762	4.500	4.464	3.000	2.979	2.700	…	2 1/4 2.250	1.495	1.140	0.031
3 1/4 — 3.2500	3.2500	3.2262	4.875	4.837	3.250	3.228	2.925	…	2 1/4 2.250	1.620	1.235	0.031
3 1/2 — 3.5000	3.5000	3.4762	5.250	5.211	3.500	3.478	3.150	…	2 3/4 2.750	1.745	1.330	0.031
3 3/4 — 3.7500	3.7500	3.7262	5.625	5.584	3.750	3.727	3.375	…	2 3/4 2.750	1.870	1.425	0.031
4 — 4.0000	4.0000	3.9762	6.000	5.958	4.000	3.976	3.600	…	3 3.000	1.995	1.520	0.031

APPENDIX E – FASTENERS (INCH SERIES) Continued

DIMENSIONS OF HEXAGON AND SPLINE SOCKET FLAT COUNTERSUNK HEAD CAP SCREWS

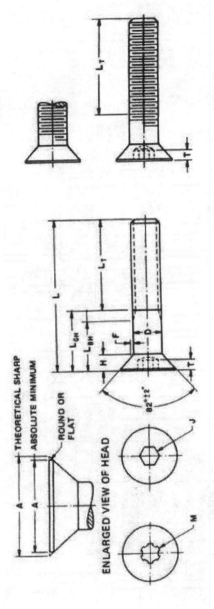

Nominal Size or Basic Screw Diameter	D Body Diameter Max	D Body Diameter Min	A Head Diameter Theoretical Sharp Max	A Head Diameter Abs. Min	H Head Reference	H Height Flushness Tolerance	M Spline Socket Size	J Hexagon Socket Size Nom	T Key Engagement Min	F Fillet Extension Above D Max
0	0.0600	0.0568	0.138	0.117	0.044	0.006	0.048	0.035	0.025	0.006
1	0.0730	0.0695	0.168	0.143	0.054	0.007	0.060	0.050	0.031	0.008
2	0.0860	0.0822	0.197	0.168	0.064	0.008	0.060	0.050	0.038	0.010
3	0.0990	0.0949	0.226	0.193	0.073	0.010	0.072	1/16 0.062	0.044	0.010
4	0.1120	0.1075	0.255	0.218	0.083	0.011	0.072	1/16 0.062	0.055	0.012
5	0.1250	0.1202	0.281	0.240	0.090	0.012	0.096	5/64 0.078	0.061	0.014
6	0.1380	0.1329	0.307	0.263	0.097	0.013	0.096	5/64 0.078	0.066	0.015
8	0.1640	0.1585	0.359	0.311	0.112	0.014	0.111	3/32 0.094	0.076	0.015
10	0.1900	0.1840	0.411	0.359	0.127	0.015	0.145	1/8 0.125	0.087	0.015
1/4	0.2500	0.2435	0.531	0.480	0.161	0.016	0.183	5/32 0.156	0.111	0.015
5/16	0.3125	0.3053	0.656	0.600	0.198	0.017	0.216	3/16 0.188	0.135	0.015
3/8	0.3750	0.3678	0.781	0.720	0.234	0.018	0.251	7/32 0.219	0.159	0.015
7/16	0.4375	0.4294	0.844	0.781	0.234	0.018	0.291	1/4 0.250	0.159	0.015
1/2	0.5000	0.4919	0.938	0.872	0.251	0.018	0.372	5/16 0.312	0.172	0.015
5/8	0.6250	0.6163	1.188	1.112	0.324	0.022	0.454	3/8 0.375	0.220	0.015
3/4	0.7500	0.7406	1.438	1.355	0.396	0.024	0.454	1/2 0.500	0.220	0.015
7/8	0.8750	0.8647	1.688	1.604	0.468	0.025	...	9/16 0.562	0.248	0.015
1	1.0000	0.9886	1.938	1.841	0.540	0.028	...	5/8 0.625	0.297	0.015
1 1/8	1.1250	1.1086	2.188	2.079	0.611	0.031	...	3/4 0.750	0.325	0.031
1 1/4	1.2500	1.2336	2.438	2.316	0.683	0.035	...	7/8 0.875	0.358	0.031
1 3/8	1.3750	1.3568	2.688	2.553	0.755	0.038	...	7/8 0.875	0.402	0.031
1 1/2	1.5000	1.4818	2.938	2.791	0.827	0.042	...	1 1.000	0.435	0.031

Enlarged view of head labels: THEORETICAL SHARP, ABSOLUTE MINIMUM, ROUND OR FLAT, ENLARGED VIEW OF HEAD, 82°±2°

APPENDIX E – FASTENERS (INCH SERIES) Continued

DIMENSIONS OF SLOTTED FLAT COUNTERSUNK HEAD CAP SCREWS

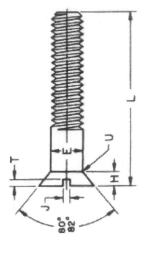

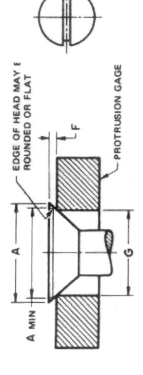

Nominal Size[1] or Basic Screw Diameter	E Body Diameter		A Head Diameter		H[2] Head Height	J Slot Width		T Slot Depth		U Fillet Radius	F[3] Protrusion Above Gaging Diameter		G[3] Gaging Diameter
	Max	Min	Max, Edge Sharp	Min, Edge Rounded or Flat	Ref	Max	Min	Max	Min	Max	Max	Min	Gaging Diameter
1/4 0.2500	0.2500	0.2450	0.500	0.452	0.140	0.075	0.064	0.068	0.045	0.100	0.046	0.030	0.424
5/16 0.3125	0.3125	0.3070	0.625	0.567	0.177	0.084	0.072	0.086	0.057	0.125	0.053	0.035	0.538
3/8 0.3750	0.3750	0.3690	0.750	0.682	0.210	0.094	0.081	0.103	0.068	0.150	0.060	0.040	0.651
7/16 0.4375	0.4375	0.4310	0.812	0.736	0.210	0.094	0.081	0.103	0.068	0.175	0.065	0.044	0.703
1/2 0.5000	0.5000	0.4930	0.875	0.791	0.210	0.106	0.091	0.103	0.068	0.200	0.071	0.049	0.756
9/16 0.5625	0.5625	0.5550	1.000	0.906	0.244	0.118	0.102	0.120	0.080	0.225	0.078	0.054	0.869
5/8 0.6250	0.6250	0.6170	1.125	1.020	0.281	0.133	0.116	0.137	0.091	0.250	0.085	0.058	0.982
3/4 0.7500	0.7500	0.7420	1.375	1.251	0.352	0.149	0.131	0.171	0.115	0.300	0.099	0.068	1.208
7/8 0.8750	0.8750	0.8660	1.625	1.480	0.423	0.167	0.147	0.206	0.138	0.350	0.113	0.077	1.435
1 1.0000	1.0000	0.9900	1.875	1.711	0.494	0.188	0.166	0.240	0.162	0.400	0.127	0.087	1.661
1 1/8 1.1250	1.1250	1.1140	2.062	1.880	0.529	0.196	0.178	0.257	0.173	0.450	0.141	0.096	1.826
1 1/4 1.2500	1.2500	1.2390	2.312	2.110	0.600	0.211	0.193	0.291	0.197	0.500	0.155	0.105	2.052
1 3/8 1.3750	1.3750	1.3630	2.562	2.340	0.665	0.226	0.208	0.326	0.220	0.550	0.169	0.115	2.279
1 1/2 1.5000	1.5000	1.4880	2.812	2.570	0.742	0.258	0.240	0.360	0.244	0.600	0.183	0.124	2.505

[1] Where specifying nominal size in decimals, zeros preceding decimal and in the fourth decimal place shall be omitted.

[2] Tabulated values determined from formula for maximum H.

[3] No tolerance for gaging diameter is given.

Reprinted from The American Society of Mechanical Engineers-ANSI 318.6.2-R1993.

APPENDIX E – FASTENERS (INCH SERIES) Continued

DIMENSIONS OF SLOTTED ROUND HEAD CAP SCREWS

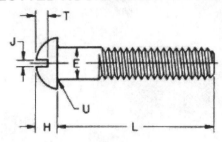

Nominal Size' or Basic Screw Diameter		E Body Diameter		A Head Diameter		H Head Height		J Slot Width		T Slot Depth		U Fillet Radius	
		Max	Min	Max	Min	Max	Min	Max	Min	Max	Min	Max	Min
1/4	0.2500	0.2500	0.2450	0.437	0.418	0.191	0.175	0.075	0.064	0.117	0.097	0.031	0.016
5/16	0.3125	0.3125	0.3070	0.562	0.540	0.245	0.226	0.084	0.072	0.151	0.126	0.031	0.016
3/8	0.3750	0.3750	0.3690	0.625	0.603	0.273	0.252	0.094	0.081	0.168	0.138	0.031	0.016
7/16	0.4375	0.4375	0.4310	0.750	0.725	0.328	0.302	0.094	0.081	0.202	0.167	0.047	0.016
1/2	0.5000	0.5000	0.4930	0.812	0.786	0.354	0.327	0.106	0.091	0.218	0.178	0.047	0.016
9/16	0.5625	0.5625	0.5550	0.937	0.909	0.409	0.378	0.118	0.102	0.252	0.207	0.047	0.016
5/8	0.6250	0.6250	0.6170	1.000	0.970	0.437	0.405	0.133	0.116	0.270	0.220	0.062	0.031
3/4	0.7500	0.7500	0.7420	1.250	1.215	0.546	0.507	0.149	0.131	0.338	0.278	0.062	0.031

'Where specifying nominal size in decimals, zeros preceding decimal and in the fourth decimal place shall be omitted.

DIMENSIONS OF SLOTTED FILLISTER HEAD CAP SCREWS

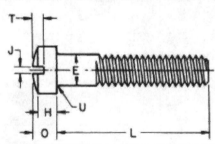

Nominal Size' or Basic Screw Diameter		E Body Diameter		A Head Diameter		H Head Side Height		0 Total Head Height		J Slot Width		T Slot Depth		U Fillet Radius	
		Max	Min	Max	Min	Max	Min	Max	Min	Max	Min	Max	Min	Max	Min
1/4	0.2500	0.2500	0.2450	0.375	0.363	0.172	0.157	0.216	0.194	0.075	0.064	0.097	0.077	0.031	0.016
5/16	0.3125	0.3125	0.3070	0.437	0.424	0.203	0.186	0.253	0.230	0.084	0.072	0.115	0.090	0.031	0.016
3/8	0.3750	0.3750	0.3690	0.562	0.547	0.250	0.229	0.314	0.284	0.094	0.081	0.142	0.112	0.031	0.016
7/16	0.4375	0.4375	0.4310	0.625	0.608	0.297	0.274	0.368	0.336	0.094	0.081	0.168	0.133	0.047	0.016
1/2	0.5000	0.5000	0.4930	0.750	0.731	0.328	0.301	0.413	0.376	0.106	0.091	0.193	0.153	0.047	0.016
9/16	0.5625	0.5625	0.5550	0.812	0.792	0.375	0.346	0.467	0.427	0.118	0.102	0.213	0.168	0.047	0.016
5/8	0.6250	0.6250	0.6170	0.875	0.853	0.422	0.391	0.521	0.478	0.133	0.116	0.239	0.189	0.062	0.031
3/4	0.7500	0.7500	0.7420	1.000	0.976	0.500	0.466	0.612	0.566	0.149	0.131	0.283	0.223	0.062	0.031
7/8	0.8750	0.8750	0.8660	1.125	1.098	0.594	0.556	0.720	0.668	0.167	0.147	0.334	0.264	0.062	0.031
1	1.0000	1.0000	0.9900	1.312	1.282	0.656	0.612	0.803	0.743	0.188	0.166	0.371	0.291	0.062	0.031

APPENDIX E – FASTENERS (INCH SERIES) Continued

DIMENSIONS OF SLOTTED FLAT COUNTERSUNK HEAD MACHINE SCREWS

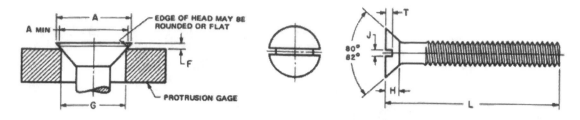

Nominal Size' or Basic Screw Diameter		L[2] These Lengths or Shorter are Undercut	A Head Diameter		H[3] Head Height	J Slot Width		T Slot Depth		F[4] Protrusion Above Gaging Diameter		G[4] Gaging Diameter
			Max. Edge Sharp	Min, Edge Rounded or Flat	Ref	Max	Min	Max	Min	Max	Min	
0000	0.0210	-	0.043	0.037	0.011	0.008	0.004	0.007	0.003	*	*	*
000	0.0340	-	0.064	0.058	0.016	0.011	0.007	0.009	0.005	*	*	*
00	0.0470	-	0.093	0.085	0.028	0.017	0.010	0.014	0.009	*	*	*
0	0.0600	1/8	0.119	0.099	0.035	0.023	0.016	0.015	0.010	0.026	0.016	0.078
1	0.0730	1/8	0.146	0.123	0.043	0.026	0.019	0.019	0.012	0.028	0.016	0.101
2	0.0860	1/8	0.172	0.147	0.051	0.031	0.023	0.023	0.015	0.029	0.017	0.124
3	0.0990	1/8	0.199	0.171	0.059	0.035	0.027	0.027	0.017	0.031	0.018	0.148
4	0.1120	3/16	0.225	0.195	0.067	0.039	0.031	0.030	0.020	0.032	0.019	0.172
5	0.1250	3/16	0.252	0.220	0.075	0.043	0.035	0.034	0.022	0.034	0.020	0.196
6	0.1380	3/16	0.279	0.244	0.083	0.048	0.039	0.038	0.024	0.036	0.021	0.220
8	0.1640	1/4	0.332	0.292	0.100	0.054	0.045	0.045	0.029	0.039	0.023	0.267
10	0.1900	5/16	0.385	0.340	0.116	0.060	0.050	0.053	0.034	0.042	0.025	0.313
12	0.2160	3/8	0.438	0.389	0.132	0.067	0.056	0.060	0.039	0.045	0.027	0.362
1/4	0.2500	7/16	0.507	0.452	0.153	0.075	0.064	0.070	0.046	0.050	0.029	0.424
5/16	0.3125	1/2	0.635	0.568	0.191	0.084	0.072	0.088	0.058	0.057	0.034	0.539
3/8	0.3750	9/16	0.762	0.685	0.230	0.094	0.081	0.106	0.070	0.065	0.039	0.653
7/16	0.4375	5/8	0.812	0.723	0.223	0.094	0.081	0.103	0.066	0.073	0.044	0.690
1/2	0.5000	3/4	0.875	0.775	0.223	0.106	0.091	0.103	0.065	0.081	0.049	0.739
9/16	0.5625	-	1.000	0.889	0.260	0.118	0.102	0.120	0.077	0.089	0.053	0.851
5/8	0.6250	-	1.125	1.002	0.298	0.133	0.116	0.137	0.088	0.097	0.058	0.962
3/4	0.7500	-	1.375	1.230	0.372	0.149	0.131	0.171	0.111	0.112	0.067	1.186

' Where specifying nominal size in decimals, zeros preceding decimal and in the fourth decimal place shall be omitted.

[2] Screws of these lengths and shorter shall have undercut heads as shown in Table 5.

[3] Tabulated values determined from formula for maximum H, Appendix V.

[4] No tolerance for gaging diameter is given.

* Not practical to gage.

Reprinted from The American Society of Mechanical Engineers-ANSI B18.6.3-R1991. All rights reserved.

APPENDIX E – FASTENERS (INCH SERIES) Continued

DIMENSIONS OF HEX NUTS AND HEX JAM NUTS

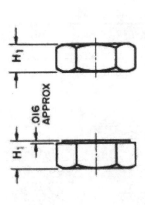
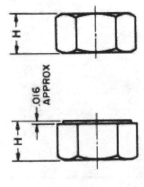

Nominal Size or Basic Major Dia of Thread		F Width Across Flats			G Width Across Corners		H Thickness Hex Nuts			H_1 Thickness Hex Jam Nuts			Hex Nuts Specified Proof Load — Runout of Bearing Face, FIR Max		Jam Nuts All Strength Levels
		Basic	Max	Min	Max	Min	Basic	Max	Min	Basic	Max	Min	Up to 150,000 psi	150,000 psi and Greater	
1/4	0.2500	7/16	0.438	0.428	0.505	0.488	7/32	0.226	0.212	5/32	0.163	0.150	0.015	0.010	0.015
5/16	0.3125	1/2	0.500	0.489	0.577	0.557	17/64	0.273	0.258	3/16	0.195	0.180	0.016	0.011	0.016
3/8	0.3750	9/16	0.562	0.551	0.650	0.628	21/64	0.337	0.320	7/32	0.227	0.210	0.017	0.012	0.017
7/16	0.4375	11/16	0.688	0.675	0.794	0.768	3/8	0.385	0.365	1/4	0.260	0.240	0.018	0.013	0.018
1/2	0.5000	3/4	0.750	0.736	0.866	0.840	7/16	0.448	0.427	5/16	0.323	0.302	0.019	0.014	0.019
9/16	0.5625	7/8	0.875	0.861	1.010	0.982	31/64	0.496	0.473	5/16	0.324	0.301	0.020	0.015	0.020
5/8	0.6250	15/16	0.938	0.922	1.083	1.051	35/64	0.559	0.535	3/8	0.387	0.363	0.021	0.016	0.021
3/4	0.7500	1 1/8	1.125	1.088	1.299	1.240	41/64	0.665	0.617	27/64	0.446	0.398	0.023	0.018	0.023
7/8	0.8750	1 5/16	1.312	1.269	1.516	1.447	3/4	0.776	0.724	31/64	0.510	0.458	0.025	0.020	0.025
1	1.0000	1 1/2	1.500	1.450	1.732	1.653	55/64	0.887	0.831	35/64	0.575	0.519	0.027	0.022	0.027
1 1/8	1.1250	1 11/16	1.688	1.631	1.949	1.859	31/32	0.999	0.939	39/64	0.639	0.579	0.030	0.025	0.030
1 1/4	1.2500	1 7/8	1.875	1.812	2.165	2.066	1 1/16	1.094	1.030	23/32	0.751	0.687	0.033	0.028	0.033
1 3/8	1.3750	2 1/16	2.062	1.994	2.382	2.273	1 11/64	1.206	1.138	25/32	0.815	0.747	0.036	0.031	0.036
1 1/2	1.5000	2 1/4	2.250	2.175	2.598	2.480	1 9/32	1.317	1.245	27/32	0.880	0.808	0.039	0.034	0.039

APPENDIX E – FASTENERS (INCH SERIES) Continued

Drill and Counterbore Sizes for Socket Head Cap Screws

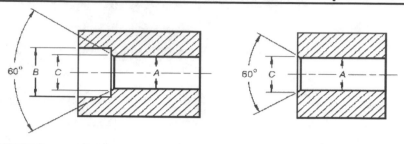

Nominal Size of Screw (D)	Nominal Drill Size (A)		Counterbore Diameter (B)	Countersink (C)
	Close Fit	**Normal Fit**		
#0 (0.0600)	(#51) 0.067	(#49) 0.073	1/8	0.074
#1 (0.0730)	(#46) 0.081	(#43) 0.089	5/32	0.087
#2 (0.0860)	3/32	(#36) 0.106	3/16	0.102
#3 (0.0990)	(#36) 0.106	(#31) 0.120	7/32	0.115
#4 (0.1120)	1/8	(#29) 0.136	7/32	0.130
#5 (0.1250)	9/64	(#23) 0.154	1/4	0.145
#6 (0.1380)	(#23) 0.154	(#18) 0.170	9/32	0.158
#8 (0.1640)	(#15) 0.180	(#10) 0.194	5/16	0.188
#10 (0.1900)	(#5) 0.206	(#2) 0.221	3/8	0.218
1/4	17/64	9/32	7/16	0.278
5/16	21/64	11/32	17/32	0.346
3/8	25/64	13/32	5/8	0.415
7/16	29/64	15/32	23/32	0.483
1/2	33/64	17/32	13/16	0.552
5/8	41/64	21/32	1	0.689
3/4	49/64	25/32	1 3/16	0.828
7/8	57/64	29/32	1 3/8	0.963
1	1 1/64	1 1/32	1 5/8	1.100
1 1/4	1 9/32	1 5/16	2	1.370
1 1/2	1 17/32	1 9/16	2 3/8	1.640
1 3/4	1 25/32	1 13/16	2 3/4	1.910
2	2 1/32	2 1/16	3 1/8	2.180

(1) Countersink. It is considered good practice to countersink or break the edges of holes that are smaller than F (max.) in parts having a hardness which approaches, equals, or exceeds the screw hardness. The countersink or corner relief, however, should not be larger than is necessary to ensure that the fillet on the screw is cleared. Normally, the diameter of countersink does not have to exceed F (max.). Countersinks or corner reliefs in excess of this diameter reduce the effective bearing area and introduce the possibility of imbedment or brinelling or flaring of the heads of the screws.

(2) Close Fit. The close fit is normally limited to holes for those lengths of screws that are threaded to the head in assemblies where only one screw is to be used or where two or more screws are to be used and the mating holes are to be produced either at assembly or by matched and coordinated tooling.

(3) Normal Fit. The normal fit is intended for screws of relatively long length or for assemblies involving two or more screws where the mating holes are to be produced by conventional tolerancing methods. It provides for the maximum allowable eccentricity of the longest standard screws and for certain variations in the parts to be fastened, such as deviations in hole straightness, angularity between the axis of the tapped hole and that of the hole for the shank, differences in center distances of the mating holes, etc.

APPENDIX E – FASTENERS (INCH SERIES) Continued

Preferred Sizes of Type A Plain Washers

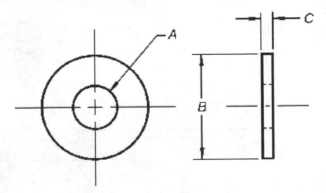

Nominal Washer Size[*]	Inside Diameter (A) Basic	Outside Diameter (B) Basic	Thickness (C)	Nominal Washer Size[*]	Inside Diameter (A) Basic	Outside Diameter (B) Basic	Thickness (C)
	0.078	0.188	0.020	1 N	1.062	2.000	0.134
	0.094	0.250	0.020	1 W	1.062	2.500	0.165
	0.125	0.312	0.032	1 1/8 N	1.250	2.250	0.134
#6 (0.138)	0.156	0.375	0.049	1 1/8 W	1.250	2.750	0.165
#8 (0.164)	0.188	0.438	0.049	1 1/4 N	1.375	2.500	0.165
#10 (0.190)	0.219	0.500	0.049	1 1/4 W	1.375	3.000	0.165
3/16	0.250	0.562	0.049	1 3/8 N	1.500	2.750	0.165
#12 (0.216)	0.250	0.562	0.065	1 3/8 W	1.500	3.250	0.180
1/4 N	0.281	0.625	0.065	1 1/2 N	1.625	3.000	0.165
1/4 W	0.312	0.734	0.065	1 1/2 W	1.625	3.500	0.180
5/16 N	0.344	0.688	0.065	1 5/8	1.750	3.750	0.180
5/16 W	0.375	0.875	0.083	1 3/4	1.875	4.000	0.180
3/8 N	0.406	0.812	0.065	1 7/8	2.000	4.250	0.180
3/8 W	0.438	1.000	0.083	2	2.125	4.500	0.180
7/16 N	0.469	0.922	0.065	2 1/4	2.375	4.750	0.220
7/16 W	0.500	1.250	0.083	2 1/2	2.625	5.000	0.238
1/2 N	0.531	1.062	0.095	2 3/4	2.875	5.250	0.259
1/2 W	0.562	1.375	0.109	3	3.125	5.500	0.284
9/16 N	0.594	1.156	0.095				
9/16 W	0.625	1.469	0.109				
5/8 N	0.656	1.312	0.095				
5/8 W	0.688	1.750	0.134				
3/4 N	0.812	1.469	0.134				
3/4 W	0.812	2.000	0.148				
7/8 N	0.938	1.750	0.134				
7/8 W	0.938	2.250	0.165				

[*]Nominal washer sizes are intended for use with comparable nominal screw or bolt sizes.
ANSI B18.22.1 - 1965 (R2003) Standard. Reprinted from the standard listed by permission of the American Society of Mechanical Engineers. All rights reserved.

APPENDIX E – FASTENERS (INCH SERIES) Continued

Regular Helical Spring-Lock Washers

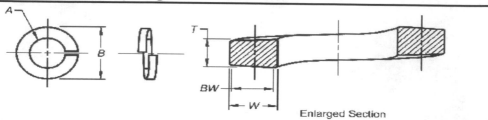

Enlarged Section

Nominal Washer Size	Min. Inside Diameter (A)	Max. Outside Diameter (B)	Mean Section Thickness (T)	Min. Section Width (W)	Min. Bearing Width (BW)
#2 (0.086)	0.088	0.172	0.020	0.035	0.024
#3 (0.099)	0.101	0.195	0.025	0.040	0.028
#4 (0.112)	0.114	0.209	0.025	0.040	0.028
#5 (0.125)	0.127	0.236	0.031	0.047	0.033
#6 (.0138)	0.141	0.250	0.031	0.047	0.033
#8 (0.164)	0.167	0.293	0.040	0.055	0.038
#10 (0.190)	0.193	0.334	0.047	0.062	0.043
#12 (0.216)	0.220	0.377	0.056	0.070	0.049
1/4	0.252	0.487	0.062	0.109	0.076
5/16	0.314	0.583	0.078	0.125	0.087
3/8	0.377	0.680	0.094	0.141	0.099
7/16	0.440	0.776	0.109	0.156	0.109
1/2	0.502	0.869	0.125	0.171	0.120
9/16	0.564	0.965	0.141	0.188	0.132
5/8	0.628	1.073	0.156	0.203	0.142
11/16	0.691	1.170	0.172	0.219	0.153
3/4	0.753	1.265	0.188	0.234	0.164
13/16	0.816	1.363	0.203	0.250	0.175
7/8	0.787	1.459	0.219	0.266	0.186
15/16	0.941	1.556	0.234	0.281	0.197
1	1.003	1.656	0.250	0.297	0.208
1 1/16	1.066	1.751	0.266	0.312	0.218
1 1/8	1.129	1.847	0.281	0.328	0.230
1 3/16	1.192	1.943	0.297	0.344	0.241
1 1/4	1.254	2.036	0.312	0.359	0.251
1 5/16	1.317	2.133	0.328	0.375	0.262
1 3/8	1.379	2.219	0.344	0.391	0.274
1 7/16	1.442	2.324	0.359	0.406	0.284
1 1/2	1.504	2.419	0.375	0.422	0.295
1 5/8	1.633	2.553	0.389	0.424	0.297
1 3/4	1.758	2.679	0.389	0.424	0.297
1 7/8	1.883	2.811	0.422	0.427	0.299
2	2.008	2.936	0.422	0.427	0.299
2 1/4	2.262	3.221	0.440	0.442	0.309
2 1/2	2.512	3.471	0.440	0.422	0.309
2 3/4	2.762	3.824	0.458	0.491	0.344
3	3.012	4.074	0.458	0.491	0.344

APPENDIX F – METRIC FASTENERS

Metric Hex Bolts

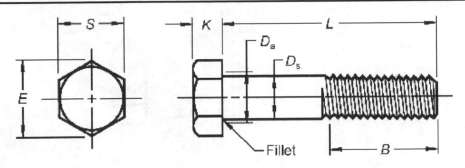

D	Ds	S	E	K	Da	Thread Length (B)		
Nominal Bolt Diameter And Thread Pitch	Max. Body Dia.	Max. Width Across Flats	Max. Width Across Corners	Max. Head Height	Fillet Transition Diameter	Bolt Lengths ≤ 125	Bolt Lengths > 125 and ≤ 200	Bolt Lengths >200
M5 x 0.8	5.48	8.00	9.24	3.88	5.7	16	22	35
M6 x 1	6.19	10.00	11.55	4.38	6.8	18	24	37
M8 x 1.25	8.58	13.00	15.01	5.68	9.2	22	28	41
M10 x 1.5	10.58	16.00	18.48	6.85	11.2	26	32	45
M12 x 1.75	12.70	18.00	20.78	7.95	13.7	30	36	49
M14 x 2	14.70	21.00	24.25	9.25	15.7	34	40	53
M16 x 2	16.70	24.00	27.71	10.75	17.7	38	44	57
M20 x 2.5	20.84	30.00	34.64	13.40	22.4	46	52	65
M24 x 3	24.84	36.00	41.57	15.90	26.4	54	60	73
M30 x 3.5	30.84	46.00	53.12	19.75	33.4	66	72	85
M36 x 4	37.00	55.00	63.51	23.55	39.4	78	84	97
M42 x 4.5	43.00	65.00	75.06	27.05	45.4	90	96	109
M48 x 5	49.00	75.00	86.60	31.07	52.0	102	108	121
M56 x 5.5	57.00	85.00	98.15	36.20	62.0		124	137
M64 x 6	65.52	95.00	109.70	41.32	70.0		140	153
M72 x 6	73.84	105.00	121.24	46.45	78.0		156	169
M80 x 6	82.16	115.00	132.79	51.58	86.0		172	185
M90 x 6	92.48	130.00	150.11	57.74	96.0		192	205
M100 x 6	102.80	145.00	167.43	63.90	107.0		212	225

APPENDIX F – METRIC FASTENERS (Continued)

Metric Hex Nuts

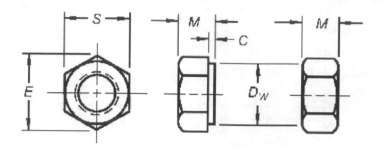

D	S	E	M	DW	C
Nominal Bolt Diameter and Thread Pitch	**Max. Width Across Flats**	**Max. Width Across Corners**	**Max. Thickness**	**Min. Bearing Face Diameter**	**Max. Washer Face Thickness**
M1.6 x 0.35	3.20	3.70	1.30	2.3	
M2 x 0.4	4.00	4.62	1.60	3.1	
M2.5 x 0.45	5.00	5.77	2.00	4.1	
M3 x 0.5	5.50	6.35	2.40	4.6	
M3.5 x 0.6	6.00	6.93	2.80	5.1	
M4 x 0.7	7.00	8.08	3.20	6.0	
M5 x 0.8	8.00	9.24	4.70	7.0	
M6 x 1	10.00	11.55	5.20	8.9	
M8 x 1.25	13.00	15.01	6.80	11.6	
M10 x 1.5	15.00	17.32	9.10	13.6	
M10 x 1.5	16.00	18.45	8.40	14.6	
M12 x 1.75	18.00	20.78	10.80	16.6	
M14 x 2	21.00	24.25	12.80	19.4	
M16 x 2	24.00	27.71	14.80	22.4	
M20 x 2.5	30.00	34.64	18.00	27.9	0.8
M24 x 3	36.00	41.57	21.50	32.5	0.8
M30 x 3.5	46.00	53.12	25.60	42.5	0.8
M36 x 4	55.00	63.51	31.00	50.8	0.8

ASME B18.2.4.1M - 2002 Standard. Reprinted from the standard listed by permission of the American Society of Mechanical Engineers. All rights reserved.

APPENDIX F – METRIC FASTENERS (Continued)

Metric Socket Head Cap Screws

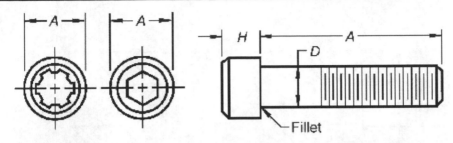

Screw Size (D)	Head Diameter (A)	Head Height (H)
1.6 through 2.5	See Table	
3 through 8	Max. A = 1.5 D + 1	Max. H = D
> 10	Max. A = 1.5 D	

Screw Size (D)	1.6	2	2.5
Max. Head Diameter (A)	3.00	3.80	4.50

ASME/ANSI B18.3.1M - 1986 (R2002) Standard. Reprinted from the standard listed by permission of the American Society of Mechanical Engineers. All rights reserved.

Metric Countersunk Socket Head Cap Screws

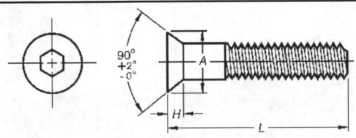

Basic Screw Diameter and Thread Pitch	Head Diameter (A) Theor. Sharp	Head Height (H)
M3 x 0.5	6.72	1.86
M4 x 0.7	8.96	2.48
M5 x 0.8	11.20	3.10
M6 x 1	13.44	3.72
M8 x 1.25	17.92	4.96
M10 x 1.5	22.40	6.20
M12 x 1.75	26.88	7.44
M14 x 2	30.24	8.12
M16 x 2	33.60	8.80
M20 x 2.5	40.32	10.16

ASME/ANSI B18.3.5M - 1986 (R2002) Standard. Reprinted from the standard listed by permission of the American Society of Mechanical Engineers. All rights reserved.

APPENDIX F – METRIC FASTENERS (Continued)

Drill and Counterbore Sizes for Socket Head Cap Screws

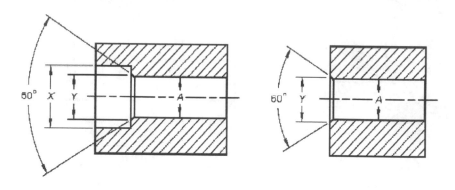

Nominal Size or Basic Screw Diameter	A		X	Y
	Nominal Drill Size		Counterbore Diameter	Countersink Diameter
	Close Fit	Normal Fit		
M1.6	1.80	1.95	3.50	2.0
M2	2.20	2.40	4.40	2.6
M2.5	2.70	3.00	5.40	3.1
M3	3.40	3.70	6.50	3.6
M4	4.40	4.80	8.25	4.7
M5	5.40	5.80	9.75	5.7
M6	6.40	6.80	11.25	6.8
M8	8.40	8.80	14.25	9.2
M10	10.50	10.80	17.25	11.2
M12	12.50	12.80	19.25	14.2
M14	14.50	14.75	22.25	16.2
M16	16.50	16.75	25.50	18.2
M20	20.50	20.75	31.50	22.4
M24	24.50	24.75	37.50	26.4
M30	30.75	31.75	47.50	33.4
M36	37.00	37.50	56.50	39.4
M42	43.00	44.00	66.00	45.6
M48	49.00	50.00	75.00	52.6

APPENDIX F – METRIC FASTENERS (Continued)

Drill and Countersink Sizes for Flat Countersunk Head Cap Screws

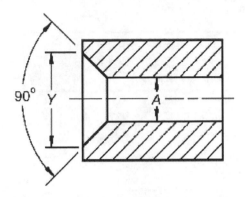

D	A	Y
Nominal Screw Size	**Nominal Hole Diameter**	**Min. Countersink Diameter**
M3	3.5	6.72
M4	4.6	8.96
M5	6.0	11.20
M6	7.0	13.44
M8	9.0	17.92
M10	11.5	22.40
M12	13.5	26.88
M14	16.0	30.24
M16	18.0	33.60
M20	22.4	40.32

ASME/ANSI B18.3.5M - 1986 (R2002) Standard. Reprinted from the standard listed by permission of the American Society of Mechanical Engineers. All rights reserved.

APPENDIX G – FASTENERS

BOLT AND SCREW CLEARANCE HOLES

(1) Inch Clearance Holes

Nominal Screw Size	Fit Classes		
	Normal	Close	Loose
	Nominal Drill Size		
#0 (0.06)	#48 (0.0760)	#51 (0.0670)	3/32
#1 (0.073)	#43 (0.0890)	#46 (0.0810)	#37 (0.1040)
#2 (0.086)	#38 (0.1015)	3/32	#32 (0.1160)
#3 (0.099)	#32 (0.1160)	#36 (0.1065)	#30 (0.1285)
#4 (0.112)	#30 (0.1285)	#31 (0.1200)	#27 (0.1440)
#5 (0.125)	5/32	9/64	11/64
#6 (0.138)	#18 (0.1695)	#23 (0.1540)	#13 (0.1850)
#8 (0.164)	#9 (0.1960)	#15 (0.1800)	#3 (0.2130)
#10 (0.190)	#2 (0.2210)	#5 (0.2055)	B (0.238)
1/4	9/32	17/64	19/64
5/16	11/32	21/64	23/64
3/8	13/32	25/64	27/64
7/16	15/32	29/64	31/64
1/2	9/16	17/32	39/64
5/8	11/16	21/32	47/64
3/4	13/16	25/32	29/32
7/8	15/16	29/32	1 1/32
1	1 3/32	1 1/32	1 5/32
1 1/8	1 7/32	1 5/32	1 5/16
1 1/4	1 11/32	1 9/32	1 7/16
1 3/8	1 1/2	1 7/16	1 39/64
1 1/2	1 5/8	1 9/16	1 47/64

APPENDIX G – FASTENERS

BOLT AND SCREW CLEARANCE HOLES (Continued)

(2) Metric Clearance Holes

Nominal Screw Size	Fit Classes		
	Normal	Close	Loose
	Nominal Drill Size		
M1.6	1.8	1.7	2
M2	2.4	2.2	2.6
M2.5	2.9	2.7	3.1
M3	3.4	3.2	3.6
M4	4.5	4.3	4.8
M5	5.5	5.3	5.8
M6	6.6	6.4	7
M8	9	8.4	10
M10	11	10.5	12
M12	13.5	13	14.5
M14	15.5	15	16.5
M16	17.5	17	18.5
M20	22	21	24
M24	26	25	28
M30	33	31	35
M36	39	37	42
M42	45	43	48
M48	52	50	56
M56	62	58	66
M64	70	66	74
M72	78	74	82
M80	86	82	91
M90	96	93	101
M100	107	104	112

APPENDIX H – REFERENCES

- ASME B1.1 - 2003: Unified Inch Screw Threads (UN and UNR Thread Form)
- ASME B1.13M - 2001: Metric Screw Threads: M Profile
- USAS/ASME B4.1 - 1967 (R2004): Preferred Limits and Fits for Cylindrical Parts
- ANSI B4.2 - 1978 (R2004): Preferred Metric Limits and Fits
- ASME B18.2.1 - 1996: Square and Hex Bolts and Screws (Inch Series)
- ASME/ANSI B18.2.2 - 1987 (R1999): Square and Hex Nuts (Inch Series)
- ANSI B18.2.3.5M - 1979 (R2001): Metric Hex Bolts
- ASME B18.2.4.1M - 2002: Metric Hex Nuts, Style 1
- ASME B18.2.8 - 1999: Clearance Holes for Bolts, Screws, and Studs
- ASME B18.3 - 2003: Socket Cap, Shoulder, and Set Screws, Hex and Spline Keys (Inch Series)
- ASME/ANSI B18.3.1M - 1986 (R2002): Socket Head Cap Screws (Metric Series)
- ASME/ANSI B18.3.5M - 1986 (R2002): Hexagon Socket Flat Countersunk Head Cap Screws (Metric Series)
- ASME 18.6.2 - 1998: Slotted Head Cap Screws, Square Head Set Screws, and Slotted Headless Set Screws (Inch Series)
- ASME B18.21.1 - 1999: Lock Washers (Inch Series)
- ANSI B18.22.1 - 1965 (R2003): Plain Washers
- ASME Y14.2M - 1992 (R2003): Line Conventions and Lettering
- ASME Y14.3 - 2003: Multiview and Sectional view Drawings
- ASME Y14.4M - 1989 (R1999): Pictorial Drawings
- ASME Y14.5M - 1994: Dimensioning and Tolerancing
- ASME Y14.6 - 2001: Screw Thread Representation
- ASME Y14.100 - 2000: Engineering Drawing Practices

INDEX

Notes: